FERROHYDRODYNAMICS

FERROHYDRODYNAMICS

R. E. Rosensweig

DOVER PUBLICATIONS, INC.
Mineola, New York

Bibliographical Note

This Dover edition, first published by Dover Publications, Inc., in 2014, is an unabridged republication of *Ferrohydrodynamics,* published by Dover Publications, Inc., in 1997, which was a slightly corrected edition of the work first published by Cambridge University Press in 1985.

International Standard Book Number

ISBN-13: 978-0-486-67834-4
ISBN-10: 0-486-67834-2

Manufactured in the United States by Courier Corporation
67834202 2014
www.doverpublications.com

To my wife
RUTH
for her love, devotion, and good spirits

CONTENTS

	Preface	*page* xiii
1	**Introduction**	**1**
1.1	Scope of ferrohydrodynamics	1
1.2	Ferromagnetic solids	4
1.3	Magnetic fluids	7
1.4	Ferromagnetic concepts and units	8
	Definition of the field	8
	External field of a dipole source	12
	Magnetic force and torque on dipolar matter	13
	Interaction energy of two dipoles	15
1.5	Concepts of fluid mechanics	16
	Continuity equation	18
	Substantial derivative	22
1.6	Generalized Bernoulli equation	23
1.7	Stress tensor and its physical meaning	25
	Force resulting from a stress tensor	29
	Addendum: equivalence of dipolar and polar representations	30
	Comments and supplemental references	31
2	**Magnetic fluids**	**33**
2.1	Stability requirements	34
	Stability in a magnetic-field gradient	34
	Stability against settling in a gravitational field	35
	Stability against magnetic agglomeration	36
	Necessity to guard against the van der Waals attractive force	37
2.2	Preparation of magnetic colloids by size reduction	38
2.3	Preparation of ferrofluids by chemical precipitation	40
	Magnetite precipitation with steric stabilization	40
	Cobalt particles in an organic carrier	43
	Charge-stabilized magnetite	43
2.4	Other magnetic fluids	44
	Paramagnetic salt solutions	44

	Metallic-base ferrofluid	45
2.5	Surface adsorption and steric stabilization	46
	Steric repulsion mechanism	46
	Net interaction curve	48
	Dispersant structural guidelines	48
2.6	Ferrofluid modification	50
	Phenomenological basis	50
	Carrier liquid exchange	51
	Surfactant exchange	54
2.7	Physical properties	54
	Equilibrium magnetization: superparamagnetism	55
	Magnetic relaxation	61
	Viscosity	63
	Concentrated suspensions	66
2.8	Correlation phenomena	67
2.9	Tabulated physical properties	70
	Comments and supplemental references	72
3	**Electromagnetism and fields**	**74**
3.1	Magnetostatic field equations	74
	Scalar potential	77
3.2	Magnetic-field boundary conditions	77
3.3	Maxwell stress tensor	81
	Portrait of the Maxwell stress tensor	83
3.4	Maxwell's equations	84
	Integral equations	87
	Differential equations	90
3.5	Energy density of the electromagnetic field	94
3.6	Transformed expression for the field energy	95
	Comments and supplemental references	98
4	**Stress tensor and the equation of motion**	**100**
4.1	Thermodynamic background	101
4.2	Formulation of the magnetic stress tensor	103
	Stress tensor of a magnetizable fluid	106
	What is the "pressure" in a magnetized fluid?	108
4.3	Magnetic body-force density	110
	Alternative general forms	111
	Alternative reduced forms	113
	Remarks concerning striction in compressible media	117
4.4	Equation of motion for magnetic fluid	119
	Alternative forms of the equation of motion	121
	Comments and supplemental references	122
5	**The ferrohydrodynamic Bernoulli equation**	**124**
5.1	Derivation	124
5.2	Boundary conditions	126
5.3	Categories of equilibrium inviscid flows	131

5.4 Applications of the FHD Bernoulli equation 132
Classical Quincke problem 132
Surface elevation in a normal field 133
Magnetic nozzle 134
Modified Gouy experiment 136
Conical meniscus 137
Origin of the radial force 141
Magnetic-fluid rotary-shaft seals 142
5.5 Earnshaw's theorem and magnetic levitation 146
Simplified treatment of the levitation of a nonmagnetic body 149
Phenomenon of self-levitation 150
Analysis of forces on an immersed body 153
5.6 Striction effect 157
Comments and supplemental references 158

6 Magnetocaloric energy conversion 161
6.1 Thermodynamics of magnetic materials 162
6.2 Mechanism for power generation 164
6.3 Linear equation of state 166
6.4 General cycle analysis 168
6.5 Cycle efficiency for a linear material 171
6.6 Cycle efficiency with regeneration 172
6.7 Implementing the cycle 172
6.8 Summary 174
Comments and supplemental references 175

7 Ferrohydrodynamic instabilities 177
7.1 Normal-field instability 178
Equation set 178
Interfacial force balance 179
Kinematics 182
Flow-field analysis 183
Perturbed magnetic field 185
Boundary conditions on the field vectors 185
Applying the magnetic boundary conditions 187
Theoretical predictions 188
Normal-field instability experiments 193
Nonlinear analysis 196
7.2 General dispersion relation for moving media with oblique magnetic field 199
7.3 Rayleigh–Taylor problem 200
Experimental confirmation 202
7.4 Kelvin–Helmholtz instability 202
Criterion for Kelvin–Helmholtz instability in a ferrofluid 204
7.5 Gradient-field stabilization 206
7.6 Labyrinthine instability 208
Labyrinth spacing 212
7.7 Stability of fluid cylinders 216
Magnetic-fluid jet in a uniform field 217

	Cylindrical column in a radial gradient field	222
7.8	Porous-medium flow: fingering instability	225
7.9	Thermoconvective stability	228
7.10	Retrospective	232
	Addendum: representation of disturbance waves	232
	Comments and supplemental references	235
8	**Magnetic fluids and asymmetric stress**	**237**
8.1	Phenomena	237
8.2	Cauchy stress principle and conservation of momentum	240
8.3	Integral equation for conservation of angular momentum for nonpolar materials	242
8.4	Reynolds' transport theorem	243
8.5	Cauchy equation of motion	245
8.6	Symmetry of the stress tensor for nonpolar fluids	246
8.7	Analysis for polar fluids	249
8.8	Summary of the basic laws of continuum mechanics	252
8.9	Constitutive relations	253
	Magnetization relaxation process	255
8.10	Analogs of the Navier–Stokes equations for fluids with internal angular momentum	257
8.11	Ferrohydrodynamic-torque-driven flow	258
8.12	Effective viscosity of a magnetized fluid	263
	Comments and supplemental references	269
9	**Magnetic two-phase flow**	**272**
9.1	Background	272
9.2	Ordinary fluidization	273
9.3	Some fundamental problems in fluidization engineering	274
9.4	Basic relationships in two-phase flow	276
	Definition of averages	276
	Averaged continuity equations	279
	Averaged magnetostatic relationships	280
	Averaged momentum balances	282
	Constitutive relations	285
9.5	Summary of the averaged equations	288
9.6	Magnetized fluidized solids	289
	Solution in the steady state with a uniform magnetic field	290
9.7	Stability of the steady-state solution	291
	Derivation of the linearized equations	291
	Form of the linearized equations for plane waves	293
	General solution for the voidage perturbation	297
	Nature of the predicted behavior	301
9.8	Experimental behavior	304
	Transition velocity	304
	Influence of the field orientation	306
	Rheology of magnetically stabilized fluidized solids	307

Directions for further study 312
Comments and supplemental references 312

Appendixes **315**
1 Vector and tensor notation 315
2 Application of the Maxwell stress tensor: analysis of a sheet jet 318

References 323

Citation index 335

Subject index 339

PREFACE

The forces of ordinary magnetism, transmitted at a distance, are apt to stir wonder in all. Separate curiosity arises concerning the motion and forms of flowing liquid, whether observed in ocean surf or a teacup. Consider then magnetic fluids, in which the features of magnetism and fluid behavior are combined in one medium. These fluids display novel and useful behavior. It is hoped that some of the fascination of such a fluid is expressed in this work.

My initial studies with my colleagues were motivated by engineering endeavors and the hope that adding a magnetic term to the equation of fluid motion would lead to interesting and useful consequences. This hope was quickly substantiated with the realization of a succession of novel equilibrium flows arising in family groups both in theory and in the laboratory. Subsequently, more subtle manifestations of magnetics in fluids made an appearance in the form of striking stability phenomena, some leading to the spontaneous organization of fluid into geometric patterns not seen before. In this context the magnetic surface force density unexpectedly emerges to play a key role. Another broad class of flow phenomena arises as the result of the antisymmetric stress produced by the magnetic body torque present in dynamic flows along with magnetic body force. Interactions in magnetic multiphase flow illustrate the most recent extension of the general principles.

Ferrohydrodynamics is an interdisciplinary topic having inherent interest of a physical and mathematical nature with applications in tribology, separations science, instrumentation, information display, printing, medicine, and other areas. Because practitioners in these pursuits have widely different backgrounds, an effort has been made to produce a work that is sufficiently self-contained to be accessible to engineers, scientists, and students from many fields. Concepts of magnetism are scarcely included in any contemporary curriculum, and then often as an afterthought based on an appeal to electrostatics, and the treatment of polarization force receives less emphasis. Accordingly, a careful preparation in the most important aspects of magnetism and introductory material of fluid dynamics are included in the initial chapters. These lay the groundwork for developing in Chapter 4 a rigorous description of stress in magnetic fluids, which leads to formulation of the equations of motion and, later, treatment of the surface boundary conditions. Numbers of equilibrium flows are then treated in Chapters 5 and 6, with emphasis on the use of the generalized Bernoulli equation as a unifying concept that is simple to apply yet powerful in the results achieved.

Chapter 7 treats the basic instability flows, a more sophisticated topic that repays the effort to understand it. The final chapters broach the more complex topics of flows dominated by asymmetric stress and behavior in magnetic multiphase flow, a topic of developing interest in the processing of chemicals and fossil fuel. In these topics the state of theoretical knowledge is less perfect, and so pains are taken to distinguish those parts of the subject that stand on firm ground from those having a tentative status.

Throughout the work, selected applications of ferrohydrodynamics are discussed, for the most part in places where the subject matter serves to illuminate the theoretical principles and where combined liquid and magnetic behavior is prominent.

The topics of this work were offered in a course presented at the University of Minnesota in 1980, when the author visited as professor in the Department of Chemical Engineering and Materials Science. O. A. Basaran recorded the lectures and assisted in producing bound notes that served as a point of departure for this work. His efforts and contributions are deeply appreciated. I am especially grateful to L. E. Scriven for inviting the visit and encouraging the work as well as for many stimulating discussions, and to H. T. Davis for his support and suggestions. Interactions with M. Zahn of MIT have been unstintingly enthusiastic and productive, and I am appreciative of his reading and commenting on the entire text.

Help in the production of this work has been given through the generosity of my employer Exxon and its Corporate Research organization. The support of F. A. Horowitz has been vital. I thank Carolyn Dupré for her sustained efforts in producing the typed manuscript. It has also been a pleasure to work with the publisher's staff and D. Tranah.

Collaboration with many other colleagues over a number of years has been enriching and enjoyable. Most will be identified in this work by citation to their published research. Finally, it is noted that international symposia devoted exclusively to aspects of the subject are being held at three-year intervals. Proceedings of the meetings together with extensive bibliographies of the scientific and patent literature are published as special issues of the *Journal of Magnetism and Magnetic Materials* (Amsterdam) and provide an important source of recent and archival information.

Ronald E. Rosensweig

FERROHYDRODYNAMICS

1

INTRODUCTION

Prior to recent years the engineering applications of fluid mechanics were restricted to systems in which electric and magnetic fields play no role. However, the interaction of electromagnetic fields and fluids has been attracting increasing attention with the promise of applications in areas as diverse as controlled nuclear fusion, chemical reactor engineering, medicine, and high-speed silent printing. The study of various field and fluid interactions may be divided into three main categories:

1. *electrohydrodynamics* (EHD), the branch of fluid mechanics concerned with electric force effects;
2. *magnetohydrodynamics* (MHD), the study of the interaction between magnetic fields and fluid conductors of electricity; and
3. *ferrohydrodynamics* (FHD), the subject of this work, which has become of interest owing to the emergence in recent years of magnetic fluids.

1.1 **Scope of ferrohydrodynamics**

Ferrohydrodynamics deals with the mechanics of fluid motion influenced by strong forces of magnetic polarization. Developing an understanding of the consequences of these forces occupies most of this book. It will be well at the outset to emphasize the difference between ferrohydrodynamics and the relatively better-known discipline of magnetohydrodynamics. In MHD the body force acting on the fluid is the Lorentz force that arises when electric current flows at an angle to the direction of an impressed magnetic field. However, in FHD there need be no electric current flowing in the fluid, and usually there is none. The body force in FHD is due to polarization force, which in turn requires material magnetization in the presence of magnetic field gradients or discontinuities. Likewise, the force interaction arising in EHD is often due to free electric charge acted upon by an electric force field. In comparison, in FHD free electric charge is normally absent, and the analog of electric charge, the monopole, has not been found in nature. An analogy between EHD and FHD arises, however, for charge-free electrically polarizable fluids exposed to a gradient electric field. A major difference from FHD is the magnitude of the effect, which is normally much smaller in the electrically polarizable media. This work is concerned exclusively with FHD; however, the reader interested in EHD or MHD will find excellent starting points in the references cited at the end of this chapter.

Ferrohydrodynamics began to be developed in the early to mid-1960s, motivated initially by the objective of converting heat to work with no mechanical parts. However, as colloidal magnetic fluids (ferrofluids) became available, many other uses of these fascinating liquids were recognized. Many of these ideas are concerned with the remote positioning and control of magnetic fluid using magnetic force fields. An aspect of this behavior is illustrated in the photograph of Figure 1.1.

Ferrohydrodynamics has inherent interest if for no other reason than the uniqueness of fluid having giant magnetic response. As a result, a number of striking phenomena are exhibited by the magnetic fluids in response to impressed magnetic fields. These responses include the normal field instability, because of which a pattern of spikes appears on the fluid surface; the spontaneous formation of intricate labyrinthine patterns in thin layers; the generation of body couple in rotary fields, which is manifested as antisymmetric stress; unusual buoyancy relationships, such as the self-levitation of an immersed magnet; and enhanced convective cooling in ferrofluids having a temperature-dependent magnetic moment. It is a major objective of this work to build a significant understanding of the subject, based on the continuum-mechanical approach as augmented, where needed, by the microscopic description.

Demonstrated applications of ferrofluids span a very wide range. Actual commercial usage presently includes novel zero-leakage rotary shaft seals used in computer disk drives (Bailey 1983), vacuum feedthroughs for semiconductor manufacturing and related uses (Moskowitz 1975), pressure seals for compressors and blowers (Rosensweig 1979a), and more. A *drop* of fluid makes these devices possible; without that strategic drop the devices would not function. Also in use are liquid-cooled loudspeakers that employ mere drops of ferrofluid to conduct heat away from the speaker coils (Hathaway 1979). This innovation

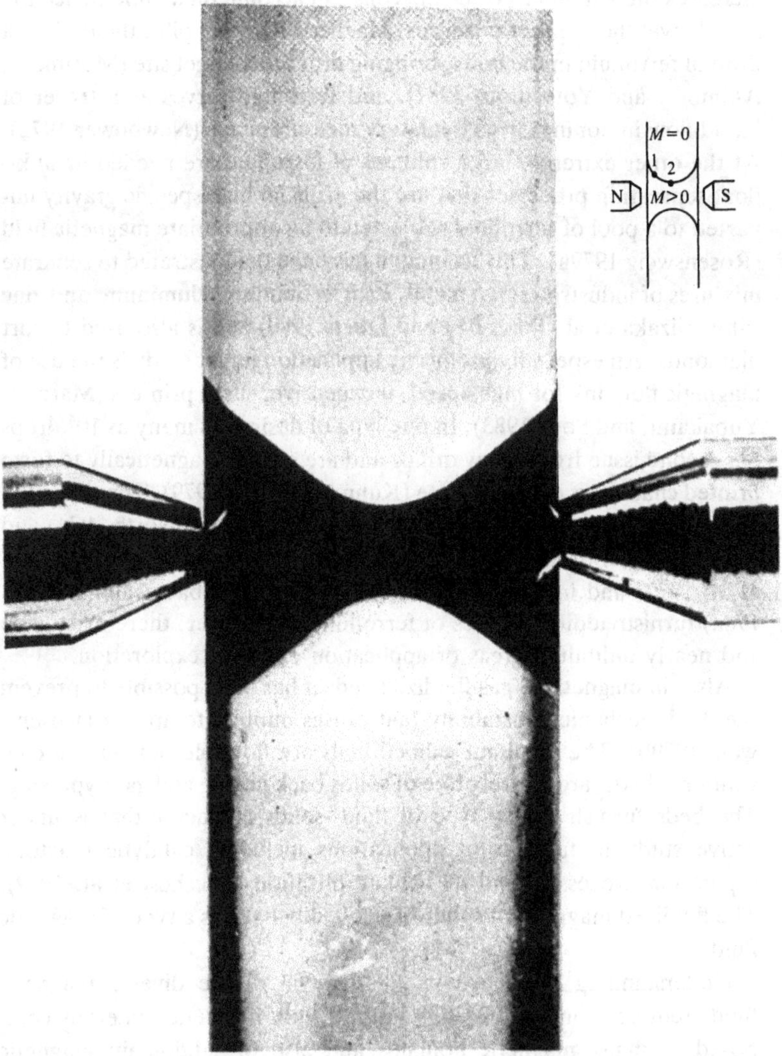

1.1 Magnetic fluid within the transparent tube, shaped and positioned by field of the permanent magnets located outside the tube. (*Courtesy of D. A. Roth.*) Inset identifies nomenclature for "the simplest example" discussed in Section 1.6.

increases the amplifier power the coils can accommodate and hence the sound level the speaker produces. Magnetic field can pilot the path of a drop of ferrofluid in the body, bringing drugs to a target site (Morimoto, Akimoto, and Yotsumoto 1982), and ferrofluid serves as a tracer of blood flow in noninvasive circulatory measurements (Newbower 1972). At the other extreme, large volumes of ferrofluid are needed in sink–float separation processes that use the artificial high specific gravity imparted to a pool of ferrofluid subjected to an appropriate magnetic field (Rosensweig 1979a). This technique has been demonstrated to separate mixtures of industrial scrap metals such as titanium, aluminum, and zinc (Shimoiizaka et al. 1980; Fay and Quets 1980) and is also used to sort diamonds. An especially promising application under study is the use of magnetic fluid ink for high-speed, inexpensive, silent printers (Maruno, Yubakami, and Soga 1983). In one type of design, as many as 10^4 drops per second issue from a tiny orifice and are guided magnetically to form printed characters on a substrate (Kuhn and Myers 1979). The detection of magnetic domains in alloys and crystals (Wolfe and North 1974) and the production of magnetically responsive enzyme supports (Adelstein et al. 1979) and immobilized microorganisms (Birnbaum and Larsson 1982) furnish additional uses of ferrofluids. Moreover, there exist wide and nearly unlimited areas of application open for exploration.

Also, in magnetized gas-fluidized beds it has been possible to prevent the fluid-mechanical instability that causes bubbles to appear (Rosensweig 1979b). The resultant calmed beds are flowable but, unlike conventional beds, are entirely free of solids back mixing and gas bypassing. The beds furnish a new type of fluid–solids contactor that is under active study in families of applications including catalytic reactors, separation processes, and particulate filtration (Lucchesi et al. 1979). The fluidized magnetized solids of the bed behave as a type of magnetic fluid.

Understanding the magnetic phenomena of the diverse magnetic fluids requires some knowledge both of bulk magnetic materials composed of many magnetic domains and also of subdomain magnetic materials. Accordingly, a few rudiments concerning the magnetic materials of interest are introduced in the next section. Subsequently, certain relationships from electromagnetism and continuum mechanics are developed as a foundation for the rest of this book.

1.2 Ferromagnetic solids

Ferromagnetic solids are composed of domains in each of which the magnetic moments of individual atoms are oriented in a fixed direc-

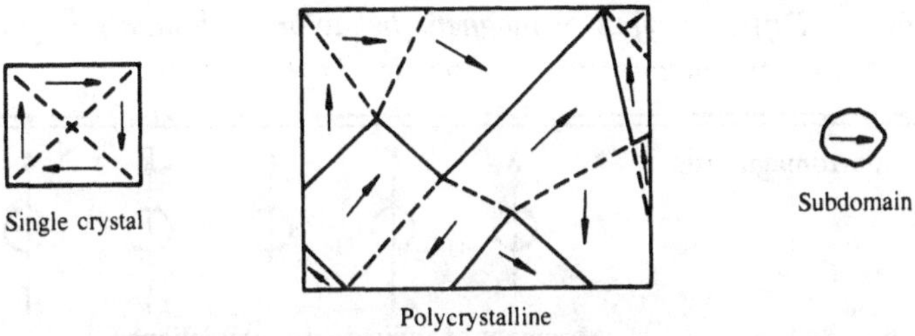

1.2 Ferromagnetic domain structures for single-crystal, polycrystalline, and subdomain samples. Crystal walls are shown solid, domain walls dashed.

tion. The existence of domains was first postulated by Pierre Weiss, in 1907. Domain structures for single-crystal and polycrystalline materials are sketched in Figure 1.2, where dotted lines represent domain walls and solid lines represent crystal boundaries. Basically, a ferromagnetic material breaks up into domains to minimize the field energy, which would be considerable if the material were magnetized in one direction. However, the material does not divide itself into domains indefinitely, for it requires energy to create the *domain walls*, which separate the domains. The domain wall structure shown in Figure 1.2 is simplified. In reality the transition in the direction of the atomic moment vector is gradual and takes place across ~100 atoms.

A fundamental theoretical understanding of ferromagnetism did not come about until 1928, when Werner Heisenberg finally explained it on the basis of the newly developed quantum theory. According to quantum mechanics, it is mainly the spin magnetic moments that contribute to the molecular field. Heisenberg showed that when the spins on neighboring atoms change from parallel alignment to antiparallel alignment there is an accompanying change in the electron charge distribution in the atoms that alters the electrostatic energy of the system. In some cases parallel alignment is energetically more favorable; this is what is known as ferromagnetism. A thorough discussion of magnetism is beyond the scope of this introduction; let it suffice merely to summarize certain relevant aspects of magnetic behavior. Table 1.1 lists for convenience some facts needed to understand the subsequent topics of this book.

Ferromagnetism is exhibited by iron, nickel, cobalt, and many of their alloys; some rare earths, such as gadolinium; and certain intermetallics, such as gold–vanadium. Ferromagnetic ordering disappears at the Curie temperature θ. Antiferromagnetic materials exhibit no net

Table 1.1 *Different types of magnetic behavior. Colloidal ferrofluids exhibit superparamagnetism, as discussed in the text.*

Spontaneous domain formation	Ferromagnetism	 Moments of individual atoms aligned
	Antiferromagnetism	 Moments alternating from atom to atom
	Ferrimagnetism	 Unequal moments alternate
No domains	Paramagnetism	No long-range order; alignment with applied field
	Diamagnetism	No long-range order; alignment opposes field

moment at any temperature. The antiferromagnetic ordering disappears at the Néel temperature. *Antiferromagnetism* is a property of MnO, FeO, NiO, $FeCl_2$, MnSe, and many other compounds. In *ferrimagnetism* the net moment is smaller than in a typical ferromagnetic material. Ferrites of the general formula $MO \cdot Fe_2O_3$ exhibit ferrimagnetism where M stands for Fe, Ni, Mn, Cu, Mg. Magnetite, having composition Fe_3O_4 and possessing cubic crystalline structure, is the best known ferrite. Hexagonal ferrites and garnets, which are cubic insulators composed of iron, other metals, and oxygen atoms, give additional examples of ferrimagnetic materials.

Paramagnetism is a behavior resulting from the tendency of molecular moments to align with the applied magnetic field but in the absence of long-range order. The property is exhibited by liquid oxygen, rare-earth salt solutions, ferromagnets above the Curie temperature, and many other substances. *Diamagnetism* represents the weakest type of magnetic behavior and is prominent only in materials with closed electron shells. Inert gases, many metals, most nonmetals, and many organic compounds are diamagnetic. A colloidal magnetic fluid consists of a collection of ferro- or ferrimagnetic single-domain particles with no long-range order between particles. The resultant behavior, termed *superparamagnetism*, is similar to paramagnetism except that the magnetization in low to moderate fields is much larger. Relationships of superparamagnetism are developed in Chapter 2.

1.3 Magnetic fluids

Several types of magnetic fluids arise with FHD; the principal type is *colloidal ferrofluid*. A colloid is a suspension of finely divided particles in a continuous medium, including suspensions that settle out slowly. However, a true ferrofluid does not settle out, even though a slight concentration gradient can become established after long exposure to a force field (gravitational or magnetic). Such ferrofluids are composed of small (3–15 nm) particles of solid, magnetic, single-domain particles coated with a molecular layer of a dispersant and suspended in a liquid carrier (see Figure 1.3). Thermal agitation keeps the particles suspended because of Brownian motion, and the coatings prevent the particles from sticking to each other.

It will be recalled that Brownian motion is named after the botanist Robert Brown. In 1827 he discovered the continuous random motion of small particles suspended in water that may be observed under a microscope. Albert Einstein developed in 1905 a theory of Brownian motion based on the assumption that translational kinetic energy is equally partitioned between the particles and the molecules of the surrounding fluid. Comparison of the theory with measurements provided the earliest and most direct experimental evidence for the reality of the atom.

The colloidal ferrofluid must be synthesized, for it is not found in nature. Also, it is far different in its properties than the "magnetic fluids" for clutches and brakes introduced in the late 1940s. Composed of micron- and larger-size iron particles slurried in an oil, the clutch fluids solidify in the presence of an applied magnetic field. In comparison, colloidal ferrofluids retain liquid flowability in the most intense applied magnetic fields. A typical ferrofluid contains 10^{23} particles per cubic meter and is opaque to visible light.

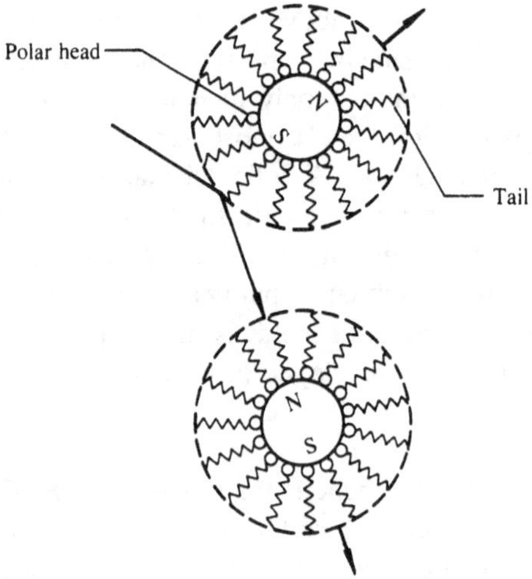

1.3 Schematic representation of coated, subdomain, magnetic particles in a colloidal ferrofluid. Collisions of suitably coated particles are elastic.

Recently there have been successful attempts to use paramagnetic solutions of rare-earth salts as a magnetic fluid; these examples of molecularly dispersed systems require the use of relatively intense magnetic fields to generate appreciable forces. Certain pure substances such as liquid oxygen are strongly attracted to a magnet and behave like a ferrofluid, but, because they exist only at cryogenic temperatures, to date they have not been used. A more recent development is the subject of magnetized fluidized solids, which has grown out of FHD thinking and employs a number of the same principles. Whereas ferrofluids are wet, these fluidized systems can be dry and yet exhibit aspects of fluid behavior.

1.4 Ferromagnetic concepts and units

Definition of the field

When a magnet is dipped into a magnetic fluid, the fluid clings to it in the manner of iron filings, especially in certain places called *poles*, which are usually located near the ends of the magnet. The concept of poles is useful even though isolated poles are unknown in nature. Charles Coulomb, in 1785, determined as the result of experimental observations that like poles repel and unlike poles attract with a force that is proportional to the product of the pole strengths and inversely proportional to the square of the distance between them. For magnetic point poles of strength p and p' separated in a vacuum by a distance r, the magnitude of the force is given by $pp'/4\pi\mu_0 r^2$, with the direction of the force along the line connecting the poles. If p' is a unit north-seeking pole, the force acting on it is defined as the *magnetic field* **H**. Thus, surrounding a point pole p the magnetic field is

$$\mathbf{H} = \frac{p\hat{\mathbf{r}}}{4\pi\mu_0 r^2} = \frac{p\mathbf{r}}{4\pi\mu_0 r^3} \tag{1.1}$$

where **r** is the position vector directed from p to p' and $\hat{\mathbf{r}} \equiv \mathbf{r}/r$ is a unit vector having the orientation of **r**.*

The value of the proportionality constant $1/4\pi\mu_0$ in equation (1.1) depends on the system of units used. Throughout this book, SI units (Système International d'Unités) are used, the base units for which are taken from the rationalized mksa system of units: Distances are measured in meters (m), mass in kilograms (kg), time in seconds (s), and electric current in amperes (A). The adjective "rationalized" is used because the factor of 4π is arbitrarily introduced into the proportionality factor of Coulomb's law. This cancels a 4π that arises in other frequently used laws to be introduced later, Maxwell's equations. The magnetic field **H** has units of amperes per meter. The parameter μ_0 is called the permeability of free space and has the value $\mu_0 = 4\pi \times 10^{-7}$ $\mathrm{H \cdot m^{-1}}$, where H stands for the henry. Force in (1.1) is measured in newtons (N).

The notion of a magnetic field **H** simplifies the detailed description of external conditions. Thus, instead of stating for a given experiment that the test conducted was performed at a particular distance and orientation with respect to a magnet constructed according to certain specifications, it may be said that the apparatus was placed at a given location in a field **H**.

In SI, an *induction field B* (in tesla) is defined such that in vacuum, $B = \mu_0 H$. According to Faraday's law, time rates of change in the induction field have fundamental importance in determining voltages a topic that will be further considered in connection with Maxwell's laws. From (1.1) and the definition of B, the induction field surrounding an SI pole of strength p is given by $B = p/4\pi r^2$. The B field may be pictured as lines

*An equation that defines a quantity is indicated throughout with the symbol $\equiv$.

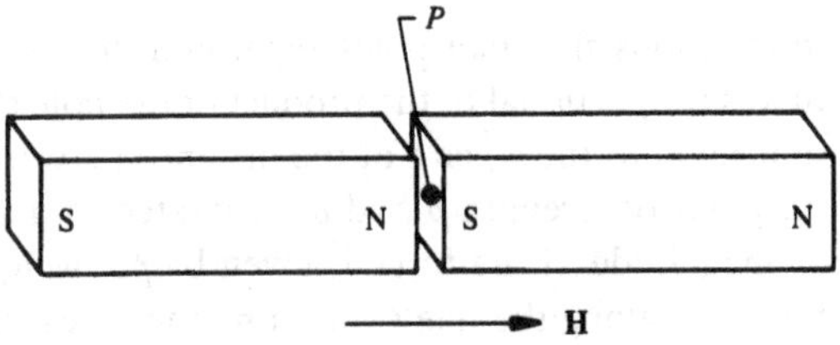

1.4 Ferromagnetic bar containing a narrow transverse gap.

of induction. In a uniform B field of unit intensity, one line (or weber, denoted Wb) is said to cross each square meter of perpendicular surface. Thus B has units of webers per square meter, also known as teslas (T). Alternative units for μ_0 are tesla-meters per ampere, and the units of pole strength are teslas per square meter. A sphere surrounding a point pole p is crossed by a total number of lines $\phi = 4\pi r^2 B = p$, so from a unit SI pole there emanates one line of magnetic induction.

The *intensity of magnetization* M denotes the state of polarization of magnetized matter. If an SI magnetic pole of uniform strength p has an area a (m^2) the intensity of magnetization is $M \equiv p/a\mu_0 = \rho_s/\mu_0$, where ρ_s is the surface density of magnetic poles. Consider, as in Figure 1.4, a ferromagnetic bar containing an extremely narrow gap with orientation transverse to the direction of magnetization. Poles appear on both faces of the gap, and the field in the gap is the superposition of the field emanating from the north poles, appearing on the gap surface to the left, and the field due to the south poles, located on the opposite surface of the gap. Because one line of induction emanates from each north pole, by symmetry $p/2a$ lines cross a unit area of the gap perpendicularly owing to the presence of poles on the north face, and an additional $p/2a$ lines are contributed by the poles on the south face, for a total of $p/a = \mu_0 M$ lines. The impressed magnetizing field H contributes an additional $\mu_0 H$ to the induction field present in the gap, so the total induction field is $B = \mu_0(H + M)$; if the narrow gap is closed, B remains the same. The lines of the B field are continuous loops that emanate from the north end of the bar and enter the south end, and within the bar the B field is directed from south to north.

Materials scientists frequently work in the *cgs system of units* in which $4\pi\mu_0$ in (1.1) is unity, r is measured in centimeters, and the force is given in dynes (dyn). Thus, a cgs unit magnetic pole is such that, when placed in a vacuum at a distance of 1 cm from a precisely similar pole, it

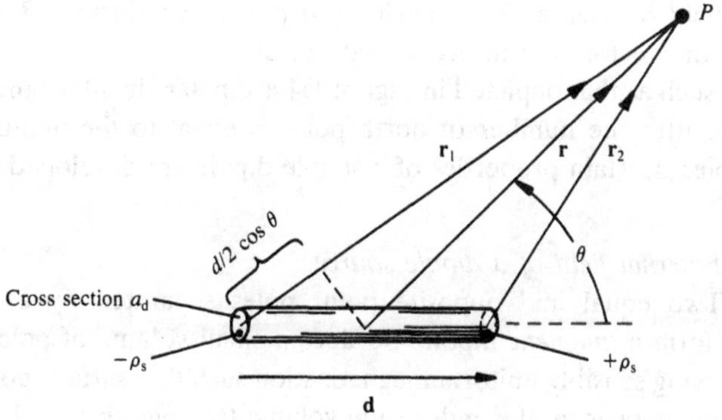

1.5 Development of the field at the point P due to a magnetic dipole.

is repelled with a force of 1 dyn. The magnetic force at a point in an arbitrary magnetic field in a vacuum is the force in dynes upon a unit north pole placed at the point; this force is termed the magnetic field H (in oersted, denoted Oe). Associating one line of force (or maxwell, denoted Mx) crossing perpendicularly a unit area (in square centimeters) with a magnetic field of unit intensity, it follows from (1.1) that a unit cgs magnetic pole emanates 4π lines of force. The intensity of magnetization is $I = p/a$, where a is the area in square centimeters. The field in a narrow gap of a uniformly magnetized ferromagnetic bar is now found from symmetry to be given by $B = H + 4\pi I$, where $4\pi I = M$ is known as the ferric induction. Thus $B = H + M$, where B and M are conventionally expressed in gauss (G). In a vacuum in the cgs system the B field (in gauss) is numerically equal to the H field (in oersteds). Since the reader will commonly encounter cgs units as well as SI nomenclature in the literature it is necessary to be familiar with both systems. The SI unit of magnetic induction, the weber per square meter or tesla, is equal to 10^4 G. In comparison, the Earth's magnetic field averages about 0.7 G in magnitude.

In this section B, H, and M have been introduced as scalar quantities. More generally, these field variables possess orientation as well as magnitude and hence are vectorial in nature. Thus, B is the magnitude of a vector $\mathbf{B}$, H that of a vector $\mathbf{H}$, and M that of a vector $\mathbf{M}$. These vectors are related by

$$\mathbf{B} \equiv \mu_0(\mathbf{H} + \mathbf{M}) \tag{1.2}$$

For example, a permanent magnet with magnetization M placed in an applied field H oriented at an angle to M produces induction B that is given as the vector sum in accord with (1.2).

A bar such as that depicted in Figure 1.4 is dipolar. In all magnetized dipolar matter the number of north poles is equal to the number of south poles. Certain properties of a simple dipole are developed next.

External field of a dipole source

Two equal and opposite point poles separated by a small distance form a *magnetic dipole*. So does a small volume of polarized matter having sensibly uniform magnetization such that surface poles of density $\pm\rho_s$ appear on the ends of the volume (see Figure 1.5). Let the orientation and length of the small volume be specified by the vector $\mathbf{d}$.

The magnetic field at an arbitrary point P can be found by application of Coulomb's law and assuming superposition of the fields. The result is

$$\mathbf{H}(\mathbf{r}) = \frac{\rho_s a_d}{4\pi\mu_0}\left(-\frac{\mathbf{r}_1}{r_1^3} + \frac{\mathbf{r}_2}{r_2^3}\right) \tag{1.3}$$

where $\mathbf{r}_1 = \frac{1}{2}\mathbf{d} + \mathbf{r}$ and $\mathbf{r}_2 = -\frac{1}{2}\mathbf{d} + \mathbf{r}$. Though this relation holds for any separation d, what is desired is a good approximation to the field when the separation d is small compared to r. When $d \ll r$, to a good approximation r_1 and r_2 are given by

$$r_1 \approx r + \frac{d}{2}\cos\theta, \qquad r_2 \approx r - \frac{d}{2}\cos\theta \tag{1.4}$$

By the binominal theorem,

$$\begin{aligned} r_{1\text{ or }2}^{-3} &\approx \left(r \pm \frac{d}{2}\cos\theta\right)^{-3} = r^{-3}\left(1 \pm \frac{d}{2r}\cos\theta\right)^{-3} \\ &\approx r^{-3}\left(1 \mp \frac{3d}{2r}\cos\theta\right) \end{aligned} \tag{1.5}$$

$$\begin{aligned} \mathbf{H}(\mathbf{r}) \approx \frac{\rho_s a_d}{4\pi\mu_0 r^3}\Bigg[&\left(-\tfrac{1}{2}\mathbf{d} - \mathbf{r}\right)\left(1 - \frac{3d}{2r}\cos\theta\right) \\ &+ \left(-\tfrac{1}{2}\mathbf{d} + \mathbf{r}\right)\left(1 + \frac{3d}{2r}\cos\theta\right)\Bigg] \end{aligned} \tag{1.6}$$

and canceling the common terms gives

$$\mathbf{H}(\mathbf{r}) \approx \frac{\rho_s a_d d}{4\pi\mu_0 r^3}\left[-\hat{\mathbf{d}} + 3\cos\theta\,\hat{\mathbf{r}}\right] \tag{1.7}$$

where $\hat{\mathbf{d}}$ is the unit vector $\mathbf{d}/d$. However, $\hat{\mathbf{d}}\cdot\hat{\mathbf{r}} = \cos\theta$, $\rho_s = \mu_0 M$, and $V = a_d d$, so this result may be written

$$\frac{\mathbf{H}(\mathbf{r})}{M} \approx \frac{V}{4\pi r^3}\left[-\hat{\mathbf{d}} + 3(\hat{\mathbf{d}}\cdot\hat{\mathbf{r}})\hat{\mathbf{r}}\right] \tag{1.8}$$

Thus the field of a point dipole falls off as $1/r^3$ and hence is of shorter range than the $1/r^2$ field of Coulomb's law.

A uniformly magnetized sphere produces an external field exactly equivalent to that of a point dipole possessing the same total magnetic moment and located at the sphere center. From (1.8) it is easily shown that the field at the north pole is oriented oppositely to the field at the equator, with the field at the pole twice as strong as the field at the equator.

Magnetic force and torque on dipolar matter

A general expression for the magnetic force on a magnetized body will now be developed. Note at the outset that this computation tells nothing whatsoever about the state of stress within the body. Evaluation of the stress distribution is a much more difficult task and requires a thermodynamic treatment, a topic returned to in Chapter 4.

To begin, consider an isolated small cylindrical volume of magnetically polarized substance with geometric axis **d** aligned with the magnetization vector **M** (Figure 1.6). The material is subjected to an applied field $\mathbf{H}_0$, and poles of density $\rho_s = \mu_0 M$ appear in equal number and opposite polarity on the ends of area a_d. A volume δV of the element is $a_d d$. The applied field $\mathbf{H}_0$ may be taken to be the force on a unit pole, and hence the force experienced by the volume element is

$$-\mathbf{H}_0\rho_s a_d + (\mathbf{H}_0 + \delta\mathbf{H}_0)\rho_s a_d = \delta\mathbf{H}_0\rho_s a_d \tag{1.9}$$

where $\delta\mathbf{H}_0$ is the change in $\mathbf{H}_0$ along the direction of **d**. Thus $\delta\mathbf{H}_0 = (\mathbf{d}\cdot\nabla)\mathbf{H}_0 = (d/M)(\mathbf{M}\cdot\nabla)\mathbf{H}_0$, and the *Kelvin force density* is given by

$$\text{force density} = \mu_0(\mathbf{M}\cdot\nabla)\mathbf{H}_0 \tag{1.10}$$

Note that $\mu_0\mathbf{M}$ represents the vector moment per unit volume because the definition of *dipole moment* **m** is

$$\mathbf{m} \equiv \rho_s a_d \mathbf{d} = \mu_0 \mathbf{M} a_d d \tag{1.11}$$

and $a_d d$ is the volume of the element.

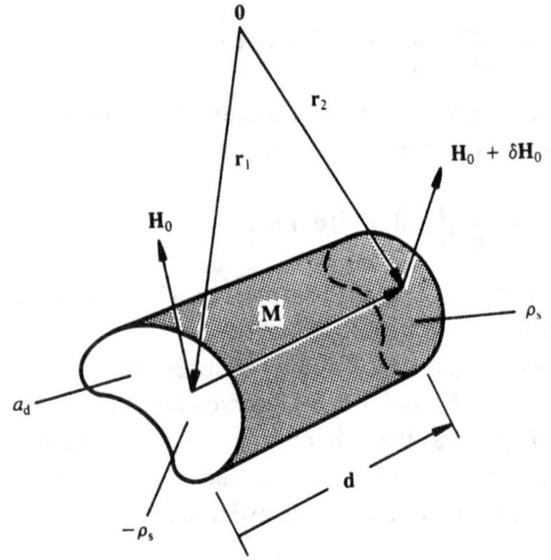

1.6 Development of the gradient-field force and the magnetic torque on a small element of magnetically polarized substance.

The torque $\delta\mathbf{T}$ acting on a small volume of magnetically polarized matter in magnetic field can also be derived with reference to Fig. 1.6 for an impressed magnetic field that is spatially uniform, i.e., for which $\delta\mathbf{H}_0 = \mathbf{0}$. Summing moments about an origin 0 with position vector $\mathbf{r}_1$ of the south face of the volume and $\mathbf{r}_2 = \mathbf{r}_1 + \mathbf{d}$, the position vector of the north face is seen to give

$$\delta\mathbf{T} = \rho_s a_d(-\mathbf{r}_1 \times \mathbf{H}_0 + \mathbf{r}_2 \times \mathbf{H}_0) = \rho_s a_d \mathbf{d} \times \mathbf{H}_0$$

Because $\rho_s a_d \mathbf{d} = \mu_0 \mathbf{M}\,\delta V$, the torque per unit volume is given by

$$\text{torque density} = \mu_0 \mathbf{M} \times \mathbf{H}_0 \tag{1.12}$$

and the expression is independent of the choice of origin.

Note that for "soft" magnetic materials $\mathbf{M}$ is parallel to $\mathbf{H}_0$. The saturation magnetization of a purely ferromagnetic sample is the domain magnetization M_d of the material. Otherwise, magnetization curves for soft and "hard" materials are quite different, as shown in Figure 1.7. The behavior depicted in Figure 1.7b is known as *hysteresis*. M_r is called the *retentivity* or *remanence*, and the magnitude of H_c is known as the *coercivity* of the material. Permanent magnets are made

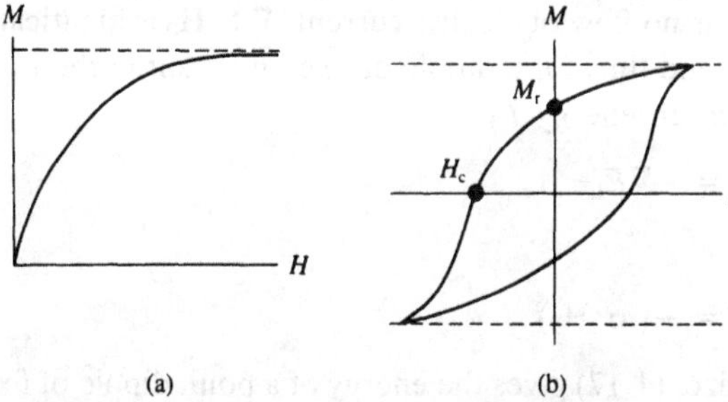

1.7 Typical magnetization curves for (a) soft and (b) hard magnetic materials.

from hard material and thus retain a magnetic moment after the applied field is removed.

For soft materials the force density of equation (1.10) reduces to the term $\mu_0 M \nabla H_0$. The reduction, requiring the use of a Maxwell relationship, is carried out in Section 4.3. Thus, the work per unit volume in a displacement $d\mathbf{s}$ having magnitude ds is $\mu_0 M(\nabla H_0)\cdot d\mathbf{s}$, which can be written $\mu_0 M\,dH$.

Thought problem: Given two metal bars of identical appearance, one a magnet and the other soft iron, how can the identities of the bars be determined using no other equipment?

Interaction energy of two dipoles

As demonstrated by equation (1.10), the force per unit volume acting on an elementary volume of dipolar matter in an external magnetic field $\mathbf{H}_0$ is $\mu_0(\mathbf{M}\cdot\nabla)\mathbf{H}_0$, or, from the definition of $\mathbf{m}$ in (1.11), the total force is

$$\mathbf{F} = (\mathbf{m}\cdot\nabla)\mathbf{H}_0 \tag{1.13}$$

With the help of a vector identity, $(\mathbf{m}\cdot\nabla)\mathbf{H}_0$ can be rewritten

$$(\mathbf{m}\cdot\nabla)\mathbf{H}_0 = \nabla(\mathbf{m}\cdot\mathbf{H}_0) - \mathbf{H}_0\cdot\nabla\mathbf{m} - \mathbf{m}\times(\nabla\times\mathbf{H}_0) - \mathbf{H}_0\times(\nabla\times\mathbf{m}) \tag{1.14}$$

For constant $\mathbf{m}$ this simplifies to

$$(\mathbf{m}\cdot\nabla)\mathbf{H}_0 = \nabla(\mathbf{m}\cdot\mathbf{H}_0) - \mathbf{m}\times(\nabla\times\mathbf{H}_0) \tag{1.15}$$

When there is no flow of electric current, $\nabla \times \mathbf{H}_0$ is identically zero (see Chapter 3), and thus for a dipole of fixed moment $\mathbf{m}$ the force $\mathbf{F}$ can be derived from an energy E_h by

$$\mathbf{F} = -\nabla E_h \tag{1.16}$$

where

$$E_h = -(\mathbf{m} \cdot \mathbf{H}_0) \tag{1.17}$$

To summarize, (1.17) gives the energy of a point dipole of fixed moment $\mathbf{m}$ as a function of its orientation and position in the applied field $\mathbf{H}_0$.

It is now possible to consider the interaction energy of two point dipoles, as shown in Figure 1.8. If, for instance, dipole (1) is regarded as the source of the magnetic field felt by (2), then, according to (1.8),

$$\mathbf{H}_0(\mathbf{r}) = \frac{M_1 V_1}{4\pi r^3}\left[-\hat{\mathbf{d}}_1 + 3(\hat{\mathbf{d}}_1 \cdot \hat{\mathbf{r}})\hat{\mathbf{r}}\right] \tag{1.18}$$

Now using (1.18) and the relation $\mathbf{m}_2 = \mu_0 \mathbf{M}_2 V_2 = \mu_0 M_2 V_2 \hat{\mathbf{d}}_2$ in (1.17) gives the desired expression for the interaction energy of two point dipoles:

$$E_{dd} = \frac{\mu_0 (M_1 V_1)(M_2 V_2)}{4\pi r^3}\left[\hat{\mathbf{d}}_1 \cdot \hat{\mathbf{d}}_2 - 3(\hat{\mathbf{d}}_1 \cdot \hat{\mathbf{r}})(\hat{\mathbf{d}}_2 \cdot \hat{\mathbf{r}})\right] \tag{1.19}$$

where E_{dd} is the particular form of E_h and may be equivalently written

$$E_{dd} = \frac{1}{4\pi\mu_0}\left[\frac{\mathbf{m}_1 \cdot \mathbf{m}_2}{r^3} - \frac{3}{r^5}(\mathbf{m}_1 \cdot \mathbf{r})(\mathbf{m}_2 \cdot \mathbf{r})\right] \tag{1.20}$$

This expression is used later to evaluate the colloidal stability of magnetic particle dispersions in liquid carriers.

1.5 Concepts of fluid mechanics

The term “fluids” includes both liquids and gases. The physical properties of ideal gases can be explained on a molecular basis by the kinetic theory of gases, but there is no generally applicable kinetic theory of liquids. It might be thought that it would be necessary for technological purposes to develop the mechanics of liquids and gases as separate subjects, but this is not the case. In fact, by treating the general fluid as a homogeneous substance or continuum and invoking general conservation laws of mass, momentum, and energy, one can describe many flow fields with penetrating accuracy.

It is assumed that the reader will be familiar with the principles of

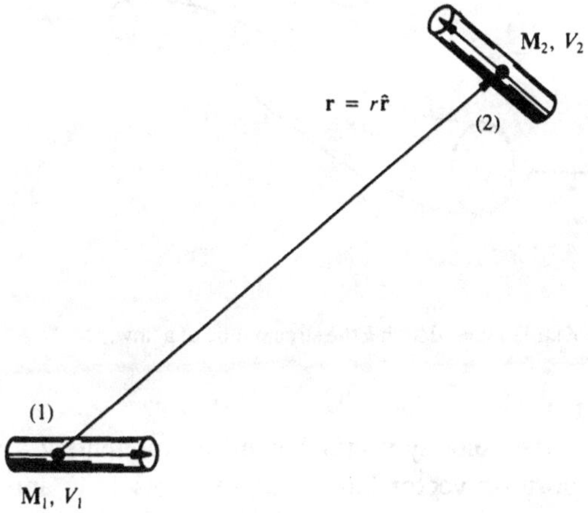

1.8 Interaction of two point dipoles with moment vectors oriented arbitrarily.

elementary hydrodynamics. For example, in a fluid at rest the pressure at a point is the same in all directions and shear stresses are absent. The existence of shearing stress in a moving fluid is associated with the physical property of viscosity. It is also an experimental fact that a fluid does not slip at a boundary. A very extensive theory has been developed, however, to describe the flow of ideal or inviscid fluids in which the action of viscosity is ignored altogether. It will be seen in this book that concepts of hydrostatics are modified in interesting and useful ways by the influence of magnetic force and that the topic of inviscid flow gains a broader relevance. Magnetically influenced flows with viscous effects are also broached, as are some more complex situations.

As usual in continuum mechanics, some mathematical notation is needed in order to describe fluid motion. It is convenient to take position coordinates x, y, z to locate a fixed point in the flow. The velocity components v_x, v_y, v_z measured in the x, y, and z directions, respectively, are functions of the coordinates x, y, z and, in general, the time t:

$$\begin{aligned} v_x &= v_x(x,y,z,t) \\ v_y &= v_y(x,y,z,t) \\ v_z &= v_z(x,y,z,t) \end{aligned} \tag{1.21}$$

Vector notation enables these relationships to be expressed more

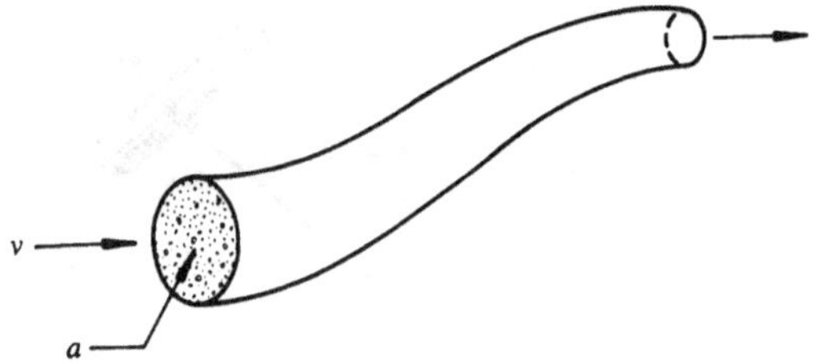

1.9 Bundle of streamlines defining the stream tube of a flow.

concisely. If **v** is the velocity vector having components v_x, v_y, and v_z, and if **r** is the position vector having components x, y, and z, then

$$\mathbf{v} = f(\mathbf{r}, t) \tag{1.22}$$

Because the velocity **v** is continuously distributed over space, **v**, like the magnetic vectors **B**, **H**, and **M**, is a field variable.

Continuity equation

It is useful to picture a set of lines drawn so as to coincide with the velocity vectors at every point. These are known as *streamlines*. By definition no flow crosses a streamline. The streamlines passing through a closed curve form a stream tube (see Figure 1.9), and, if the curve is drawn small enough, the velocity can be considered uniform over any cross section of the stream tube. Conservation of mass requires that along the stream tube

$$\rho a v = \text{const} \tag{1.23}$$

where ρ is the density of the fluid, a is the cross-sectional area of the stream tube, and $v = |\mathbf{v}|$ is the speed of the fluid. However, a more general equation of mass conservation, the differential equation of continuity, is required in the analysis. Streamlines of the velocity field are analogous to lines of the magnetic **B** field, with mass conserved along a stream tube in a flow field and magnetic flux conserved in a magnetic field. In fact, historically, the analogy arose in the reverse order.

In setting up the equation of continuity as well as differential equations governing the dynamics of fluid motion, there is a choice of employing coordinates fixed in space or moving with the fluid. The two procedures are known, respectively, as the Eulerian and Lagrangian specifications, and both are employed in this book. Furthermore, one

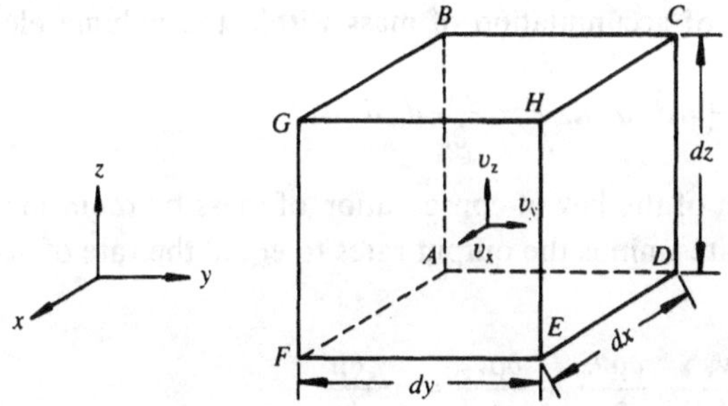

1.10 Eulerian differential control volume.

may choose to focus on a differential or a macroscopic (integral) control volume. The continuity equation will be developed in alternative formulations to illustrate the various procedures.

Eulerian differential formulation: Consider the rectangular parallelepiped of Figure 1.10, whose sides are of length dx, dy, and dz relative to a rectangular coordinate system. Now consider the face $ABCD$. The rate of input of matter across this face and into the element will be the product of the area $dy\,dz$ and the mean value of the density and velocity in the x direction, ρv_x. This mean mass flux can be evaluated as the value at the center of the face and is

$$\rho v_x - \frac{\partial \rho v_x}{\partial x}\frac{dx}{2}$$

The rate of input of mass across face $ABCD$ into the volume element is therefore

$$\left(\rho v_x - \frac{\partial \rho v_x}{\partial x}\frac{dx}{2}\right) dy\,dz$$

and the rate of output of mass across face $EFGH$ is

$$\left(\rho v_x + \frac{\partial \rho v_x}{\partial x}\frac{dx}{2}\right) dy\,dz$$

Similar expressions can be written for the inputs across faces $ABGF$ and $ADEF$ and the outputs across faces $CDEH$ and $CBGH$.

The rate of accumulation of mass within the volume element is

$$\frac{\partial}{\partial t}\left(\rho\, dx\, dy\, dz\right) = \frac{\partial \rho}{\partial t} dx\, dy\, dz \tag{1.24}$$

Application of the law of conservation of mass by requiring the sum of the input rates minus the output rates to equal the rate of accumulation yields

$$\frac{\partial \rho v_x}{\partial x} + \frac{\partial \rho v_y}{\partial y} + \frac{\partial \rho v_z}{\partial z} = -\frac{\partial \rho}{\partial t} \tag{1.25}$$

or, in vector notation,

$$\frac{\partial \rho}{\partial t} + \nabla \cdot (\rho \mathbf{v}) = 0 \tag{1.26}$$

which is the continuity equation of fluid mechanics. For an incompressible liquid, as a ferrofluid, to an excellent approximation, can usually be assumed to be, ρ is a constant and the continuity equation reduces to

$$\nabla \cdot \mathbf{v} = 0 \tag{1.27}$$

This method of derivation, employing a differential Eulerian control volume, is used in Section 1.7 to equate the divergence of any surface stress tensor with the body force density associated with the stress distribution.

A method of deriving the equation of continuity by introducing directly the definition of derivative is found in Bird, Stewart, and Lightfoot (1960).

Eulerian integral formulation: Starting afresh, consider an arbitrarily shaped control volume of macroscopic size fixed in space, as in Figure 1.11. Fluid moves into or out of this volume at points over its surface. An element of the surface, denoted $d\mathbf{S}$, has area given by the magnitude dS of $d\mathbf{S}$ and direction given by the outward normal. The component of the velocity $\mathbf{v}$ parallel to $d\mathbf{S}$ transfers fluid out of V. Hence

$$\text{rate of depletion of mass from } V = \int_S \rho \mathbf{v} \cdot d\mathbf{S} \tag{1.28}$$

Also,

$$\text{total mass in volume } V = \int_V \rho\, dV \tag{1.29}$$

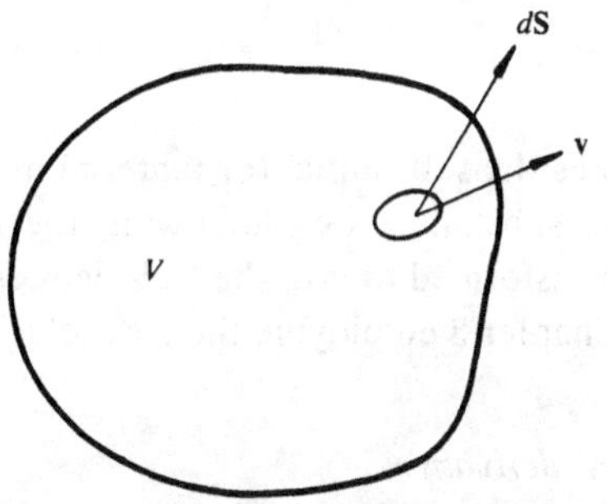

1.11 Eulerian integral control volume.

Hence

$$\frac{d}{dt}\int_V \rho\, dV = \int_V \frac{\partial \rho}{\partial t}\, dV = -\int_S \rho \mathbf{v}\cdot d\mathbf{S} \tag{1.30}$$

The surface element $d\mathbf{S} = \mathbf{n}\, dS$, where $\mathbf{n}$ is the unit outward normal vector. Thus, from Gauss's theorem the surface integral appearing in (1.30) can be transformed to an integral over the contained volume.

$$\int_S \rho \mathbf{v}\cdot \mathbf{n}\, dS = \int_V \nabla\cdot(\rho \mathbf{v})\, dV \tag{1.31}$$

Therefore, combining (1.30) and (1.31) gives

$$\int_V \frac{\partial \rho}{\partial t}\, dV + \int_V \nabla\cdot(\rho \mathbf{v})\, dV = 0 \tag{1.32}$$

Because this relationship must be true no matter the location, shape, or size of the volume, the integral signs may be dropped to give

$$\frac{\partial \rho}{\partial t} + \nabla\cdot(\rho \mathbf{v}) = 0 \tag{1.33}$$

which is the continuity equation obtained previously as (1.26).

Lagrangian formulation: In a Lagrangian formulation the control volume contains a constant mass, and as the volume of fluid flows, its size and shape change. For an integral volume of fluid, the statement of mass conservation is formulated

$$\frac{D}{Dt}\int_V \rho\, dV = 0 \tag{1.34}$$

where D/Dt denotes the substantial (or material or convective) derivative following the substance, i.e., following the mass motion. This equation can be transformed to give the Eulerian result. The procedure is carried out in Chapter 8 employing the Reynolds transport theorem.

Substantial derivative

The nature of the substantial derivative bears further discussion. Let $f = f(x, y, z, t)$ be any scalar field variable such as density or magnetic permeability of the fluid. During the time dt the change in f, from differential calculus, is

$$df = \frac{\partial f}{\partial t}\,dt + \frac{\partial f}{\partial x}\,dx + \frac{\partial f}{\partial y}\,dy + \frac{\partial f}{\partial z}\,dz$$

where the various derivatives are expressed in terms of Eulerian coordinates; e.g., $\partial/\partial t$ is evaluated at constant x, y, z and $\partial/\partial x$ at constant t, y, z.

The change df in the Lagrangian framework may be divided by dt to give

$$\frac{df}{dt} = \frac{\partial f}{\partial t} + \frac{\partial f}{\partial x}\frac{dx}{dt} + \frac{\partial f}{\partial y}\frac{dy}{dt} + \frac{\partial f}{\partial z}\frac{dz}{dt}$$

Now, f could have been detected by an observer moving at any velocity. As a special case of great importance, if the variation of f is noted by an observer traveling with the fluid, then dx/dt is the velocity component v_x of the fluid, $dy/dt = v_y$, and $dz/dt = v_z$. For this special case the time derivative d/dt is denoted D/Dt. Thus, in vector form the equation can be written

$$\frac{Df}{Dt} = \frac{\partial f}{\partial t} + \mathbf{v}\cdot\nabla f = \left(\frac{\partial}{\partial t} + \mathbf{v}\cdot\nabla\right)f \tag{1.35}$$

so that

$$\frac{D}{Dt} = \frac{\partial}{\partial t} + \mathbf{v}\cdot\nabla \tag{1.36}$$

is an operator representing the substantial derivative (follows the substance). The term $\mathbf{v}\cdot\nabla f$ expresses the fact that, in a time-independent flow field in which the fluid properties depend only upon the

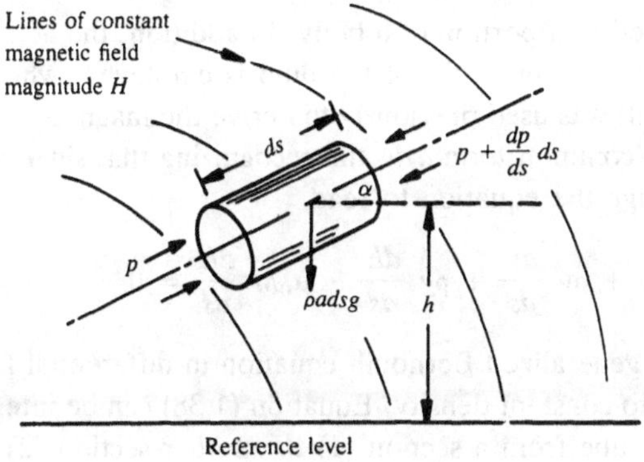

1.12 Derivation of the generalized Bernoulli equation.

spatial coordinates, there is a change in f due to the changing spatial position of a given fluid element as it flows. The term $\partial f/\partial t$ is the familiar Eulerian derivative evaluated at a fixed point in space. Equation (1.35) expresses the Lagrangian rate of change for a given fluid element Df/Dt in terms of the Eulerian derivatives on the right side of the equation.

1.6 **Generalized Bernoulli equation**

An augmented form of the well-known Bernoulli equation that includes the effect of magnetic forces can now be derived. The result applies to the idealized case for the flow of an inviscid or frictionless fluid along a stream tube. Although the derivation at this stage is lacking in rigor and the result is incomplete, the insight obtained makes the effort worthwhile.

With reference to Figure 1.12, let s be distance measured along the direction of the stream tube, v the speed of a fluid element, and h the height measured vertically above a fixed reference level. Consider an element of fluid of instantaneous length ds and of mass $\rho a\, ds$ moving along the tube. For *steady flow* ($\partial/\partial t = 0$), by (1.35) with $f = v$, $x = s$, and $dx/dt = v$, the acceleration of the element is $v\, dv/ds$, and an equation of motion expressing Newton's law may be formulated:

$$\underset{\text{mass}\times\text{acceleration}}{\rho a\, ds\, v \frac{dv}{ds}} = \underset{\text{net force due to pressure}}{-a \frac{dp}{ds} ds} - \underset{\text{force due to gravity}}{\rho a\, ds\, g \sin\alpha} + \underset{\text{force due to magnetism}}{\mu_0 M \frac{dH}{ds} a\, ds} \tag{1.37}$$

An uncertainty in this relationship is the use of the expression for magnetic force, which was derived for a whole body but which is now being applied to a portion of a body. In addition, the actual field H as influenced by the presence of the fluid is employed, whereas H_0, the applied field, was used previously to derive the magnetic force. Factoring out the common term $a\,ds$ and recognizing that $\sin\alpha = dh/ds$, one can rearrange the equation to read

$$\frac{dp}{ds} + \rho v \frac{dv}{ds} + \rho g \frac{dh}{ds} - \mu_0 M \frac{dH}{ds} = 0 \tag{1.38}$$

This is the generalized Bernoulli equation in differential form without restriction to constant density. Equation (1.38) can be integrated along the stream tube from a section (1) to another section (2) to give

$$\int_1^2 \frac{dp}{\rho} + \frac{v_2^2 - v_1^2}{2} + g(h_2 - h_1) - \mu_0 \int_1^2 \frac{M}{\rho}\, dH = 0 \tag{1.39}$$

With ferrofluids taken to be incompressible, ρ is constant, and the augmented Bernoulli equation can be expressed

$$p_1 + \rho\frac{v_1^2}{2} + \rho g h_1 - \mu_0 \int_0^{H_1} M\,dH$$

$$= p_2 + \rho\frac{v_2^2}{2} + \rho g h_2 - \mu_0 \int_0^{H_2} M\,dH \tag{1.40}$$

The dimensions of each term are energy per unit volume (i.e., $N \cdot m \cdot m^{-3}$) or pressure.*

It might at first be thought that the Bernoulli equation would be of very limited application because of the restriction to situations where viscous effects are relatively unimportant. However, for nonmagnetic flows it has surprisingly wide application in, for example, the theory of flow-measuring devices – such as the venturi tube, the sharp edge orifice, and the pitot tube – and determining the reaction force on a horizontal pipe bend or in flow past a cascade of vanes. When, in addition, it is considered that each combination of the magnetic term with one of the ordinary terms in the Bernoulli equation describes a different category of fluid flow and response, a glimpse can be obtained of the richness that magnetics brings to hydrodynamics.

The simplest example: The inset to Figure 1.1 illustrates a cylindrical transparent tube containing a layer of ferrofluid located

*μ_0 has numerically equivalent units of henry per meter ($H \cdot m^{-1}$), tesla-meter per ampere ($T \cdot m \cdot A^{-1}$), newtons per square ampere ($N \cdot A^{-2}$) or $kg \cdot m \cdot s^{-2} \cdot A^{-2}$, and $m^2 \cdot T^2 \cdot N^{-1}$.

between regions of immiscible, transparent fluid having nearly the same mass density. Permanent magnets brought up to opposite sides of the tube position the fluid and shape both the top and bottom interfaces. Analyze the meniscus shape in terms of the generalized Bernoulli equation.

Analysis: Choose any two points on a given meniscus and call them points 1 and 2. The Bernoulli relationship, equation (1.40), can be applied to the liquid along either side of the meniscus. On the transparent liquid side,

$$\begin{aligned} v_1 &= v_2 = 0 \\ M_1 &= M_2 = 0 \\ p_1 + \rho g h_1 &= p_2 + \rho g h_2 \end{aligned} \tag{1.41}$$

On the magnetic liquid side,

$$v_1 = v_2 = 0$$

$$p_1 + \rho g h_1 - \mu_0 \int_0^{H_1} M\,dH = p_2 + \rho g h_2 - \mu_0 \int_0^{H_2} M\,dH \tag{1.42}$$

Subtracting (1.41) from (1.42) and rearranging gives

$$\int_{H_1}^{H_2} M\,dH = 0 \tag{1.43}$$

Since the magnetization magnitude M acquires only positive values, (1.43) can be valid only if the field values are equal:

$$H_1 = H_2 \tag{1.44}$$

To the approximation studied here, this indicates that the meniscus shape maps a contour of constant magnetic field magnitude. The ferrofluid behaves as a novel gaussmeter.

An interaction that is consistent with the Bernoulli relationship but not predicted by it is illustrated in Figure 1.13. A spherical droplet of magnetic fluid subjected to a uniform applied magnetic field elongates along the direction of the field. Understanding this phenomenon requires the more advanced treatment of Chapter 5, which considers the magnetic and surface interfacial force densities arising at a surface of discontinuity.

1.7 Stress tensor and its physical meaning

The notion of stress and its tensorial description arise in magnetic as well as hydrodynamic studies. Consider an arbitrary volume V

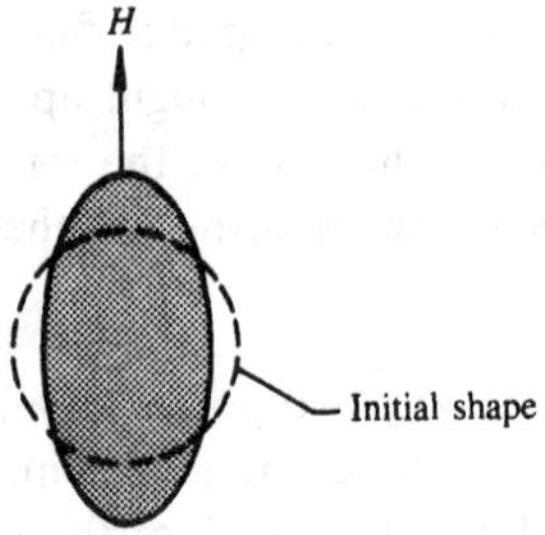

1.13 A ferrofluid droplet exposed to a uniform magnetic field elongates.

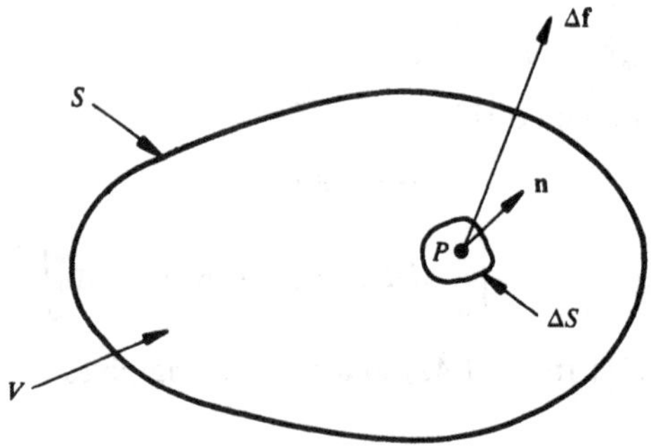

1.14 Definition of the stress vector.

enclosed by the surface S (Figure 1.14). Let $\Delta\mathbf{f}$ be the resultant force exerted across ΔS by material outside on matter just inside. Clearly the $\Delta\mathbf{f}$ will depend both on the choice of ΔS and on $\mathbf{n}$, the outward-pointing unit normal to ΔS. The average surface or contact force per unit area on ΔS is given by $\Delta\mathbf{f}/\Delta S$. According to the *Cauchy stress principle*, this ratio tends to a definite limit $d\mathbf{f}/dS$ as $\Delta S \to 0$ at the point P located by a positional vector $\mathbf{r}$. This represents the continuum point of view usual in mechanics and corresponds to using an expectation value. The quantity $d\mathbf{f}/dS$ is called the stress vector $\mathbf{t}_n(\mathbf{r})$:

$$\lim_{\Delta S \to 0} \frac{\Delta\mathbf{f}}{\Delta S} = \frac{d\mathbf{f}}{dS} \equiv \mathbf{t}_n(\mathbf{r}) \tag{1.45}$$

It should be kept in mind that the stress vector acting on a surface element varies from point to point in the fluid and also depends on the orientation of the element. The stress vector is often called the *traction vector*.

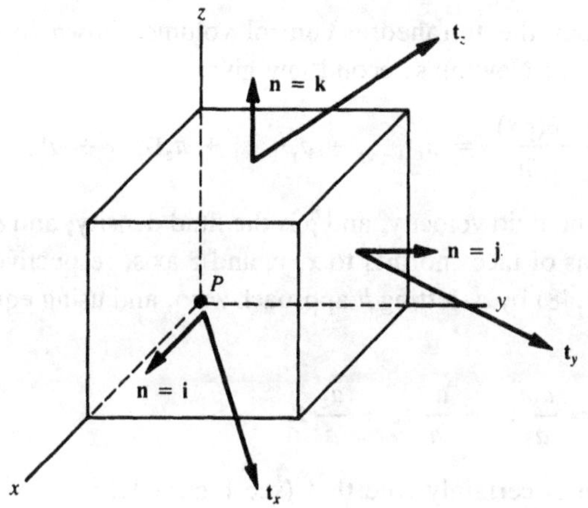

1.15 Cube to introduce the definition of the stress components.

In order to describe the state of stress at a given point P, it suffices to know the traction vector on three mutually perpendicular planes at P.* If these three planes are the coordinate planes in a rectangular Cartesian coordinate system, then reference to Figure 1.15 shows that

$$\mathbf{t}_x = T_{xx}\mathbf{i} + T_{xy}\mathbf{j} + T_{xz}\mathbf{k} \tag{1.46a}$$
$$\mathbf{t}_y = T_{yx}\mathbf{i} + T_{yy}\mathbf{j} + T_{yz}\mathbf{k} \tag{1.46b}$$
$$\mathbf{t}_z = T_{zx}\mathbf{i} + T_{zy}\mathbf{j} + T_{zz}\mathbf{k} \tag{1.46c}$$

where T_{xx} is the x component of the stress vector acting on an element of surface whose unit normal is $\mathbf{i}$, T_{yx} is the x component of the stress vector acting on an element of surface whose unit normal is $\mathbf{j}$, and so on. The general rule is that the component T_{ij} acts in the direction of the jth coordinate axis and on the plane whose outward normal is parallel to the ith coordinate axis.

The traction $\mathbf{t}_n$ as defined above is force exerted per unit area on matter on the $-\mathbf{n}$ side of a surface element by material on the $+\mathbf{n}$ side of the surface. Thus $\mathbf{t}_{(-n)}$ is the force exerted on matter on the $+\mathbf{n}$ side by material on the $-\mathbf{n}$ side. From Newton's third law or from application of Newton's second law to a surface element dS (of negligible mass) it follows that

$$\mathbf{t}_n = -\mathbf{t}_{(-n)} \tag{1.47}$$

*This assertion is justified in Section 8.2.

Now consider the tetrahedral control volume shown in Figure 1.16. Application of Newton's second law gives

$$\frac{ha}{3}\frac{\partial(\rho\mathbf{v})}{\partial t} = a_x\mathbf{t}_{(-x)} + a_y\mathbf{t}_{(-y)} + a_z\mathbf{t}_{(-z)} + a\mathbf{t}_{\mathrm{n}} \tag{1.48}$$

where $\mathbf{v}$ is the fluid velocity, and ρ is the fluid density, and a_x, a_y, and a_z are the areas of faces normal to x, y, and z axis, respectively. Dividing equation (1.48) by a, letting h approach zero, and using equation (1.47) gives

$$\mathbf{t}_{\mathrm{n}} = \frac{a_x}{a}\mathbf{t}_x + \frac{a_y}{a}\mathbf{t}_y + \frac{a_z}{a}\mathbf{t}_z \tag{1.49}$$

However, it is certainly true that (see Figure 1.16)

$$\frac{a_x}{a} = \mathbf{n}\cdot\mathbf{i}, \qquad \frac{a_y}{a} = \mathbf{n}\cdot\mathbf{j}, \qquad \frac{a_z}{a} = \mathbf{n}\cdot\mathbf{k} \tag{1.50}$$

and (1.49) becomes

$$\mathbf{t}_{\mathrm{n}} = \mathbf{n}\cdot(\mathbf{i}\mathbf{t}_x + \mathbf{j}\mathbf{t}_y + \mathbf{k}\mathbf{t}_z) \tag{1.51}$$

The expression within the parenthesis is a dyadic and it is called the *stress dyadic* or *stress tensor* $\mathbf{T}$:

$$\mathbf{T} = \mathbf{i}\mathbf{t}_x + \mathbf{j}\mathbf{t}_y + \mathbf{k}\mathbf{t}_z \tag{1.52}$$

The stress tensor may be written out by substituting (1.46) into (1.52) to give

$$\begin{aligned}\mathbf{T} = {} & T_{xx}\mathbf{ii} + T_{xy}\mathbf{ij} + T_{xz}\mathbf{ik} + T_{yx}\mathbf{ji} + T_{yy}\mathbf{jj} + T_{yz}\mathbf{jk} \\ & + T_{zx}\mathbf{ki} + T_{zy}\mathbf{kj} + T_{zz}\mathbf{kk}\end{aligned} \tag{1.53}$$

Thus, the state of stress at a point is not a hopelessly complicated concept but depends on just nine numbers.

Introducing $\mathbf{T}$ into (1.51) gives the compact relationship for the stress vector in symbolic notation:*

$$\mathbf{t}_{\mathrm{n}} = \mathbf{n}\cdot\mathbf{T} \tag{1.54}$$

It is verified below for any tensor $\mathbf{T}$ that $\mathbf{n}\cdot\mathbf{T} = \mathbf{T}^{\mathrm{T}}\cdot\mathbf{n}$, where $\mathbf{T}^{\mathrm{T}}$ is the transpose of $\mathbf{T}$.

$\mathbf{n}\ \mathbf{T} = \mathbf{T}^{\mathrm{T}}\cdot\mathbf{n}$	poses the question in symbolic notation
$n_k\mathbf{e}_k\cdot T_{ij}\mathbf{e}_i\mathbf{e}_j = T_{ji}\mathbf{e}_i\mathbf{e}_j\cdot n_k\mathbf{e}_k$	rewrite in indicial notation

*The symbolic notation is summarized in Appendix 1.

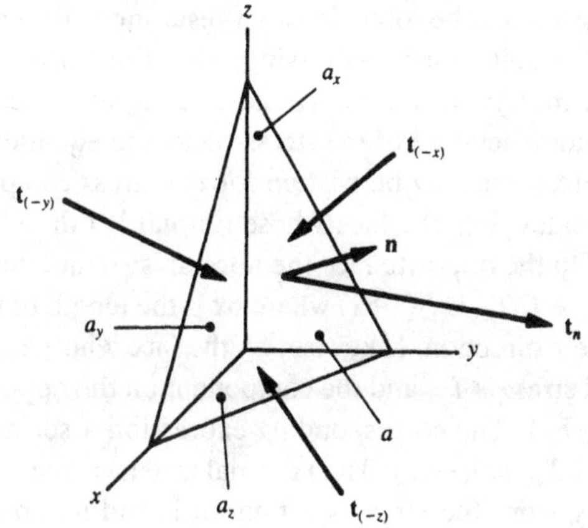

1.16 Tetrahedron used in the derivation of the stress dyadic.

$n_i T_{ij} \mathbf{e}_j = T_{ji} \mathbf{e}_i n_j$	disappearance of the orthogonal vectors
$n_i T_{ij} \mathbf{e}_j = T_{ij} \mathbf{e}_j n_i$	rearrangement of dummy subscripts
$\mathbf{n} \cdot \mathbf{T} = \mathbf{T}^{\mathrm{T}} \cdot \mathbf{n}$	QED

Note that only when the tensor is symmetric is it permissible to write $\mathbf{n} \cdot \mathbf{T} = \mathbf{T} \cdot \mathbf{n}$.

Force resulting from a stress tensor

As shown in equation (1.54), the traction vector $\mathbf{t}_n$ on any surface element dS is obtained by dotting the unit normal $\mathbf{n}$ to dS into the stress tensor $\mathbf{T}$ to give $\mathbf{n} \cdot \mathbf{T}$. Summing the traction vectors acting on differential elements of the surface area over the whole surface of the body gives the net surface force acting on the body:

$$\mathbf{F} = \oint_S \mathbf{t}_n \, dS = \oint_S \mathbf{n} \cdot \mathbf{T} \, dS \tag{1.55}$$

From the divergence theorem for a tensor, the net force on the body is also given by the following expression, which results from transforming the rightmost integral appearing in (1.55):

$$\mathbf{F} = \int_V \nabla \cdot \mathbf{T} \, dV \tag{1.56}$$

Thus the integrand $\nabla \cdot \mathbf{T}$ can be taken to represent the force density, or force per unit volume $\mathbf{f}$, which is compactly represented as

$$\mathbf{f} = \nabla \cdot \mathbf{T} \tag{1.57}$$

This expression for the force density **f** resulting from a stress tensor is an important result worth rederiving, this time using the Eulerian differential formulation. Refer once more to the stress "cube" of Figure 1.15 and the nomenclature of the stress vectors in equations (1.46). The following expressions may be written for the stress components on the various cube sides: For the face whose normal is **i** the x component of stress is T_{xx}. On the opposite face the normal is $-\mathbf{i}$ and the x component of stress is $T_{xx} + (\partial T_{xx}/\partial x)(-\delta x)$ where δx is the length of the side of the cube along the x direction. Likewise, on the face whose normal is **j** the x component of stress is T_{yx} and the component on the opposite face is $T_{yx} + (\partial T_{yx}/\partial y)(-\delta y)$. The corresponding expression associated with the **k** face is $T_{zx} + (\partial T_{zx}/\partial z)(-\delta z)$. The **i** normal face has area $\delta y\, \delta z$ and so the force resulting from the stresses acting on it and its opposing face is

$$\left\{T_{xx} - \left[T_{xx} + \frac{\partial T_{xx}}{\partial x}\left(-\delta x\right)\right]\right\}\delta y\, \delta z = \frac{\partial T_{xx}}{\partial x}\delta V$$

where $\delta V = \delta x\, \delta y\, \delta z$ is the volume of the fluid cube. The corresponding contributions resulting from the **j** and **k** normal faces are $(\partial T_{yx}/\partial y)\, \delta V$ and $(\partial T_{zx}/\partial z)\, \delta V$, respectively. Thus, the sum of the x-direction forces acting over the entire surface is represented as $\mathbf{i}(\partial T_{xx}/\partial x + \partial T_{yx}/\partial y + \partial T_{zx}/\partial z)\, \delta V$.

When similar expressions are developed for y and z directions, the result is that the vector force density **f** acting on the cube is given by

$$\begin{aligned}\mathbf{f} = &\left(\frac{\partial T_{xx}}{\partial x} + \frac{\partial T_{yx}}{\partial y} + \frac{\partial T_{zx}}{\partial z}\right)\mathbf{i} \\ &+ \left(\frac{\partial T_{xy}}{\partial x} + \frac{\partial T_{yy}}{\partial y} + \frac{\partial T_{zy}}{\partial z}\right)\mathbf{j} \\ &+ \left(\frac{\partial T_{xz}}{\partial x} + \frac{\partial T_{yz}}{\partial y} + \frac{\partial T_{zz}}{\partial z}\right)\mathbf{k}\end{aligned} \tag{1.58}$$

The right side of this equation is precisely the Cartesian equivalent of the right side of (1.57), giving again $\mathbf{f} = \nabla \cdot \mathbf{T}$.

Addendum: equivalence of dipolar and polar representations

The treatment in this chapter implies that the concept of magnetic poles has utility and validity, which indeed is true. Now it is desired to illustrate a formal mathematical equivalence that exists between the dipolar description of magnetized matter and the polar description. The relationship can be established in the following manner. From (1.10), giving the magnetic force density as $\mu_0(\mathbf{M}\cdot\nabla)\mathbf{H}_0$, and the vector identity $\nabla\cdot(\mathbf{M}\mathbf{H}_0) = \mathbf{H}_0(\nabla\cdot\mathbf{M}) + (\mathbf{M}\cdot\nabla)\mathbf{H}_0$, the net force acting on a whole magnetized body becomes

$$\mathbf{f} = \int_V \mu_0(\mathbf{M}\cdot\nabla)\mathbf{H}_0\,dV$$
$$= \int_V \nabla\cdot(\mu_0\mathbf{M}\mathbf{H}_0)\,dV + \int_V (-\mu_0\nabla\cdot\mathbf{M})\mathbf{H}_0\,dV \tag{1.59}$$

The first term on the right side, with the aid of the divergence theorem, can be rewritten

$$\int_V \nabla\cdot(\mu_0\mathbf{M}\mathbf{H}_0)\,dV = \oint_S \mathbf{n}\cdot\mu_0\mathbf{M}\mathbf{H}_0\,dS = \oint_S \mu_0(\mathbf{n}\cdot\mathbf{M})\mathbf{H}_0\,dS \tag{1.60}$$

Combining these relationships yields the transformed force expression:

$$\mathbf{f} = \int_V \mu_0(\mathbf{M}\cdot\nabla)\mathbf{H}_0\,dV$$
$$= \oint_S (\mu_0\mathbf{M}\cdot\mathbf{n})\mathbf{H}_0\,dS + \int_V (-\mu_0\nabla\cdot\mathbf{M})\mathbf{H}_0\,dV \tag{1.61}$$

Because $\mathbf{H}_0$ is the force per unit pole and $\mu_0\mathbf{M}\cdot\mathbf{n}$ and $-\mu_0\nabla\cdot\mathbf{M}$ have units of poles per unit area and poles per unit volume, respectively, in view of (1.61) it is appropriate to make the association that

$$\rho_S = \mu_0\mathbf{M}\cdot\mathbf{n} \quad \equiv \text{surface density of poles} \tag{1.62}$$
$$\rho_V = -\mu_0\nabla\cdot\mathbf{M} \equiv \text{volume density of poles} \tag{1.63}$$

which establishes the equivalence. This result was stated by Brown (1951), who also gives corresponding equations for the torque on a whole body. A wealth of related theorems, including an extension to quadrupolar matter, is given by Dahler and Scriven (1963).

Comments and supplemental references

The topic of ferrohydrodynamics (FHD) was introduced by

Neuringer and Rosensweig (1964)

Overviews of ferrohydrodynamics are presented by

Shaposhnikov and Shliomis (1975)

Rosensweig (1982a)

Numerous applications are described by

Rosensweig (1979a)

Electrohydrodynamics is the topic of an excellent review authored by

Melcher and Taylor (1969)

and suggested introductory works dealing with MHD are those of

Hughes and Young (1966)

Cabannes (1970)

Modern electromagnetics texts develop the concept of electric charge and dipoles but neglect to treat magnetics on an equal footing. The treatment of magnetic phenomena in terms of the pole description has fallen into disfavor relative to models of current loops. An older, well written text developed around the pole model is that of

Bitter (1937)

Recommended introductory texts covering the broader field of electromagnetism are those by

Zahn (1979)

Reitz, Milford, and Christy (1979)

Every student of magnetics should be acquainted with the classical work of

Bozorth (1951)

which contains a great deal of data concerning magnetic materials available up to that time, as well as informative discussions of magnetic theory and behavior.

Many good texts on fluid mechanics are available. An authoritative introduction is given by

Batchelor (1970)

and a book with topics chosen to provide insight into the behavior of fluids in motion is that of

Tritton (1977)

For an elementary but clear account of stress in a continuum see

Mase (1970), Chapter 2

A lucid discussion of surface stress and its historical contribution as a conceptual aid to logical thinking is given by

Serrin (1959)

2

MAGNETIC FLUIDS

Theory admits the possibility of ferromagnetism in homogeneous liquid, but no substance is known whose Curie point exceeds its melting point.

A magnetic ferrofluid consists of a stable colloidal dispersion of subdomain magnetic particles in a liquid carrier. The properties of the ferrofluid are profoundly affected by the thermal Brownian motion of the suspended particles and the circumstance that each subdomain particle is permanently magnetized. In addition to the particles and the carrier liquid, the third essential ingredient of a colloidal ferrofluid is the presence of an adsorbed long-chain molecular species on the particle surface that prevents agglomeration of the particles to each other.

Stability as a colloid is an important property that ensures the investigator of a well-defined material suitable for fluid applications as well as for scientific studies. Early interest in magnetic colloids appears in the literature, but stabilization was not achieved with concentrated colloids, and the maximum magnetizations were smaller than in ferrofluids by several orders of magnitude. Thus, Elmore (1938a) developed a chemical method of preparing a magnetic colloid that was used to visualize the magnetic domains at the surface of a ferromagnetic solid by the technique of Bitter (1931). The method depends on the particles concentrating near the magnetic domain walls, where the field is strongest, rather than maintaining their homogeneity of dispersion throughout the liquid carrier.

The magnetic ferrofluids of the type in general use today are an outgrowth of discoveries made in the early 1960s.

2.1 Stability requirements

It is instructive to consider the simple physics of certain mechanisms that are responsible for the existence of the ferrofluids. Dimensional reasoning may be used to arrive at criteria for physicochemical stability. To begin it is useful to write expressions for various energy terms. These energies per particle are

$$\begin{aligned} \text{thermal energy} &= kT \\ \text{magnetic energy} &= \mu_0 MHV \\ \text{gravitational energy} &= \Delta\rho\, VgL \end{aligned}$$

where k is Boltzmann's constant and equals $1.38 \times 10^{-23}\ \mathrm{N\cdot m\cdot K^{-1}}$, T is the absolute temperature in degrees Kelvin, μ_0 is the permeability of free space and has the value $4\pi \times 10^{-7}\ \mathrm{H\cdot m}$, volume $V = \pi d^3/6m^3$ for a spherical particle of diameter d, and L is the elevation in the gravitational field. Ratios of one term to another yield dimensionless quantities that can alter the stability of the ferrofluids, as will be seen.

Stability in a magnetic-field gradient

Consider the stability against settling of particles in a field gradient due to an external magnetic source. Particles are attracted to the higher-intensity regions of a magnetic field, while thermal motion counteracts the field force and provides statistical motions that allow the particle to sample all portions of the fluid volume. The magnetic energy $\mu_0 MHV$ represents the reversible work in removing a magnetized particle from a point in the fluid, where the field is H, to a point in the fluid that is outside the field:

$$W = -\int_H^0 \left(\mu_0 M \frac{dH}{ds} V \right) ds \approx \mu_0 MHV$$

Provided some part of the fluid volume is located in a field-free region, then stability against segregation is favored by a high ratio of the thermal energy to the magnetic energy:

$$\frac{\text{thermal energy}}{\text{magnetic energy}} = \frac{kT}{\mu_0 MHV} \geq 1 \tag{2.1}$$

Rearranging and substituting for the volume of a sphere gives an expression for the maximum particle size:

$$d \leq (6kT/\pi\mu_0 MH)^{1/3} \tag{2.2}$$

Consider the conditions existing in a beaker of magnetic fluid containing

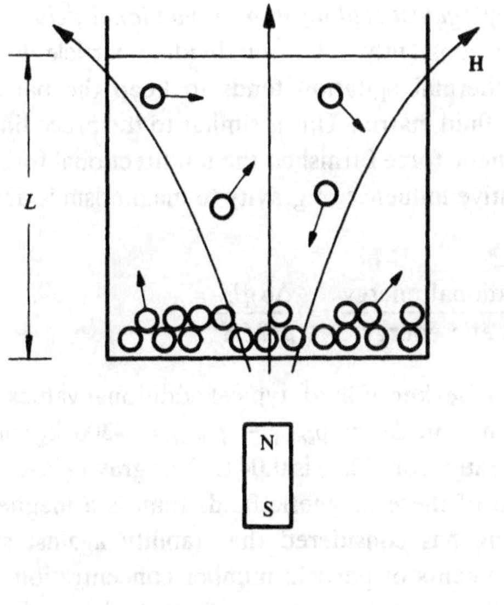

2.1 The concentrating of magnetic particles in a ferrofluid subjected to a gradient magnetic field is limited by diffusion due to particle thermal motion and steric hindrance due to finite particle size.

magnetite (Fe_3O_4) particles subject to the magnetic gradient field of a typical hand-held permanent magnet:

$$H = 8 \times 10^4 \text{ A} \cdot \text{m}^{-1}$$
$$M = 4.46 \times 10^5 \text{ A} \cdot \text{m}^{-1}$$
$$T = 298 \text{ K}$$

The value of H stated here in SI units corresponds to an induction field in air of 1,000 G in cgs units; the value of M given corresponds to 5,600 G. The particle size, computed from (2.2), is $d \leqslant 8.1 \times 10^{-9}$ m or 8.1 nm. Actual particle sizes of stable colloids range up to about 10 nm. From the calculation it is seen that particle size could scarcely be any larger and remain stable against segregation in a field gradient.

Another physical feature limits the concentration of particles in regions of more intense magnetic fields. As seen from Figure 2.1, *steric hindrance* puts an upper limit on the particle number concentration. Although concentration gradients can be established in situations like this, when the field is removed particles of a well-stabilized ferrofluid spontaneously redistribute throughout the fluid volume over a period of time.

Stability against settling in a gravitational field

Gravity constantly pulls an individual particle downward in the beaker while thermal agitation tends to keep the particle dispersed throughout the fluid matrix. This is similar to the preceding situation, in which the magnetic force furnished the unidirectional force at any given point. The relative influence of gravity to magnetism is described by the ratio

$$\frac{\text{gravitational energy}}{\text{magnetic energy}} = \frac{\Delta\rho gL}{\mu_0 MH} \tag{2.3}$$

Again for a little beaker of fluid, typical additional values of parameters are $L = 0.05$ m and $\Delta\rho = \rho_{\text{solid}} - \rho_{\text{fluid}} = 4300\ \text{kg}\cdot\text{m}^{-3}$; with $g = 9.8\ \text{m}\cdot\text{s}^{-2}$, the ratio from (2.3) is 0.047. Thus gravity is less of a threat to the segregation of these magnetic fluids than is a magnetic field.

The foregoing has considered the stability against segregating or forming big gradients of particle number concentration. The particles are assumed to be monodisperse, or not attached to each other. It is not a trivial task to achieve monodispersion, and this topic is examined next.

Stability against magnetic agglomeration

A typical colloidal magnetic fluid contains on the order of 10^{23} particles per cubic meter, and collisions between particles are frequent. Hence, if the particles adhere together, agglomeration will be rapid. Each particle is permanently magnetized, so the maximum energy to separate a pair of particles of diameter d is needed when the particles are aligned. This energy, given by (1.20) with $\mathbf{m}_1 \cdot \mathbf{m}_2 = m^2$, $(\mathbf{m}_1 \cdot \mathbf{r})(\mathbf{m}_2 \cdot \mathbf{r}) = m^2 r^2$, and $m = \mu_0 M \pi d^3/6$, is the dipole–dipole pair energy E_{dd}:

$$E_{\text{dd}} = \frac{\pi}{9} \frac{\mu_0 M^2 d^3}{(l + 2)^3} \tag{2.4}$$

where $l = 2s/d$, with s the surface-to-surface separation distance. Thus, when particles are in contact, (2.4) reduces to the following form, giving the dipole–dipole contact energy:

$$E_{\text{dd}} = \tfrac{1}{12} \mu_0 M^2 V \qquad (l = 0) \tag{2.5}$$

Again thermal agitation is available to disrupt the agglomerates, with the effectiveness of the disruption governed by the ratio

$$\frac{\text{thermal energy}}{\text{dipole-dipole contact energy}} = \frac{12kT}{\mu_0 M^2 V} \tag{2.6}$$

Accordingly, for particles to escape agglomeration, this ratio must be greater than unity, so the particle size is given by

$$d \leq (72kT/\pi\mu_0 M^2)^{1/3} \tag{2.7}$$

For the magnetite particles at room temperature, (2.7) requires $d \leq 7.8$ nm. This estimate shows that normal magnetic fluids having particle size in the range up to 10 nm are on the threshold of agglomerating but manage to escape this fate. However, there is an additional problem to be overcome, as discussed next.

Necessity to guard against the van der Waals attractive force

The van der Waals forces arise spontaneously between neutral particles because of the fluctuating electric dipole–dipole forces that are always present. The force represents the quantum-mechanical interaction due to fluctuating orbital electrons in one particle inducing oscillating dipoles in the other. Because the dipolar field varies as r^{-3}, the field gradient is r^{-4}. If the polarization is proportional to the field, and hence to r^{-3}, then the force varies as r^{-7} and the energy as r^{-6}. It turns out that the *London model* similarly predicts an inverse sixth power law between point particles. Hamaker extended the theory to apply to equal spheres and obtained the following expression:

$$\begin{pmatrix}\text{dipole fluctuation}\\ \text{energy}\end{pmatrix} = -\frac{A}{6}\left[\frac{2}{l^2+4l} + \frac{2}{(l+2)^2} + \ln\frac{l^2+4l}{(l+2)^2}\right] \tag{2.8}$$

A, the *Hamaker constant*, is roughly calculable from the ultraviolet (or optical) dielectric properties of particles and the medium. For Fe, Fe_2O_3, or Fe_3O_4 in hydrocarbon, $A = 10^{-19}$ N·m is representative (within a factor of 3). Note that the Hamaker energy is proportional to l^{-1} for close spheres and l^{-6} for distant spheres. The l^{-6} dependence is in agreement with the London result, as it must be. Reduction of (2.8) for large separation is a mathematical exercise in retaining higher-order terms in expansions of the functions. It should be noted especially that the l^{-1} dependence indicates that infinite energy is required to separate a particle pair, and it will be recalled that the magnetic pair energy was finite for contacting pairs. Thus prevention of contact is another necessity if a stable colloid is to be obtained. Before discussing how this may be accomplished, it will be helpful to consider preparative techniques in some detail.

Exercise: By expanding the logarithmic term in (2.8) in an appropriate way as $l \to \infty$, show that the dipole energy goes as l^{-6}. [Hint: Use the Taylor series expansion

$$\ln(1 + x) = x - \frac{x^2}{2} + \frac{x^3}{3} - \frac{x^4}{4} + \cdots$$

valid for $x \ll 1$.]

2.2 Preparation of magnetic colloids by size reduction

There are two broad methods of preparing a magnetic colloid – size reduction and precipitation, i.e., making little particles out of big ones and producing little ones from solution initially. It is a remarkable fact that size reduction by grinding can succeed in reducing bulk (micron-size) material to the order of 100 Å in size. The method was discovered by S. Papell (1965), yet the basic materials and equipment have been available since antiquity. The "secret" of the process is to wet grind in the presence of a surfactant (grinding aid) for long periods of time (~1000 h). A flowchart of the process is shown in Figure 2.2. Magnetic powder (most commonly magnetite) of several microns in size is mixed with carrier solvent and a surfactant dispersing agent. About 10–20 vol % of dispersing agent, based on solvent volume, and 0.2 $kg \cdot l^{-1}$ of magnetite may be used. The proportion of dispersant to solid corresponds closely to a monolayer coating on the particles in the product, in that it is found that an average particle size of about 10 nm is automatically produced in the process as found by Rosensweig, Nestor, and Timmins (1965). They also developed the concept of the monolayer coating, principles for selection of a dispersant, definitive measurement of the colloid particle size, models for viscosity coefficient, and factors affecting the maximum concentration of solids. Near-quantitative conversion of the feed solids is possible, although the net rate of production is increased by earlier removal of oversize solids and starting a new run. Rosensweig and Kaiser (1967) further developed the process and succeeded in dispersing the magnetic particles into other families of solvents, including water, hydrocarbons, aromatics, and esters. Through systematic study and application of the physicochemical principles, the saturation magnetization of fluids was increased tenfold (an increase in $\mu_0 M$ from 0.01 T to 0.1 T).

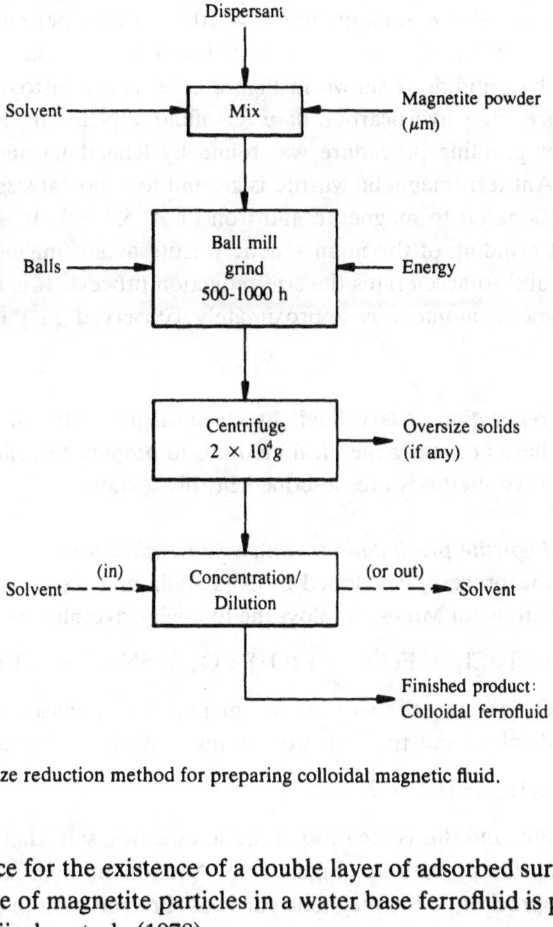

2.2 Size reduction method for preparing colloidal magnetic fluid.

Evidence for the existence of a double layer of adsorbed surfactant at the surface of magnetite particles in a water base ferrofluid is presented by Shimoiizaka et al. (1978).

The number concentration n of particles in a ferrofluid can be enormous. Letting ϕ denote the volume fraction of magnetic solid and d the particle diameter, as previously defined, gives $n = 6\phi/\pi d^3$. The saturation magnetization M_s of the dispersion is proportional to the domain magnetization M_d of the solid particles by $M_s = \phi M_d$. Eliminating ϕ between these expressions gives for the number concentration

$$n = \frac{6}{\pi} \frac{M_s}{M_d} \frac{1}{d^3} \tag{2.9}$$

For a ferrofluid having magnetic saturation $\mu_0 M_s = 0.02$ T containing magnetite particles of 10-nm diameter with $\mu_0 M_d = 0.56$ T, the calculated value of n amounts to 6.8×10^{22} particles per cubic meter.

An electron micrograph of the particles in a water base ferrofluid prepared by grinding is shown in Figure 2.3a, and a histogram of the particle sizes in a hydrocarbon base ferrofluid appears in Figure 2.3b.

A faster grinding procedure was found by Khalafalla and Reimers (1973b). Antiferromagnetic wustite is ground to colloidal size and then disproportionated to magnetite and iron ($T < 530\,°C$). It is hypothesized that grinding of the nonmagnetic wustite avoids magnetic agglomeration and so accelerates the size-reduction process. It is found that total magnetic moment is approximately conserved in the grinding process.

2.3 Preparation of ferrofluids by chemical precipitation

There are many chemical methods to prepare ferrofluids. Four representative methods are described in this section.

Magnetite precipitation with steric stabilization

One process, developed by Khalafalla and Reimers (1974) at the U.S. Bureau of Mines, employs the following overall stoichiometry:

$$5NaOH + 2FeCl_3 + FeCl_2 = FeO \cdot Fe_2O_3 + 5NaCl + 4H_2O \quad (2.10a)$$

$FeO \cdot Fe_2O_3$ is magnetite written in a manner to emphasize the occurrence of divalent and trivalent iron atoms in it. Thus the ratio

$$Fe(III)/Fe(II) = 2 \quad (2.10b)$$

in magnetite, and this is the ratio of the soluble iron salts that are fed as reactants to the process (see Figure 2.4). Ammonium hydroxide can be used to coprecipitate the magnetic particles. Precipitation of iron oxide is a common procedure in wet-chemical analysis. The usual aim is to create conditions under which the precipitated particles will agglomerate to permit filtration for gravimetric analysis. In this process, however, the objective is just the opposite, as it is desired to maintain the particle size in the small-colloidal range. To achieve this goal, a

2.3 (a) Electron micrograph of Fe_3O_4 magnetic particles prepared by grinding in aqueous carrier fluid. (*Ferrofluid A01 Ferrofluidics Corporation. Photo Courtesy of Exxon Research and Engineering Company.*) (b) Particle-size histogram with Gaussian fit (solid line). Open circles represent the number of particles. The bars in each interval represent the statistical uncertainty in sampling. (*From McNab et al. 1968.*)

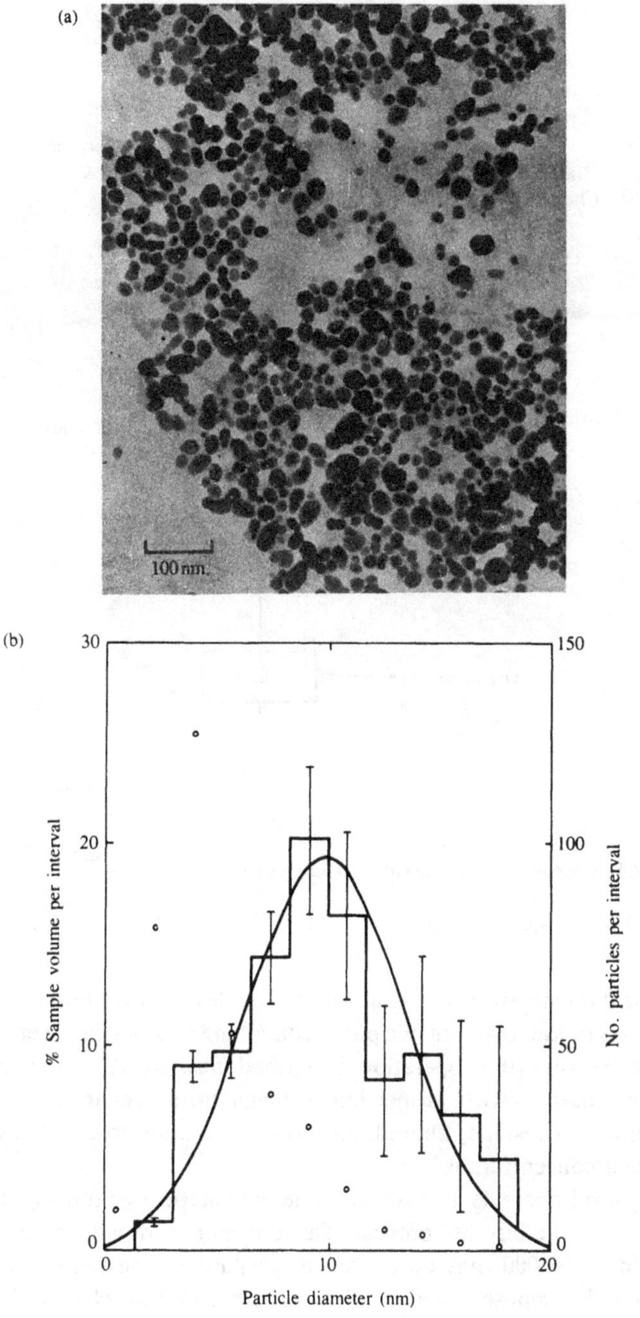
(a)
100 nm
(b)
% Sample volume per interval
No. particles per interval
Particle diameter (nm)
0
10
20
30
0
50
100
150
0
10
20

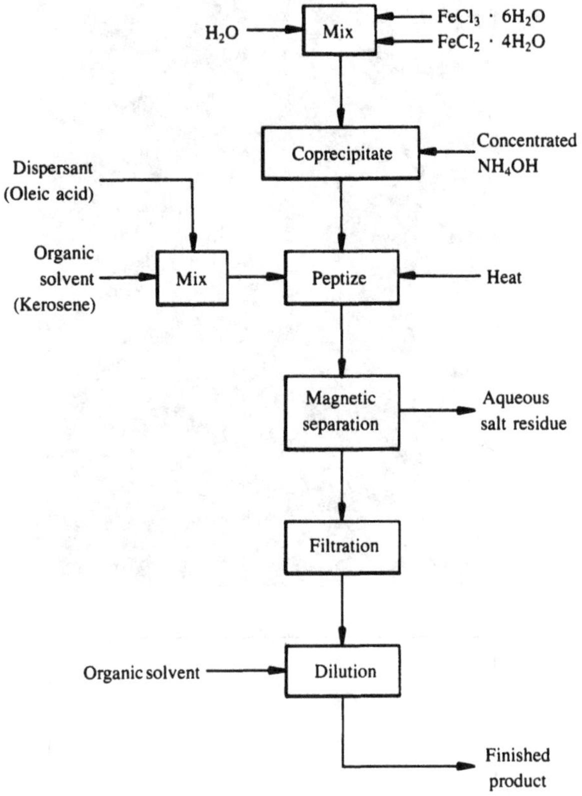

2.4 Chemical precipitation method for preparing colloidal magnetic fluid.

peptization step is included in which particles are transferred from the aqueous phase to an organic phase containing a dispersing agent. In the paint industry this operation is termed *flushing*. Subsequently the organic base particle dispersion is magnetically separated from the aqueous salt residue, filtered, and solvent adjusted to give the desired product concentration.

Lignosulfonate is a water–soluble polymeric by-product of wood pulping in the bisulfite process. This polymer, with average molecular weight of 35,000, has been used to prepare an inexpensive aqueous ferrofluid composed of the polymer anchored onto chemically precipitated magnetite microcrystals of about 10 nm in size (Neal 1977). Physical characterization of the material's susceptibility and Mossbauer spectra indicate superparamagnetic behavior, i.e., the large magnetization of a ferromagnetic material, but with the rapidly relaxing magnetization vector of a simple paramagnet (Hasset, Stecher, and Hendrickson 1980).

Cobalt particles in an organic carrier

Thomas (1966) and Hess and Parker (1966) have described a method of preparing a ferrofluid containing particles of elemental cobalt. The method consists in dissolving dicobalt-octacarbonyl [$Co_2(CO_8)$] in the solvent in the presence of the surfactant. The surfactant concentration is typically between 0.2 and 0.7 gram per gram of octacarbonyl. When the mixture is heated, the octacarbonyl decomposes, and a stable cobalt colloid results. However, the reaction is complex, and the final product depends on many parameters that are not always well controlled. The method has been further developed by Mailfert and Martinet (1973) and by Chantrell, Popplewell, and Charles (1978).

A typical composition consists of grains of cobalt dispersed in toluene with a high-molecular-weight (≥100,000) terpolymer surfactant – (methyl) methacrylate (49.5%) + ethyl acrylate (49.5%) + vinylpyrrolidone (1%). The polymer comprises a hydrocarbon chain with at least one polar adsorptive group spaced every 200 backbone atoms. Other surfactants can be used, such as Aerosol OT, which is a sodium sulfosuccinate. It appears that the rate of heating is a critical parameter. Dicobalt-octacarbonyl is not a very stable compound and decomposes slowly at room temperature. On heating to 100 °C, decomposition with the generation of carbon monoxide is observed, but the cobalt particles fail to nucleate. Study of the decomposition kinetics for temperatures up to 130 °C reveals different stages, with the formation of $Co_4(CO)_{12}$ as an important intermediate product. It may be noted that the reaction temperature exceeds the 111 °C boiling point of toluene. Ferrofluid containing cobalt particles and having saturation magnetization of 29 $kA \cdot m^{-1}$ (370 G) with viscosity of a few centipoise has been prepared.

Charge-stabilized magnetite

Massart (1981) describes the preparation of aqueous ferrofluid without the use of an organic stabilizing agent. An aqueous mixture of ferric chloride and ferrous chloride is added to a solution of ammonia, resulting in a gelatinous precipitate. The precipitate is mechanically separated from the solution and treated in one of two ways. An alkaline ferrofluid is made by peptizing the precipitate with aqueous tetramethylammonium hydroxide. An acidic sol is obtained by stirring the precipitate with aqueous perchloric acid, followed by centrifugation. Peptization occurs merely upon the addition of water. Negative charges of the colloids in alkaline solution and positive charges of the colloids in acidic medium are assumed to be adsorbed OH^- ions and adsorbed H_3O^+ ions, respectively. The stabilization is produced by the electrostatic repulsion between the magnetic particles. From pH-5 to pH-9 the colloids coagulate owing to insufficient surface charge. In this range, stable aqueous sols apparently cannot be obtained without the use of a surfactant. The average particle size is reported to be 12 nm.

The interested reader is referred to Kruyt (1952) for a general discussion of the electric-double-layer mechanism of colloid stabilization.

The chemical methods are complementary to the mechanical ones: They are fast, the times involved being minutes or hours in place of weeks or months, and they can be cheap, but these methods are restricted to a few particular compounds. Special fluids are produced mainly by size reduction.

2.4 **Other magnetic fluids**

Paramagnetic salt solutions

Of the over 100 known elements, 55 are paramagnetic. The property is most pronounced in elements having unfilled inner subshells containing unpaired electrons. This property is found in the iron group (Cr, Mn, Fe, Ni, and Co) and in the alkaline rare-earth group (Gd, Tb, Dy, Ho, and Er). Manganese in the first group and holmium, dysprosium, or erbium in the second group are especially representative. Andres (1976) has shown that aqueous solutions of paramagnetic salts of these elements provide low-susceptibility, high-saturation magnetic fluids. The water solubility of the salts exhibits the following order:

$$\text{sulfide} < \text{sulphate} < \text{perchlorate} < \text{chloride} < \text{nitrate}$$

Thus, chlorides and nitrates are the most useful compounds. Table 2.1 below lists physical properties, including the susceptibility $\chi \equiv M/H$, for a few of the more magnetic solutions obtainable in this manner. A paramagnetic salt solution is employed in sink–float separation of minerals

Table 2.1 *Properties of paramagnetic salt solutions (25 °C)*

Paramagnetic salt	Solubility limit ($kg \cdot m^{-3}$)	Susceptibility ($\chi = M/H$)	Magnetization ($\mu_0 M$ T) $\mu_0 H = 2$ T	$\mu_0 H = 20$ T	Viscosity ($N \cdot s \cdot m^{-2}$)
$FeCl_3$	1,320	0.00046	0.0009	0.009	0.0080
$MnCl_2$	1,470	0.00090	0.0018	0.018	0.0074
$Ho(NO_3)_3$	1,980	0.00276	0.0055	0.055	0.0125

employing high-field intensity (2 T) iron yoke magnets. The higher field intensity (20 T) in Table 2.1 would require a superconducting field source for realization in any steady-state process.

The susceptibility values in Table 2.1 may be scaled to other operating temperatures using the Curie law, $\chi = C/T$, where T is the absolute temperature and C is a constant. In extremely high fields or at very low temperatures, the paramagnetic salts may approach saturation, and the simple relationship no longer applies.

Metallic-base ferrofluid

A ferrofluid in a liquid-metal carrier has long been desired for uses that would benefit from the high thermal conductivity, electrical conductivity, and other properties of a liquid metal. Representative uses include electrical slip rings for rotary shafts, high-surface-speed rotary seals and bearings, and magnetocaloric energy conversion systems.

Candidate liquid metals include mercury, gallium, tin, and low-melting alloys with indium or bismuth. A convenient way of producing the small single-domain ferromagnetic metal particles and introducing them into the liquid metallic carrier involves the use of electrodeposition techniques. Salts of the ferromagnetic metals dissolved in water or alcohol are used with the metallic liquid as the cathode in an electrolytic cell. To maintain small particle size and prevent dendritic growth, the liquid metal carrier must be agitated throughout the deposition. Iron, nickel, cobalt, and their alloys have been deposited into mercury and mercury alloys.

The steric stabilization mechanism is not available in liquid metals, and to date no truly stable dispersions have been produced (Rosensweig and Kaiser 1967; Charles and Popplewell 1980). Although fine ferromagnetic particles produced, for example, by electrodeposition in mercury are well wetted by the mercury, in a magnetic field gradient a ferrous-particle-free portion of mercury is expelled from the mixture. The concentrated magnetic portion that remains is stiff and exhibits viscoplastic behavior. Trace addition of a metallic additive that is soluble in the carrier but that forms an insoluble intermetallic compound with the ferromagnetic particle – e.g., traces of tin added to iron particles in mercury carrier – improves the rheology and gives other benefits owing to the formation of an insoluble nonmagnetic surface layer on the particle. There is need for further investigations of the surface and colloidal physics of metal-in-metal colloidal systems, with special interest in the study of electronic charge transfer between the dissimilar metals and the attendant repulsive forces that might develop between charged particles of the dispersion.

2.5 Surface adsorption and steric stabilization

As discussed in Section 2.1, van der Waals attraction presents a formidable problem in obtaining colloid particles that are stable with respect to mutual agglomeration. Fortunately a mechanism exists to prevent particles from approaching so close to one another that van der Waals attractions prevail. This mechanism is the *steric repulsion* due to the presence of long chain molecules adsorbed onto the particle surface (see Figure 2.5). The polar groups of the adsorbed species associate with the particle surface either physically or chemically and disguise the surface to look like part of the fluid. Thus, the tails are chosen with properties similar to the surrounding fluid matrix or carrier fluid. Intuitively it can be appreciated that two particles approaching each other closely will tend to compress the surface-adsorbed layers surrounding each, so that the layers act as elastic bumpers. A model discussed next produces a semiquantitative description of the repulsion energies developed by these compressed layers.

Steric repulsion mechanism

Mackor (1951) treated steric repulsion as the statistical mechanics of a rigid rod attached onto a universal hinge. The "head" polar group of the adsorbed molecule is assumed to be dilute on the surface, so the "tail" rod can take up a hemisphere of orientations under the influence of thermal motions (see Figure 2.6). The expression for the repulsive energy obtained by Mackor for flat surfaces is

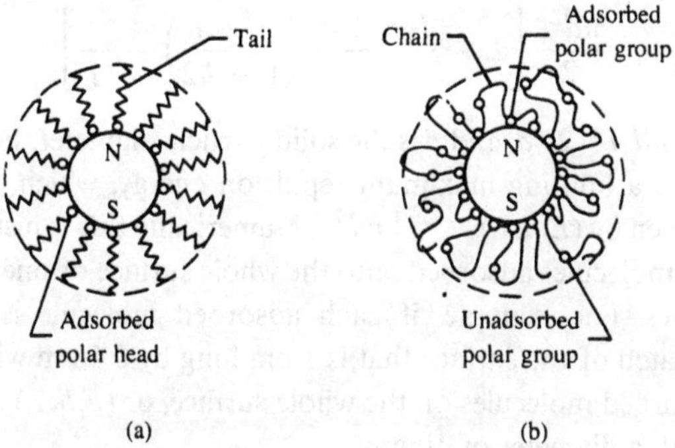

2.5 Species adsorbed onto surfaces of colloidal magnetic particles: (a) Surfactants having a polar head group and long chain tail; (b) multisite adsorption of polymer molecules containing polar groups along the length of the chain.

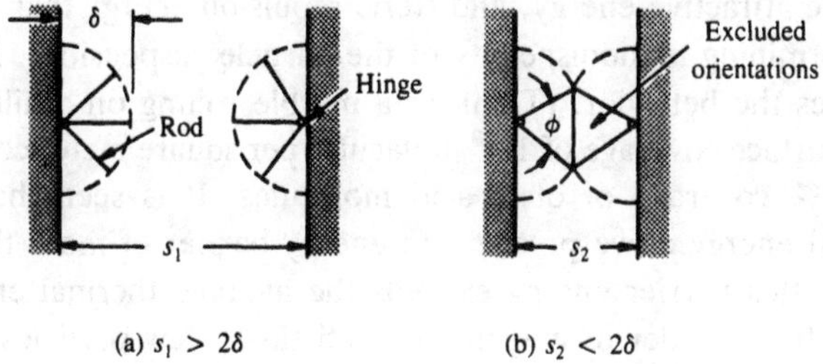

2.6 Geometric illustration of steric repulsion. (a) There can be no repulsion of surfaces that are separated by more than twice the length of the tails. (b) Closer approach of the surfaces truncates the configuration, reducing the volume available to the tails and producing a gaslike repulsion pressure.

$$\frac{E_f}{kT} = \begin{cases} \xi(1 - s/2\delta), & (s/2\delta) \le 1 \\ 0 & (s/2\delta) > 1 \end{cases} \tag{2.11}$$

where ξ is the surface concentration of adsorbed molecules, s is the surface-to-surface separation, k is the Boltzmann constant, T is the temperature, E_f is the repulsion energy per unit area of the surface, and δ is the length of the adsorbed molecule. Equation (2.11), when integrated for a neighboring pair of spheres (Rosensweig, Nestor, and Timmins 1965), gives

$$\frac{E_r}{kT} = \frac{\pi d^2 \xi}{2}\left[2 - \frac{l+2}{t}\ln\left(\frac{1+t}{1+l/2}\right) - \frac{l}{t}\right] \tag{2.12}$$

where $l = 2s/d$, $t = 2\delta/d$, and d is the solid particle diameter. For thick coatings there is a limiting maximum repulsion energy, which, according to (2.12), is given by $(E_r/kT)_{max} = \frac{1}{2}\pi d^2 \xi$. Numerically this is just one half the number of molecules adsorbed onto the whole surface of one of the spherical particles. For example, if each adsorbed molecule is assumed to occupy a patch of the surface that is 1 nm long by 0.5 nm wide, there will be 314 adsorbed molecules on the whole surface, or $(E_r/kT)_{max} = 314$ for a particle with a diameter of 10 nm.

The mechanism described here is also referred to as *entropic repulsion.*

Net interaction curve

It is the algebraic sum of the van der Waals attractive energy, magnetic attractive energy, and steric repulsion energy that is decisive for determining monodispersity of the particle suspensions. Figure 2.7 illustrates the behavior. (Think of a marble rolling on a hill.)

The surface coverage of 10^{18} molecules per square meter corresponds to a 50% coverage of oleic acid molecules. It is seen that the net potential energy curve presents an energy barrier of more than $20kT$. Because this barrier energy exceeds the average thermal energy of a particle by an order of magnitude, statistically few particles will cross the barrier, and the rate of agglomeration should be negligible. A sealed sample of ferrofluid prepared in kerosene has remained fluid and magnetic for 18 years with no sign of deterioration.

Dispersant structural guidelines

The polar adsorbing group can be carboxyl, sulfosuccinate, phosphate, phosphoric acid, polyphosphoric acid. The polar reactive group may be located at the head of the dispersant or distributed over the length of a polymer. The tails (or loops) must 1–2 nm in minimum length. These coatings take up a good deal of volume and limit the concentration of magnetic solid in the colloidal magnetic fluids, as illustrated by

$$\frac{\phi_m}{\phi_t} = \frac{1}{(1 + 2\delta/d)^3} \tag{2.13}$$

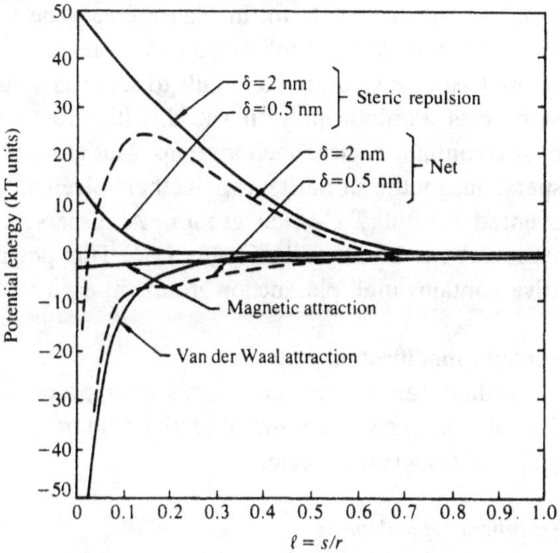

2.7 Potential energy versus surface-to-surface separation of sterically protected colloidal particles (d = 10 nm; $\xi = 10^{18}$ molecules · m^{-2}; magnetite particle).

where ϕ_m is the volume fraction of magnetic solid and ϕ_t the volume fraction of coated particle. For δ = 2 nm, d = 10 nm, and ϕ_t = 0.74, corresponding to close-packed sphere, ϕ_m = 0.27. Because an increase in particle packing leads to a rapid increase of colloid viscosity, the useful fluids are presently constrained to a solids fraction of less than 20 vol %. This corresponds for magnetite to a saturated moment of $\mu_0 M \sim 0.1$ T.

The tails of the dispersant molecules must fit well into the carrier fluid. A kink in the tail prevents crystallization (association with own species) and is desirable. Solvation indicated by exothermic mixing is considered to help. Conversely, colloids are adversely sensitive to foreign solvents; agglomeration occurs if loops or tails associate with each other better than with the solvent.

Solubility parameters (Barton 1983) should be useful in predicting compatibility of anchored chains with the carrier fluid. Leads for new dispersants can be gleaned from the technology of paints (pigment dispersion) and lube oils (sludge dispersants). Oleic acid (18 carbons) is the classic example of an agent for dispersing magnetite particles into hydrocarbons. Its structure is shown in Figure 2.8a. The kink is provided by the double bond at the ninth carbon position. Stearic acid, also an 18-carbon molecule, is saturated and fails to serve as a dispersant in the grinding process. Perfluoropolyether acid, with structure shown in Figure 2.8b, also contains an acid anchor group. This molecule has been used to disperse magnetite in perfluoropolyether solvents. Crystallization is prevented by bulky C_3 side groups. A dispersant useful in aromatic hydrocarbons is shown in Figure 2.8c. This polyphosphoric acid derivative contains multiple anchor groups in each molecule.

2.6 **Ferrofluid modification**

A ferrofluid can be refined to tailor its properties by, for example, changing the type or amount of carrier fluid or, in some cases, by exchanging the dispersant species.

Phenomenological basis

In the presence of a foreign solvent, the protective adsorbed layer on the magnetic particle responds in different ways. Consider an initially stable colloidal dispersion of magnetic particles in a fluid carrier to which a soluble second fluid is added, say, in the amount of 10 vol %. If the second fluid is similar to the first, the mixing process has no effect on the stability of the colloidal particles (Figure 2.9a). The molecules remain adsorbed, and the tails are in motion. This is the situation, for example, when decane is added to oleic-acid-protected particles in heptane.

If a polar solvent such as acetone is added to oleic-acid-stabilized ferrofluid in heptane, the magnetite particles *flocculate*, i.e., rapidly agglomerate and settle out in an external force field. It is subsequently found that the particles redisperse spontaneously when added to fresh heptane. This behavior is taken as strong evidence that the adsorbed layer remains attached to the particles' surface (Rosensweig 1970). The possible state of particles that are flocculated in this manner, depicted in Figure 2.9b, consists in their having retracted tails. Another possibility is the association of tails from neighboring particles. Silicone fluids (e.g., polydimethyl-siloxane) cause flocculation, just as acetone does. Long-chain polymers may cause flocculation by entangling a number of particles in one molecule of the polymer.

In contrast to the above phenomenology, certain systems desorb the surface species (Figure 2.9c). Evidently in these instances the surface layer is only weakly physically adsorbed and dissolves off into the new

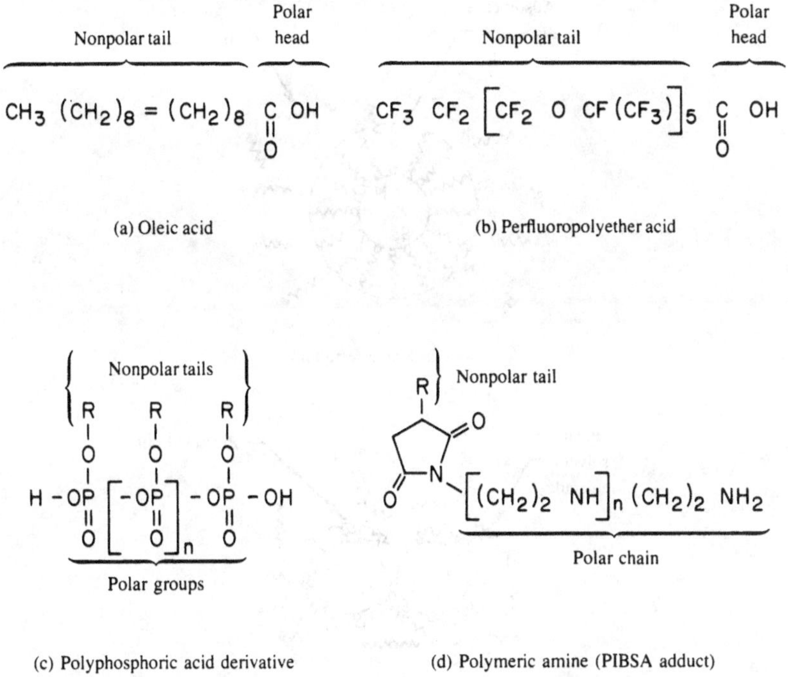

2.8 Chemical structure of molecular dispersant species used in ferrofluids.

solvent environment (Rosensweig 1975). An example is aqueous ferrofluid with an ethanolated alkylguanidine amine complex (Aerosol C61) that has been flocculated with acetone or a lower alcohol. Contacting the settled out particles with fresh solvent fails totally to produce a stable dispersion. The flocculation is *irreversible* in this case.

Carrier liquid exchange

One modification process utilizes the *reversible flocculation phenomenon* described above. As shown in the flowchart in Figure 2.10, the initial grind proceeds in the usual way, producing a stable ferrofluid. Next a dissimilar solvent is added, causing reversible flocculation, and the supernatant liquid containing the mixed solvents is removed. Magnetic separation is very useful in separating the phases at this step. The flocculated particles are washed to remove all traces of the offending foreign solvent. This is followed with an optional drying step,

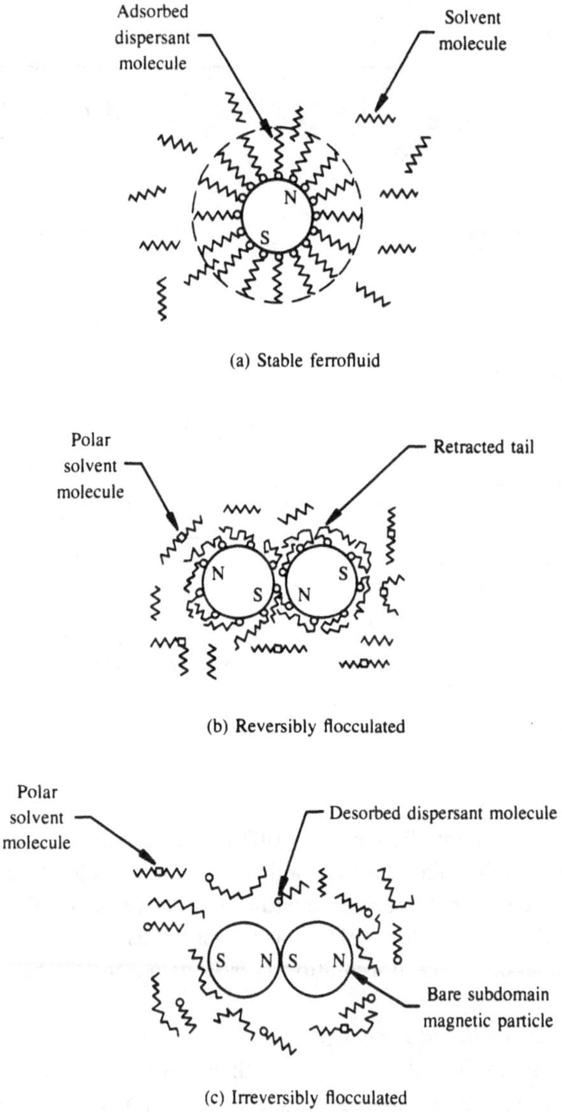

2.9 Comparative behavior of dispersant molecules in ferrofluids with added solvents. Polymeric additives can reversibly flocculate particles through entanglement.

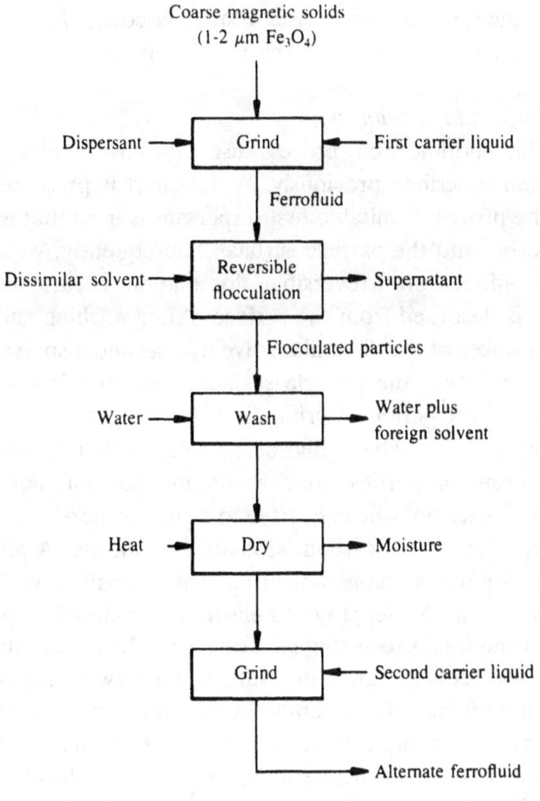

2.10 Modifying a ferrofluid by the transfer of particles to another solvent. There is no substitution of the dispersant species, which is permanently attached to the particle.

but usually such a step is not desirable. Finally, a second carrier liquid is placed in contact with the particles to effect a redispersion. The still-coated particles yield an "instant ferrofluid."

In this manner it is possible to transfer particles from a low-viscosity fluid well suited for grinding to a low-volatility fluid well suited for long-term exposure to the atmosphere. As another illustration of the utility of the process, it is well suited for concentrating particles in a given solvent while avoiding evaporation and heating. It is also possible by virtue of the process to remove excess surfactant to yield a stable colloidal mixture that can more readily be concentrated yet, because the dispersant species nearly always exhibits a higher viscosity than the carrier solvent, at the same time possesses a lower viscosity. In this process the new solvent must be similar to the initial solvent.

Surfactant exchange

This modification process uses the *irreversible flocculation phenomenon* described previously. A flowchart is presented in Figure 2.11. In the process a suitable first dispersant is used that merely physically adsorbs onto the particle surface. Subsequently, when dissimilar solvent is added, an irreversible flocculation occurs in which the dispersant is desorbed from the surface. After washing and heating to remove all traces of the dissimilar solvent, a second dispersant is chemically adsorbed onto the particle surface. This step is usually accompanied by a short period of grinding to break up particle clusters.

The new carrier may be similar or, as is most often desired, may have totally different properties from the original solvent. For example, a dispersion in water may be converted to a good dispersion in an alcohol, hydrocarbon, ester, halocarbon, aromatic, or silicone. A prerequisite is the availability of a suitable new dispersant compatible with the choice of the new solvent. No lengthy and uncertain grinding is required. Some grinding in the final processing step of Figure 2.11 is usually beneficial to effect total redispersion. Interestingly, the new surfactant need *not* qualify as a grinding aid. Evidently the two functions represent separate capabilities. An example is furnished by the surfactant Aerosol 22, tetrasodium N-(1,2 dicarboxyethyl) *n*-octadecylsulfosuccinate. This species redisperses material into aqueous media but fails to serve as a grinding aid.

The process considered here is very flexible and leads to cheaper, faster production of new families of magnetic fluids. Another benefit is the somewhat larger particle size present in the final product compared to that produced by direct grinding in the second dispersant. The larger particle size at this step is a direct result of the larger particle size present in the initial aqueous dispersion. The resultant product fluids concentrate better and produce more highly magnetic fluids of lower viscosity than are found in a parent grind.

2.7 Physical properties

The physical properties of a ferrofluid – such as its equilibrium magnetization in a steady applied field, magnetization change in a field of shifting orientation or intensity, or shear stress versus rate of strain – have obvious importance in properly formulating and interpreting the

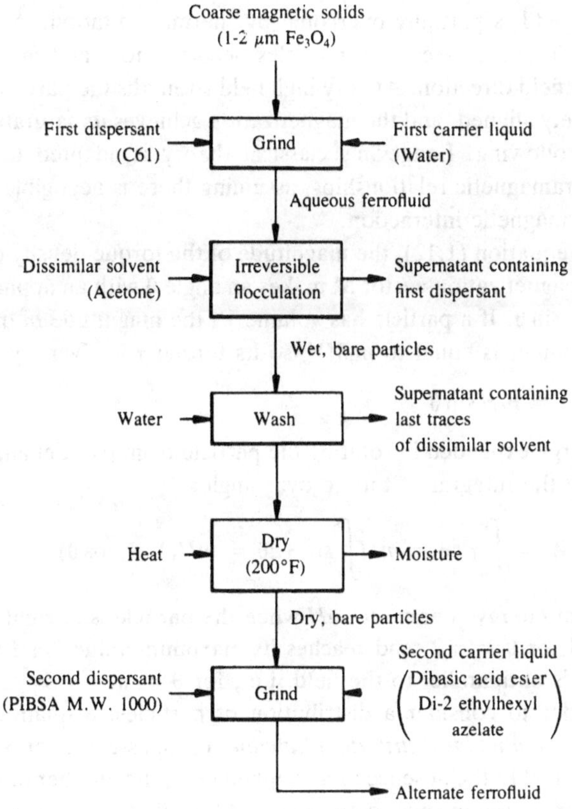

2.11 Modifying a ferrofluid by changing both the dispersant species and the solvent. Radical changes in properties are achieved. (*After Rosensweig 1975.*)

ferrohydrodynamic description of the fluids. In this section treatments are given for developing certain fundamental relationships, and other important information is reviewed.

Equilibrium magnetization: superparamagnetism

The particles in a colloidal ferrofluid, each with its embedded magnetic moment **m**, are analogous to the molecules of a paramagnetic gas. In the absence of an applied field, the particles are randomly oriented, and the fluid has no net magnetization. However, for ordinary field strengths the tendency of the dipole moments to align with the applied field is partially overcome by thermal agitation. As the field magnitude is increased, the particles become more and more aligned with the field direction. At very high field strengths the particles may be completely aligned, and the magnetization achieves its saturation value. In the following, Langèvin's classical theory is adapted to give the superparamagnetic relationships, assuming there is negligible particle–particle magnetic interaction.

From equation (1.12), the magnitude of the torque density on matter whose magnetization vector **M** makes an angle θ with an applied field **H** is $\mu_0 MH \sin\theta$. If a particle has volume V, the magnitude m of its magnetic moment is equal to $\mu_0 MV$, so its torque τ is given by

$$\tau = mH\sin\theta \tag{2.14}$$

The energy expended in rotating the particle from parallel alignment is given by the integral of torque over angle:

$$W = \int_0^\theta \tau\, d\theta = mH\int_0^\theta \sin\theta\, d\theta = mH(1 - \cos\theta) \tag{2.15}$$

Note that energy is equal to mH when the particle is at right angles to the field, or $\theta = \pi/2$, and reaches its maximum value $2mH$ when the particle is antiparallel to the field, i.e., for $\theta = \pi$.

In order to consider a distribution of particles, a quantity $n(\theta)$ is defined, the *angular distribution function* for an assembly of N independent "rods." In the absence of an applied field, the number of rods lying in the *configuration space* between θ and $\theta + d\theta$ is given by $n(\theta)\, d\theta$ (see Figure 2.12):

$$n(\theta)\, d\theta = N\frac{(2\pi\sin\theta)(d\theta)}{4\pi(1)^2} = \frac{N}{2}\sin\theta\, d\theta \tag{2.16}$$

Statistical mechanics teaches that in the presence of an applied field and at a given absolute temperature T the probability of finding a given orientation becomes proportional to

$$\text{Boltzmann factor} = e^{-W/kT} \tag{2.17}$$

Thus the number of rods lying in the configuration space between θ and $\theta + d\theta$ is proportional to

$$n(\theta)\, d\theta \propto \frac{\mathrm{N}}{4\pi}e^{-W/kT}\, 2\pi\sin\theta\, d\theta = \frac{N}{2}e^{-W/kT}\sin\theta\, d\theta \tag{2.18}$$

The proportionality constant can be calculated by requiring the number of rods to be equal to N:

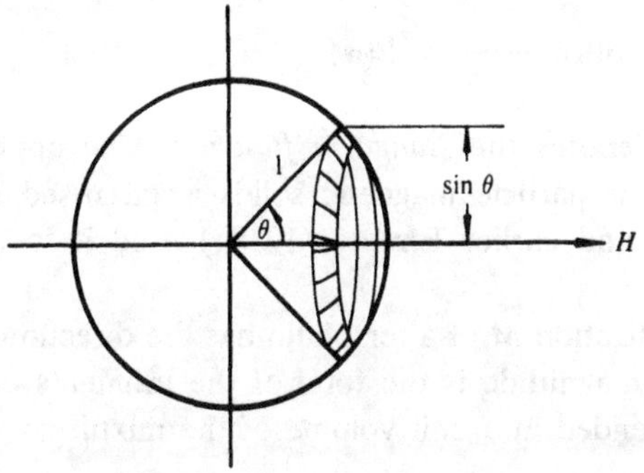

2.12 Orientational configuration space for N independent particles.

$$\int_0^{\pi} n(\theta)\, d\theta = N \tag{2.19}$$

The effective dipole moment of a particle is its component along the field direction, i.e., $m \cos\theta$. In terms of the distribution function $n(\theta)$, the average value of $m\cos\theta$, from (2.19), is given by

$$\bar{m} = \langle m\cos\theta\rangle = \int_0^{\pi} m\cos\theta\, n(\theta)\, d\theta \bigg/ \int_0^{\pi} n(\theta)\, d\theta \tag{2.20}$$

Substituting for W from (2.15) in (2.18) and putting the resultant expression for $n(\theta)$ into (2.20) reduces it, after cancellation of a proportionality constant, to

$$\bar{m} = \int_0^{\pi} m\cos\theta \exp(mH\cos\theta/kT)\sin\theta\, d\theta \bigg/ \int_0^{\pi} \exp(mH\cos\theta/kT)\sin\theta\, d\theta \tag{2.21}$$

It is convenient to introduce the energy ratio $\alpha = mH/kT$, so that (2.21) may be rewritten

$$\begin{aligned}\bar{m} &= \int_0^{\pi} m\cos\theta\, e^{\alpha\cos\theta}\, d\cos\theta \bigg/ \int_0^{\pi} e^{\alpha\cos\theta}\, d\cos\theta \\ &= \frac{m}{\alpha}\int_{-\alpha}^{\alpha} xe^x\, dx \bigg/ \int_{-\alpha}^{\alpha} e^x\, dx\end{aligned} \tag{2.22}$$

where $x = \alpha\cos\theta$. When the integrations are carried out, the result is simply

$$\frac{\bar{m}}{m} = \coth\alpha - \frac{1}{\alpha} \equiv L(\alpha) \tag{2.23}$$

where $L(\alpha)$ denotes the *Langevin function*. The application of this equation to fine-particle magnetic solids is discussed by Jacobs and Bean (1963), and earlier Elmore (1938b) used it to describe Bitter colloids.

The magnetization M of a ferrofluid has the direction of the applied field, and its magnitude is the total of the moments of the magnetic particles suspended in a unit volume of the mixture:

$$\mu_0 M = n\bar{m} \tag{2.24}$$

where $\bar{m}$ is the component of the mean magnetic moment per particle along the field direction. By the same token, the saturation magnetization M_s of the fluid is given in terms of the particle moment magnitude m by

$$\mu_0 M_s = nm \tag{2.25}$$

In addition, the saturation moment M_s of the ferrofluid is related to the saturation moment M_d of the bulk magnetic solid through the volume fraction ϕ of solid present:

$$M_s = \phi M_d$$

Eliminating M_s and n from these expressions gives

$$M/\phi M_d = \bar{m}/m \tag{2.26}$$

Combining equations (2.23) and (2.26) yields the superparamagnetic magnetization law for a monodisperse, colloidal ferrofluid.

$$\frac{M}{\phi M_d} = \coth\alpha - \frac{1}{\alpha} \equiv L(\alpha),$$

$$\alpha = \frac{\pi}{6}\,\frac{\mu_0 M_d H d^3}{kT} = \frac{mH}{kT} \tag{2.27}$$

This is the main result of the analysis.

Figure 2.13 displays magnetization curves computed from (2.27) for various particle sizes. All the curves saturate at high field, but the initial susceptibility is small when particle size is small; however, the initial susceptibility grows rapidly with increasing particle size. In addition to its importance to ferrofluids, equation (2.27) is used to determine the particle size of iron, nickel, and cobalt catalyst dispersed on a support material, and in the analysis of geological rocks.

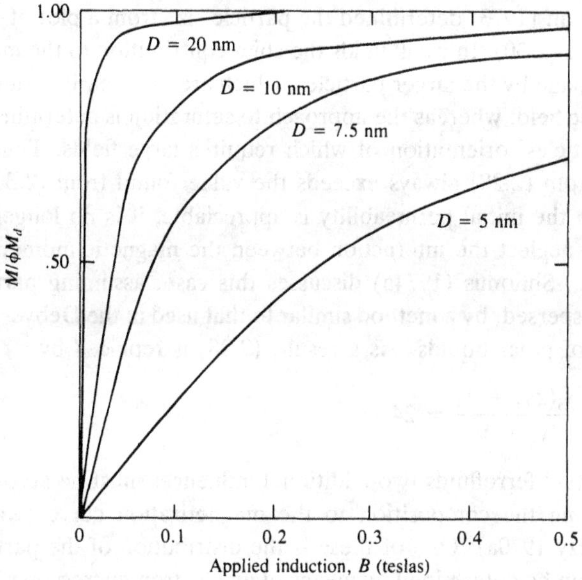

2.13 Calculated magnetization curves for monodisperse spherical particles with the domain magnetization of magnetite (4.46×10^5 A·m^{-1}).

It is of interest to examine the asymptotic values of the Langevin function for small and large values of the parameter α. A Taylor series expansion valid for small α shows that

$$L(\alpha) = \alpha/3$$

and thus the initial susceptibility $\chi_i = M/H$ is given by

$$\chi_i = \frac{\pi}{18}\phi\mu_0\frac{M_d^2 d^3}{kT} = 8\phi\lambda \qquad (\alpha \ll 1) \tag{2.28}$$

where ϕ, as before, is the volume fraction of magnetic solids, and λ is the *coupling coefficient*, which also arises in the study of correlations, as will be seen in Section 2.8.

$$\lambda \equiv m^2/4\pi d^3\mu_0 kT = \mu_0 M_d^2 V/24kT \tag{2.29}$$

The approach to saturation corresponds to α growing large, and then the asymptotic form of equation (2.27) is

$$M = \phi M_d\left(1 - \frac{6}{\pi}\frac{kT}{\mu_0 M_d H d^3}\right) \qquad (\alpha \gg 1) \tag{2.30}$$

Bibik et al. (1973) determined the particle size from a plot of M versus $1/H$ using (2.30). In weak fields the chief contribution to the magnetization is made by the larger particles, which are more easily oriented by a magnetic field, whereas the approach to saturation is determined by the fine particles, orientation of which requires large fields. Thus d computed from (2.28) always exceeds the value found from (2.30).

When the initial permeability is appreciable, it is no longer permissible to neglect the interaction between the magnetic moments of the particles. Shliomis (1974a) discusses this case, assuming particles are monodispersed, by a method similar to that used in the Debye–Onsager theory of polar liquids. As a result, (2.28) is replaced by

$$\frac{\chi_i(2\chi_i + 3)}{\chi_i + 1} = 24\phi\lambda \tag{2.31}$$

In actual ferrofluids two additional influences must be accounted for in relating the composition to the magnetization curve (Kaiser and Miskolczy 1970a). One of these is the distribution of the particle size, which can be determined by means of an electron microscope. Martinet (1983) discusses the inverse problem, namely, the use of distribution functions to derive particle sizes from the magnetization curve. The other influence is the decrease of magnetic diameter of each particle by the amount d_s, where $d_s/2$ is the thickness of a nonmagnetic surface layer formed by chemical reaction with the adsorbed dispersing agent; for example, for magnetite one takes $d_s/2 = 0.83$ nm, corresponding to the lattice constant of the cubic structure. The dispersant may be oleic acid that enters into a reaction with the Fe_3O_4; iron oleate is formed, possessing negligible magnetic properties. For a solid particle of 10-nm diameter, the volume fraction of magnetic solids is 0.58.

Thus, a better agreement is observed between the experimental magnetization curves and theoretical curves calculated with the formula

$$\frac{M}{\phi M_d} = \sum_i n_i(d_i - d_s)^3 L\left[\frac{\mu_0 M_d H}{kT}\frac{\pi}{6}(d_i - d_s)^3\right]\Big/\sum_i n_i d_i^3 \tag{2.32}$$

where d_i is the outer diameter of the particle and $d_i - d_s$ is the diameter of the magnetic core. The surface dead-layer mechanism is consistent with experiments of various authors, discussed by Bean and Livingston (1959), that detect no decrease in the spontaneous magnetization of subdomain particles having no sorbed layers for diameters down to at least 2 nm.

Mossbauer measurement, i.e., the resonant absorption of γ rays emitted by a source nucleus, and magnetic data on Fe_3O_4, $CoFe_2O_4$, and $NiFe_2O_4$ dispersions indicate that loss of magnetization due to sorption of dispersing agent on the surface layer is not due to a magnetic "dead" layer as such: The cations at the particle surface are magnetically ordered but pinned in orientation such that their net moment vanishes (Berkowitz et al. 1975).

Magnetic relaxation

There are two mechanisms by which the magnetization of a colloidal ferrofluid can relax after the applied field has been changed. In the first mechanism the relaxation occurs by particle rotation in the liquid. In the second mechanism the relaxation is due to rotation of the magnetic vector within the particle. If a ferrofluid is solidified – by freezing, for example – only the second mechanism is operative.

The particle rotation mechanism is characterized by a Brownian rotational diffusion time τ_B having hydrodynamic origin (Frenkel 1955) and given by

$$\tau_B = 3V\eta_0/kT \tag{2.33}$$

where V is the particle volume and η_0 the viscosity of the carrier liquid. The rate of Brownian rotational relaxation when a field is present is determined from solutions to the Fokker–Planck equation (Martsenyuk, Raikher, and Shliomis 1974).

For a single-domain uniaxial ferromagnetic particle in the absence of a field, the magnetization has two possible orientations, the opposite directions along the easy axis of magnetization. An energy barrier must be overcome in order to move from the one orientation to the other. The height of this energy barrier is given by KV, where K is the *anisotropy constant* of the material. When the condition $KV \ll kT$ is fulfilled, the thermal energy is large enough to induce fluctuations of the magnetization inside the grain with a characteristic time τ_N, first conceived by Néel (1949, 1953) and given as

$$\tau_N = \frac{1}{f_0} \exp\left(\frac{KV}{kT}\right) \tag{2.34}$$

where f_0 is a frequency having the approximate value 10^9 Hz. This value was confirmed for a ferrofluid composed of magnetite in kerosene by McNab, Fox, and Boyle (1968) using Mossbauer measurement.

When $\tau_N \ll \tau_B$, relaxation occurs by the Néel mechanism, and the material is said to possess *intrinsic superparamagnetism*. When $\tau_B \ll \tau_N$, the Brownian mechanism is determining and the material exhibits *extrinsic superparamagnetism*. However, if the smaller time constant is much greater than the time scale of the experiment, then the sample may be regarded as ferromagnetic.

In a ferrofluid the Brownian or Néel mechanisms both lead to an apparent superparamagnetic behavior, described by the same Langevin's law, as the Langevin relationship describes the equilibrium response and both mechanisms permit relaxation to the equilibrium state. However, in considering dynamic aspects of magnetic flows, the rate of relaxation can become an important variable. The equilibrium formulation of ferrohydrodynamics developed in Chapters 4 and 5 corresponds to relaxation times τ_B and τ_N both small compared to a characteristic flow time. In the treatment of flow with antisymmetric stress in Chapter 8 it is assumed that $\tau_N \gg \tau_B$, in which case the magnetic moment is locked to the crystal lattice of the particle.

Values of τ_B are readily calculated from equation (2.33). For a ferrofluid in kerosene with $\eta = 0.002\ \mathrm{kg \cdot m^{-1} \cdot s^{-1}}$ containing 10-nm particles at 25 °C,

$$\tau_B = 3(\pi/6)(10^{-8})^3(2 \times 10^{-3})/(1.38 \times 10^{-23})(298) = 7.6 \times 10^{-7}\ \mathrm{s}$$

In general, a value of the crystal anisotropy constant K can be determined from experiments in which magnetization curves of a single crystal are determined for field applied along the different directions of the crystallographic axes (Bozorth 1951). The Néel time τ_N can then be estimated from (2.34). For an oleic-acid-stabilized ferrofluid of magnetite in kerosene, τ_N is estimated to be $\sim 10^{-9}$ s by Martinet (1978) in a study of the magnetic anisotropy of mechanically rotated ferrofluid samples. Table 2.2 presents scaled relaxation time constants; it may be seen that the superparamagnetism of a ferrofluid with particle size above the dashed line is of the intrinsic type, whereas below the line it is extrinsic. Because of its exponential dependence on V, the numerical value of τ_N is sharply dependent on the particle size.

The transition from Néel to Brownian relaxation may be considered to take place for particles with a size D_s obtained by equating τ_B and τ_N, which gives the relationship $\kappa^{-1} \exp \kappa = 3\eta f_0/K$ where $\kappa = KV/kT$. Shliomis (1974a) calculated $D_s = 8.5$ nm for iron and 4 nm for cobalt. Charles and Popplewell (1980) discuss uncertainties in the estimate due to lack of knowledge of the crystal structure in the case of cobalt and to shape anistropy in the case of iron.

Water-base ferrofluids usually contain particles of somewhat larger

Table 2.2 *Relaxation time constants τ_N and τ_B for a ferrofluid composed of magnetite particles in a kerosene carrier (T = 25 °C)*

Particle size (nm)	τ_N (s)	τ_B (s)
8	10^{-8}	3.8×10^{-7}
10	10^{-7}	7.6×10^{-7}
16.5	1	3.4×10^{-6}

size than the particles in organic carrier. This is in accord with the finding of Sharma and Waldner (1977) that water-base ferrofluids exhibit no intrinsic superparamagnetism in Mossbauer experiments.

Viscosity

A ferrofluid remains flowable in the presence of magnetic field, even when magnetized to saturation. Nonetheless, the rheology is affected by the presence of a field. Next we discuss the viscosity of ferrofluids first in the absence, then in the presence, of an applied magnetic field.

No external field: The viscosity of a ferromagnetic suspension is greater than that of the carrier liquid and accompanies an increased rate of energy dissipation during viscous flow due to presence of the suspended particles. This situation is the same as for nonmagnetic colloids of solid particles suspended in liquid. Thus, theoretical models are available for determining the viscosity, the earliest being that of Einstein (1906, 1911), derived from the flow field of a pure strain perturbed by the presence of a sphere. The resulting formula relates the mixture viscosity η to the carrier fluid viscosity η_0 and solids fraction ϕ (assuming for the moment that the particles are bare of coatings):

$$\eta/\eta_0 = 1 + \tfrac{5}{2}\phi \tag{2.35}$$

This relationship is valid only for small concentrations. For higher concentrations a two-constant expression may be assumed (Rosensweig, Nestor, and Timmins 1965):

$$\eta/\eta_0 = 1/(1 + a\phi + b\phi^2) \tag{2.36}$$

It is insisted that this expression reduce to (2.35) for small values of ϕ so that $a = -\frac{5}{2}$. At a concentration ϕ_c the suspension becomes effectively rigid, and so the ratio η_0/η goes to zero. Thus the second constant is $b = (\frac{5}{2}\phi_c - 1)/\phi_c^2$. A value of $\phi_c = 0.74$ corresponds to close packing of spheres. Uncoated spherical particles of radius r, when present in a ferrofluid at volume fraction ϕ, will, when coated with a uniform layer of dispersing agent having thickness δ, occupy a fractional volume in the fluid of $\phi(1 + \delta/r)^3$. Combining these relationships gives

$$\frac{\eta - \eta_0}{\phi\eta} = \frac{5}{2}\left(1 + \frac{\delta}{r}\right)^3 - \left(\frac{\frac{5}{2}\phi_c - 1}{\phi_c^2}\right)\left(1 + \frac{\delta}{r}\right)^6 \phi \tag{2.37}$$

A plot of measured values of $(\eta - \eta_0)/\phi\eta$ versus ϕ yields a straight line (see Figure 2.14). Values of δ/r determined from the intercept at $\phi = 0$ and from the slope using $\phi_c = 0.74$ are in good agreement, yielding $\delta/r = 0.84$. For $\delta = 2$ nm, the particle diameter found in this manner is 4.8 nm. This is less than the mean size determined from electron microscopy. This directional variance is thought to be due to the presence of a particle size distribution; small particles with their sorbed coatings tie up a disproportionate share of the total volume in the dispersion.

If particles have a tendency to cluster and form a gel network, the value of ϕ_c for an effectively rigid suspension is reduced from 0.74. For $\phi_c < 0.4$ the slope of the curve changes sign in Figure 2.14, a phenomenon that has been observed for certain ferrofluids (Raj 1981).

When the suspended particles are nonspherical, theory predicts an increase in the coefficient of ϕ in (2.35); owing to Brownian rotation, a larger volume is swept out by a particle of given size. As an example, an axial ratio of 5 increases the coefficient from 2.5 to 6.0 (Kruyt 1952).

From these considerations it follows that highly concentrated (high-saturation-moment) ferrofluids of greatest possible fluidity are favored by particles of small coating thickness δ, large particle radius r, and spherical shape. These desired trends for δ and r are opposite to the conditions favoring stabilization as a colloid, so in any actual ferrofluid compromises must be made using intermediate values of these parameters.

External field present: When a magnetic field is applied to a sample of magnetic fluid subjected to shear deformation, the magnetic particles in the fluid tend to remain rigidly aligned with the direction of the orienting field. As a result, larger gradients in the velocity field surrounding a particle are to be expected than if the particle were not present, and viscous dissipation increases in the sample as a whole.

A theoretical treatment from first principles giving the viscosity of

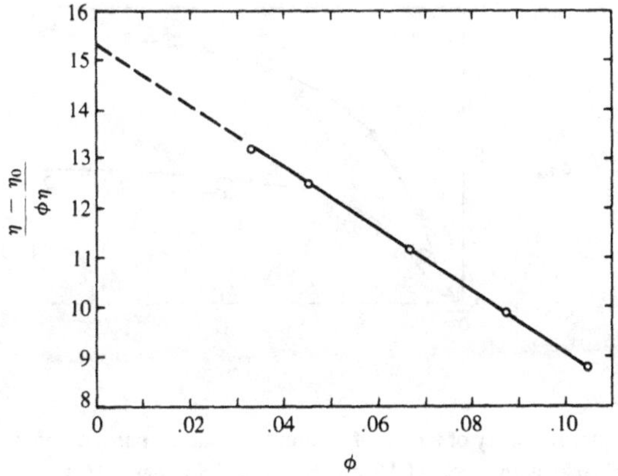

2.14 Experimental viscosity of oleic-acid-stabilized ferrite dispersions plotted to test equation (2.37). (*After Rosensweig, Nestor, and Timmins 1965.*)

dilute suspensions of single-domain spheres with large Néel time constant and accounting for Brownian rotational motion was developed by Shliomis (1972). The work represents an advance over the earlier treatment of Hall and Busenberg (1969), in which Brownian motion was neglected, leading to the prediction of saturation in a very small field. When the fluid vorticity and the magnetic field are parallel, the particles can rotate freely and magnetism exerts no influence on the viscosity, which is then given by the Einstein relationship. When the directions are perpendicular to each other, the magnetic contribution to the viscosity is maximized. The theory is discussed in detail in Chapter 8.

McTague (1969) measured the viscosity of a dilute suspension of polymer-stabilized cobalt particles in toluene in a magnetic field using a straight cylindrical capillary tube (Poiseuille flow). Measurements were made of the effect of the magnetic field both parallel and perpendicular to the direction of flow. The results of these measurements, given by the circles in Figure 2.15 for $\Delta\eta = \eta - \eta_0$, show that in both situations the viscosity increases with the Langevin argument α, finally reaching saturation as the magnetization saturates. The mean diameter of the particles used by McTague is greater than the critical diameter for intrinsic superparamagnetism. Thus, the theoretical result of Shliomis (1972), developed as (8.101), should apply. In Poiseulle flow the vorticity magnitude is constant on concentric circles in a cross-section surface of the tube. For β defined as in Chapter 8, in the parallel case $\sin^2 \beta = 1$

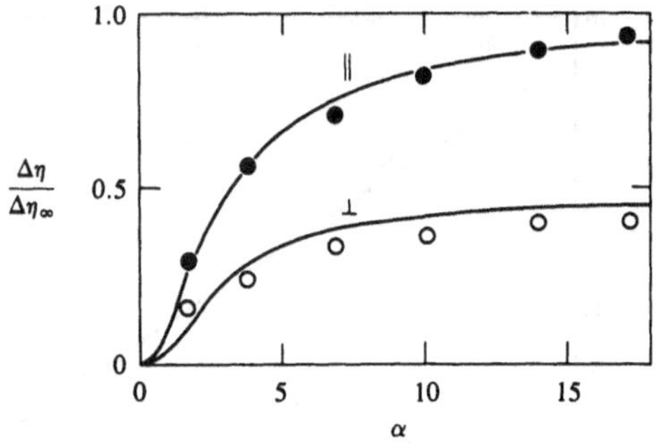

2.15 Magnetoviscosity of a dilute ferrofluid containing particles of elemental cobalt, compared with theory. (*After Shliomis and Raikher 1980.*)

because $\beta = \pi/2$ at each point of the flow. In the perpendicular orientation, averaging with respect to β gives

$$\langle \sin^2 \beta \rangle \equiv \frac{1}{2\pi} \int_0^{2\pi} \sin^2 \beta \, d\beta = \frac{1}{2} \tag{2.38}$$

Hence, for any applied field intensity

$$\Delta\eta_{\parallel} = 2 \, \Delta\eta_{\perp} \tag{2.39}$$

As shown in Figure 2.15, the theoretical curves (solid lines) agree well with the experimental data. Brenner (1970), Brenner and Weissman (1972), and Levi, Hobson, and McCourt (1973) have carried out similar analyses.

Concentrated suspensions

Rosensweig, Kaiser, and Miskoloczy (1969) measured the effect of a vertically oriented magnetic field on the viscosity of thin horizontal layers of a concentrated ferrofluid subjected to uniform shear in a horizontal plane. In some cases $\Delta\eta_{\perp}$ exceeded 2η, where η is the ferrofluid viscosity in the absence of the field. Typically, data for concentrated colloids exhibit a strain-rate dependence in the presence of magnetic field, as indicated in Table 2.3. As seen in the table, increasing values of shear, indicated by increasing values of Ω, yield shear thinning. Here Ω, the vorticity, is equal to the shear strain rate. Because in the monodisperse theory of magnetoviscosity the shear stress has negligible effect on the viscosity, the behavior evidenced in Table 2.3

Table 2.3 *Field- and shear-dependent viscosity of a fluorocarbon-base ferrofluid at 30°C*[a]

Ω (s^{-1})	B (T)	$\Delta\eta/\eta$	Comment
23	0.06	0.25	Low field reference value
46	0.06	0.13	—
115	0.06	0.06	—
230	0.06	0.05	Fivefold reduction
23	0.40	1.95	High field reference value
46	0.40	0.72	—
115	0.40	0.30	—
230	0.40	0.19	Tenfold reduction

[a] $\eta = 0.004\ \mathrm{N \cdot s \cdot m^{-2}}$; saturation $\mu_0 M = 0.0087$ T.

may be attributed to clustering effects in the concentrated colloids. Conditions for the occurrence of clustering are considered below, and concentrated colloid data that are free of shear thinning are analyzed in Chapter 8.

Calugaru, Badescu, and Luca (1976a) report similar measurements of viscosity in concentrated ferrofluids, showing values of $\Delta\eta_\perp$ exceeding 2η without saturation of the effect.

2.8 Correlation phenomena

Clustering and chaining of the particles of a ferrofluid may be regarded as steps in the direction of gelation or sedimentation. These correlation or association phenomena have been seen by electron microscopy in certain ferrofluids (Hayes 1975; Hess and Parker 1966). Such correlation behavior is undesirable in many ferrofluid uses. The behavior is undesirable in uses as a sealant, lubricant, coolant, sink–float medium, damping fluid, or printing ink and in other applications where invariancy of magnetic and fluid properties is of paramount importance. Other uses, such as in magnetic domain detection, optical shutters based on polarization of light, and the tagging of surfaces with magnetic particles, benefit from the nonideal behavior. In any case, it is helpful to understand the limitations imposed by particle association and to have knowledge of the mechanisms responsible for this behavior.

Theoretical expressions pertaining to the formation of chains of colloidal magnetic particles in uniform magnetic field have been obtained by de Gennes and Pincus (1970) and Jordan (1973) based on properties of the equation of state of a "rarified gas" of ferromagnetic particles suspended in an inert liquid. Allowance is made for the departure of the gas from ideality so far as this results from the magnetic attraction between the particles, neglecting any other forces that may be present. By considering pair correlations between particle positions, one finds that in strong external fields the ferromagnetic grains tend to form chains parallel to the field direction. The mean number of particles n_∞ in the chain is

$$n_\infty = [1 - \tfrac{2}{3}(\phi/\lambda^2)e^{2\lambda}]^{-1} \tag{2.40}$$

where λ is the (dimensionless) coupling coefficient defined in equation (2.29). It will be recalled that λ measures the strength of the grain–grain interaction in terms of m, the total moment $\mu_0 MV$ of a particle, where V is the spherical particle volume. The form of (2.40) yields two values of λ for which n_∞ approaches infinity, one <1 and the other >1. By the theoretical assumptions, only values $\lambda > 1$ have physical significance. Also, when the second term in the brackets on the right side of (2.40) exceeds unity, the approximations break down. It may be that clusters rather than chains then form in the liquid. At zero external field and $\lambda \gg 1$, there also, according to the prediction, exists a certain number of chains. Their mean length is given by

$$n_0 = [1 - \tfrac{2}{3}(\phi/\lambda^3)e^{2\lambda}]^{-1} \tag{2.41}$$

which is smaller than that in a strong field, and they are oriented in a random manner.

For a magnetite particle with magnetic diameter 10 nm with M = 446 $\mathrm{kA \cdot m^{-1}}$ and T = 298 K the coupling coefficient from (2.29) is $\lambda = 1.3$. With $\phi = 0.05$ in (2.40), $n_\infty = 1.36$, and from (2.41) $n_0 = 1.26$. The particles are essentially monodisperse, and the colloid has little clustering or agglomeration. In contrast, magnetite particles of 13-nm magnetic diameter yield $\lambda = 2.69$ and form a chain with n_∞ of infinite length. A particle of elemental iron must be smaller than a magnetite particle to avoid chaining. The domain magnetization of iron is about four times that of magnetite, so from (2.29) the magnetic diameter yielding $\lambda = 1.3$ is reduced to 4.0 nm.

A model based on Monte Carlo methods has been used to investigate the effects of particle interactions on the clustering and chaining of particles in a magnetic fluid (Chantrell et al. 1982). The energy E_i of a uniformly magnetized spherical particle in a collection of N particles is the sum

$$E_i = E_s + E_r + E_h \tag{2.42}$$

where E_s represents the magnetic dipole–dipole energy, given by E_{dd} of (2.4) summed over a neighborhood of the other particles, E_r is the steric repulsion energy of (2.12), and E_h is the energy of the particle resulting from its orientation in the external magnetic field, given by (1.17). The influence of thermal energy is simulated by moving particles to a randomly selected new location one at a time, allowing any move that gives a reduction in E_i and disallowing any move that increases E_i unless the Boltzmann factor $\exp(-\Delta E_i/kT)$ exceeds a random number X whose value ranges from 0 to 1. The procedure is carried out for all N particles in turn and is repeated for the whole system until the magnetism converges to a statistically steady value. The procedure can be shown statistically to yield a thermodynamic equilibrium configuration.

Results of the study for cobalt particles that are large enough (15 nm) to exhibit appreciable clustering and chaining are shown in Figure 2.16. In zero applied field the particles form open-loop structures with no particular spatial orientation. In large applied field of 1 T, the particles form long chains oriented along the direction of the applied field. The results are consistent with the behavior seen in electron micrographic studies of cobalt dispersions (Hess and Parker 1966; Martinet 1974). The coupling coefficient λ for 15-nm cobalt particles with domain magnetization of 1.7 T at 298 K, from (2.29), is 41, so from (2.40) even a dilute colloid is dominated by clustering, consistent with the simulation.

Monte Carlo studies of O'Grady et al. (1983) show that clustering causes the initial susceptibility of a magnetic fluid to display a Curie–Weiss type behavior:

$$\chi_i = \frac{C}{T - T_0} \tag{2.43}$$

where T_0 is an ordering temperature. This expression may be compared with (2.28) for a monodispersion, in which the parameter T_0 is absent.

Peterson and Krueger (1977) studied in situ particle clustering of ferrofluids in a vertical tube subject to an applied magnetic field. The clusters redistribute under gravity by sedimentation in the tube, with concentration detected by a Colpitts oscillator circuit. The clustering was pronounced for a water-base ferrofluid and nearly absent for many other compositions, such as well-stabilized dispersions in a diester or

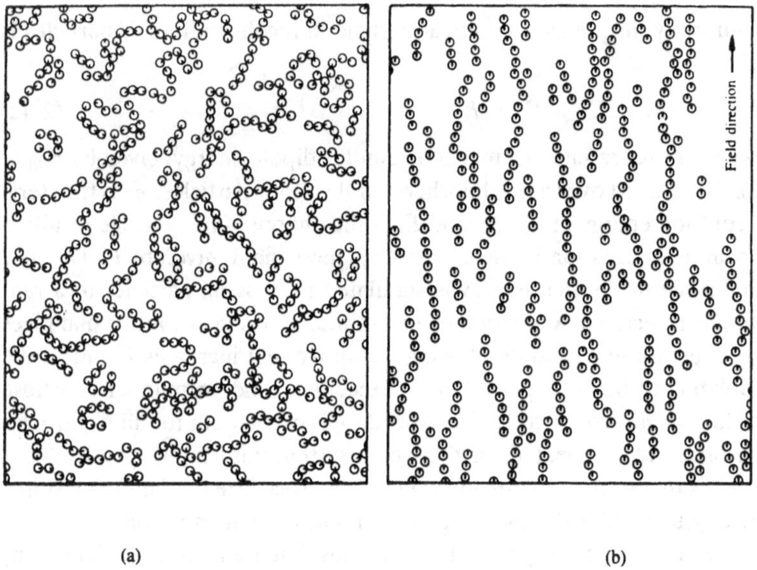

2.16 Monte Carlo simulation in two dimensions: (a) clustering in absence of external field; (b) chaining in presence of field. (*After Chantrell et. al. 1982.*)

hydrocarbon carrier fluid. Clustering, when it occurs, is reversible upon removal of the field, as thermal agitation is effective in redispersing the agglomerates. The technique has found use in evaluating the stability of new ferrofluid compositions.

Particle clustering will affect the transmission of light in a ferrofluid. Goldberg et al. (1971) report the polarization of light by magnetic fluids. Hayes (1975) related the transmission and scattering of light to particle clustering. Mehta (1978) reported additional transmission data. Taketomi (1983) reported anomalously enhanced magnetobirefringence in thin layers of concentrated ferrofluid, in which the effect is amplified by the proximity of the containing wall.

2.9 Tabulated physical properties

Table 2.4 lists fluid and thermal physical properties of numerous types and concentrations of commercially available magnetic fluids.

Table 2.4 *Nominal properties of ferrofluids (298 K)*

Carrier fluid	Magnetic saturation ($A \cdot m^{-1}$)	Density ($kg \cdot m^{-3}$)	Viscosity[a] ($N \cdot s \cdot m^{-2}$)	Pour point[b] (K)	Boiling point[c] (K)	Surface tension ($mN \cdot m^{-1}$)	Thermal conductivity ($W \cdot m^{-1} \cdot K^{-1}$)	Specific heat ($kJ \cdot m^{-3} \cdot K^{-1}$)	Thermal expansion coefficient[d] ($m^3 \cdot m^{-3} \cdot K^{-1}$)
Diester	15,900	1185	0.075	236	422	32	0.16		
Hydrocarbon	15,900	1050	0.003	278	350	28	0.15	1715	9.0×10^{-4}
	31,800	1250	0.006	281	350	28	0.19	1840	8.6×10^{-4}
Fluorocarbon	7,960	2050	2.50	239	456	18	0.20	1966	10.6×10^{-4}
Ester	15,900	1150	0.014	217	422	26	0.31	3724	8.1×10^{-4}
	31,800	1300	0.030	217	422	26		3724	8.1×10^{-4}
	47,700	1400	0.035	211	422	21		3724	8.1×10^{-4}
Water	15,900	1180	0.007	273[e]	299[f]	26	0.59	4184	5.2×10^{-4}
	31,800	1380	0.010	273[e]	299[f]	26		4184	5.0×10^{-4}
Polyphenylether	7,960	2050	7.50	283	533				

[a] Measured in the absence of a magnetic field at shear rate $>10\ s^{-1}$.
[b] Viscosity $100\ N \cdot s \cdot m^{-2}$.
[c] Under a pressure of 133 Pa (1 Torr).
[d] Average over the range 298–367 K.
[e] Freezing point.
[f] At 3.2 kPa.

The magnetic saturation can be proportionately varied by dilution with the carrier fluid. Diester-base fluid has low vapor pressure and hence can be exposed to the environment for long periods of time at normal temperatures with negligible evaporation. The hydrocarbon-base fluids have an electrical resistivity of $10^6\ \Omega\cdot m$ at 60 Hz and a relative dielectric constant of 20 at 1 kHz. The sonic velocity in a hydrocarbon-base ferrofluid was $1.201 \times 10^3\ m\cdot s^{-1}$, compared to $1.275 \times 10^3\ m\cdot s^{-1}$ in the carrier liquid (Chung and Isler 1977). A fluoroalkylpolyether constitutes the base carrier of the fluorocarbon fluids. The ester fluids are based on silicate esters and provide fluidity to low temperatures. Because of their composition, they are susceptible to hydrolytic decomposition. The carrier fluid polyphenyl ether has a radiation resistance in excess of 10^8 rad.

Thermal conductivity measurements on ferrofluids are reported by Popplewell et al. (1982). The thermal conductivity varies with particle loading in a systematic manner. No magnetic field dependence was observed in fields of intensity up to 0.1 T. The variation in thermal conductivity with particle concentration is found to be well described by the theoretical equation of Tareef. This enabled the ratio of physical to magnetic size to be determined.

The pH of water-base ferrofluids can be adjusted over a range of acidic and alkaline values. The electrical conductivity of a particular sample having saturation magnetization of $16{,}000\ A\cdot m^{-1}$ was found to be about constant at $0.2\ S\cdot m^{-1}$ at ac frequencies (Kaplan and Jacobson 1976). The same investigators report some new magnetoelectric effects on capacitance.

Any of the ferrofluids in Table 2.4 can be freeze-thawed without damage.

Comments and supplemental references

Theoretical conditions for the appearance of ferromagnetism in a liquid have been developed by a number of investigators. One of the earliest studies is that of

Gubanov (1960)

The earliest paper to discuss the preparation and physical chemistry of concentrated colloidal ferrofluid based on the Papell technique is by

Rosensweig, Nestor, and Timmins (1965)

Reviews including later information may be found in

Rosensweig (1979a)

Charles and Popplewell (1980)

Martinet (1983)
Scholten (1978, 1983)

A group of papers detail the preparation of cobalt ferrofluid by decomposition of dicobalt octacarbonyl; see

Papirer et al. (1983)

Numerous citations dealing with preparation of ferrofluids including references to the patent art are given in the bibliographies prepared by

Zahn and Shenton (1980)
Charles and Rosensweig (1983)

The formation of particle clusters in magnetic fluids may be regarded as a type of gas–liquid phase transition induced by the magnetic field. Stimulating treatments of the phenomenon using lattice-gas statistical models are developed by

Ausloos et al. (1980)
Sano and Doi (1983)

The latter treatment yields complete predictions of the coexistence curves between the condensed and dilute phases as represented on the magnetic field and concentration state plane.

The general area of ferrofluid technology is far from mature, and there is opportunity to apply ingenuity and scientific insight for the refinement and upgrading of as-produced ferrofluids.

3

ELECTROMAGNETISM AND FIELDS

Which is more real, fields or matter? This question often arises, but it is clear that the two views complement and reinforce each other. The discussion so far has emphasized the material point of view. In this chapter the field description is used, and partial differential equations are obtained governing the distribution of magnetic field in space. The field description then becomes the basis for deriving an expression for the state of stress in vacuum, the remarkable *Maxwell stress tensor*. Subsequently, with this understanding it will be possible to advance in Chapter 4 to the analysis of stress in magnetized media. For that purpose the complete Maxwell's equations of the electromagnetic field will be needed, as well as certain expressions for energy density that are derivable from them. This treatment begins with another look at Coulomb's law.

3.1 Magnetostatic field equations

Coulomb's law of magnetism was introduced as equation (1.1). The **H** field generated at position **r** by a point pole p located at the origin of coordinates can be written

$$\mathbf{H} = \frac{p}{4\pi\mu_0}\frac{\mathbf{r}}{|\mathbf{r}|^3} \tag{3.1}$$

Figure 3.1 represents the field piercing a surface S of arbitrary shape surrounding the pole.

Because **H** can be regarded as the density of field lines, the flux of field lines through a surface patch of area $d\mathbf{S}$ is given by

$$\mathbf{H}\cdot d\mathbf{S} = \frac{1}{4\pi\mu_0}\frac{p}{r^3}\mathbf{r}\cdot d\mathbf{S} = \frac{p}{4\pi\mu_0}\,d\Omega$$

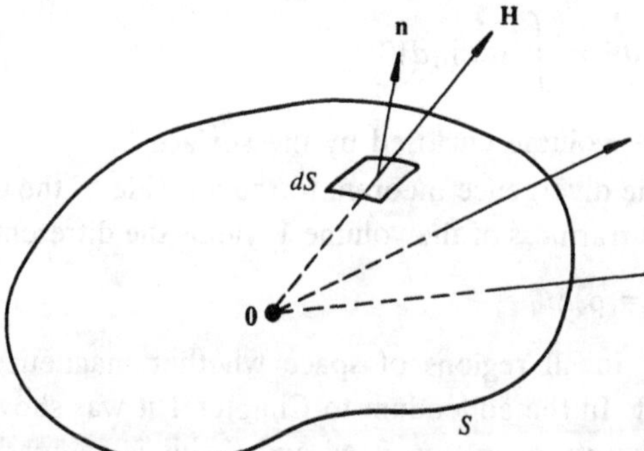

3.1 An imaginary closed surface S that encloses a point pole at the origin. This diagram is used with Coulomb's law to derive the equation $\nabla\cdot\mathbf{B} = 0$.

where $d\Omega = \mathbf{r}\cdot d\mathbf{S}/r^3$ is the solid angle in steradians subtended by $d\mathbf{S}$. Note that only the solid angle appears in the rightmost term, and not the radius, so the enclosing surface may be of any shape. The flux of field out of a closed surface is then

$$\oint_S \mathbf{H}\cdot d\mathbf{S} = \int_0^{4\pi} \frac{p}{4\pi\mu_0}\, d\Omega = \frac{p}{\mu_0}$$

because the solid angle subtended by a whole sphere is 4π steradians. For N poles inside S, where the ith pole contributes an amount $\mathbf{H}_i$ to the total field, a sum of such equations gives

$$\sum_{i=1}^{N} \oint_S \mathbf{H}_i\cdot d\mathbf{S} = \sum_{i=1}^{N} p_i/\mu_0$$

and bringing the summation inside of the integral on the left side results in

$$\oint_S \mathbf{H}\cdot d\mathbf{S} = \sum_{i=1}^{N} p_i/\mu_0$$

where it is recognized that the resultant magnetic field $\mathbf{H}$ is the vector sum of the individual contributions, and p_i can take on positive or negative values. This result can be immediately generalized to the case of a continuous distribution of poles characterized by a pole density ρ_v. The number of poles in the element dV is $\rho_v\, dV$, so

$$\oint_S \mathbf{H}\cdot d\mathbf{S} = \int_V \rho_v/\mu_0 \, dV$$

where V is the volume enclosed by the surface S.

Applying the divergence theorem to the left side of the equation and noting the arbitrariness of the volume V yields the differential equation

$$\nabla\cdot\mathbf{H} = \rho_v/\mu_0$$

which applies in all regions of space whether magnetized matter is present or not. In the addendum to Chapter 1 it was shown that volumetric pole density $\rho_v = -\mu_0 \nabla\cdot\mathbf{M}$ and if this is compatible with the expression given here for $\nabla\cdot\mathbf{H}$, it must be true that $\nabla\cdot\mathbf{H} = -\nabla\cdot\mathbf{M}$. Previously, the vector $\mathbf{B}$ was defined such that

$$\mathbf{B} = \mu_0(\mathbf{H} + \mathbf{M}) \tag{3.2}$$

and so it follows that $\mathbf{B}$ satisfies the relationship

$$\nabla\cdot\mathbf{B} = 0 \tag{3.3}$$

Equation (3.3) is one of Maxwell's equations, and the vector $\mathbf{B}$, previously introduced in Chapter 1, is known as the magnetic induction.

From the form of (3.3) it follows that $\mathbf{B}$ is analogous to the velocity vector of an incompressible fluid. Thus, the $\mathbf{B}$ field can be pictured as the flow of an incompressible fluid. The amount of the fluid entering an arbitrary volume equals the amount flowing out – none accumulates. Also, the lines of $\mathbf{B}$ cannot terminate but must form closed loops or extend indefinitely far.

Additional information can be obtained from Coulomb's law. From (3.1) it may be inferred that the field at a position $\mathbf{r}$ due to a *collection* of poles p_i located at positions $\mathbf{r}_i$ can be written

$$\mathbf{H} = \sum_{i=1}^{N} \frac{p_i}{4\pi\mu_0} \frac{\mathbf{r} - \mathbf{r}_i}{|\mathbf{r} - \mathbf{r}_i|^3}$$

Next it is desired to find the curl of $\mathbf{H}$, which involves evaluating a sum of terms of the type

$$\nabla \times \frac{\mathbf{r} - \mathbf{r}_i}{|\mathbf{r} - \mathbf{r}_i|^3} = \frac{1}{|\mathbf{r} - \mathbf{r}_i|^3} \nabla \times (\mathbf{r} - \mathbf{r}_i) + \nabla\left(\frac{1}{|\mathbf{r} - \mathbf{r}_i|^3}\right) \times (\mathbf{r} - \mathbf{r}_i)$$

The right side of this equation results from applying the vector theorem for the curl of the product of a vector and a scalar function (see Appendix 1). By direct calculation,

$$\nabla \times (\mathbf{r} - \mathbf{r}_i) = \mathbf{0} \qquad \text{and} \qquad \nabla\left(\frac{1}{|\mathbf{r} - \mathbf{r}_i|^3}\right) = -3\frac{\mathbf{r} - \mathbf{r}_i}{|\mathbf{r} - \mathbf{r}_i|^5}$$

Because the cross product of parallel vectors is zero, it follows that

$$\nabla \times \mathbf{H} = \mathbf{0} \tag{3.4}$$

This result represents the *magnetostatic* form of *Ampère's law*, which is applicable when there is no current flow.

Thus, it has been shown that the magnetostatic relations $\nabla \cdot \mathbf{B} = 0$ and $\nabla \times \mathbf{H} = \mathbf{0}$ are mathematical consequences of Coulomb's law.

Scalar potential

Because the curl of **H** is zero, a scalar magnetic potential Ψ can be defined such that

$$\mathbf{H} \equiv -\nabla\Psi \tag{3.5}$$

and (3.4) is satisfied identically. With **B** proportional to **H**, or for uniform **M**, the divergence of **H** is zero, which follows from (3.2) and (3.3). Thus, taking the divergence of both sides of (3.5) shows that the scalar magnetic potential obeys Laplace's equation,

$$\nabla^2\Psi = 0 \tag{3.6}$$

for which many solutions are known from problems in electrostatics, thermal conduction, molecular diffusion, and elasticity. An illustrative solution of equation (3.6) will be developed in Section 3.2, after the general boundary conditions that the fields must satisfy have been developed.

3.2 Magnetic-field boundary conditions

The divergence theorem gives an integral representation that is equivalent to the relationship $\nabla \cdot \mathbf{B} = 0$.

$$\int_V \nabla \cdot \mathbf{B}\, dV = \oint_S \mathbf{B} \cdot d\mathbf{S} = 0 \tag{3.7a}$$

Similarly, Stokes's theorem applied to $\nabla \times \mathbf{H} = \mathbf{0}$ results in Ampère's circuital law:

$$\int_D (\nabla \times \mathbf{H}) \cdot d\mathbf{S} = \oint_L \mathbf{H} \cdot d\mathbf{l} = 0 \tag{3.7b}$$

which is valid when there is no current flow.

At interfacial boundaries separating materials of different properties, the magnetic fields on either side of the boundary obey conditions that are determined by (3.7a) and (3.7b) applied to a differential volume straddling the interface. Figure 3.2 shows at the left a small volume (the *Gaussian pillbox*) whose upper and lower surfaces are parallel and are located on either side of the interface. The short cylindrical side, being of zero length, offers no contribution in the first integral relation, (3.7a), which thus reduces to

$$\oint_S \mathbf{B}\cdot d\mathbf{S} = (B_{2n} - B_{1n})\, dS = 0$$

where the subscript n denotes the normal component. This result states that the component of **B** normal to the interface is continuous. The result may also be expressed

$$\mathbf{n}\cdot(\mathbf{B}_1 - \mathbf{B}_2) = 0 \tag{3.8}$$

Ampère's circuital law may be applied to the contour of differential size enclosing the interface, as shown on the right in Figure 3.2. Because the sides labeled BC and AD approach zero length, they offer no contribution to the integral. The remaining two sides yield

$$\oint_L \mathbf{H}\cdot d\mathbf{l} = (H_{1t} - H_{2t})\, dl = 0$$

Thus, the tangential component of magnetic field is continuous. This may be written

$$\mathbf{n} \times (\mathbf{H}_2 - \mathbf{H}_1) = \mathbf{0} \tag{3.9}$$

The use of these relationships will now be illustrated.

Example – magnetic slab within a uniform, applied magnetic field: A slab of infinite extent in the x and y directions is placed within a uniform magnetic field $\mathbf{H}_0$ having orientation normal to the slab boundaries, as shown in Figure 3.3. Find **B**, **H**, and **M** within the slab when it is

(a) permanently magnetized with magnetization $\mathbf{M}_0 = M_0\mathbf{k}$, and
(b) a permeable material with constant permeability $\mu = B/H$.

Solution: For both situations, (3.8) requires that the normal component of the magnetic induction $\mathbf{B} = \mu_0\mathbf{H}_0$ be continuous across the boundaries.

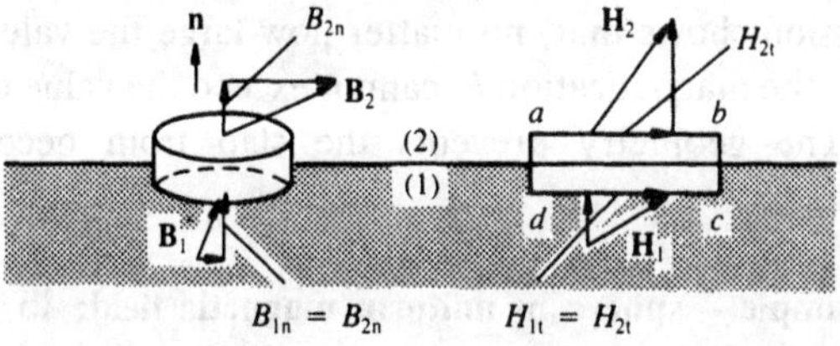

3.2 The normal component of **B** and the tangential component of **H** are continuous across the interface between dissimilar regions.

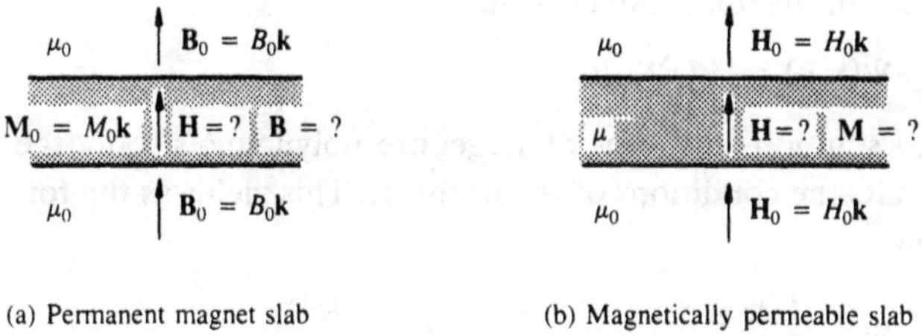

3.3 Uniform magnetic field applied to slabs of hard and soft magnetic materials.

(a) Thus, for the permanently magnetized slab, the continuity of **B** across the boundaries requires that

$$B_0 = B$$

Therefore,

$$\mu_0 H_0 = \mu_0(H + M_0) \qquad \text{or} \qquad H = H_0 - M_0$$

By definition, M_0 is constant. Note that when there is no externally applied field ($H_0 = 0$), the resultant field within the slab is oppositely directed to the magnetization and equal to it; i.e., $H = -M_0$. The field within the slab disappears when $H_0 = M_0$.

(b) For a linearly permeable medium, (3.8) requires that

$$\mu H = \mu_0(H + M) = \mu_0 H_0$$

Eliminating H gives the magnetization induced in the slab:

$$M = H_0(1 - \mu_0/\mu)$$

This expression shows that, no matter how large the value of the permeability μ, the magnetization M cannot exceed the value of the applied field H_0. The geometry prevents the slab from becoming highly magnetized.

Example – sphere in uniform magnetic field: In spherical coordinates with axisymmetric symmetry, Laplace's equation for the potential has the form

$$\frac{\partial}{\partial r}\left(r^2 \frac{\partial \Psi}{\partial r}\right) + \frac{1}{\sin\theta}\frac{\partial}{\partial \theta}\left(\sin\theta \frac{\partial \Psi}{\partial \theta}\right) = 0$$

where r is the radial distance from the origin and θ is the polar angle. A trial solution in the product form

$$\Psi(r, \theta) = R(r)\Psi(\theta)$$

leads to solutions in terms of Legendre polynomials, only the first of which suits the conditions of the problem. This yields as the form of the solution,

$$\Psi = \begin{cases} Ar\cos\theta, & r < R \\ (Cr + D/r^2)\cos\theta, & r > R \end{cases}$$

where R is the sphere radius. The associated magnetic field is then

$$\mathbf{H} = -\nabla\Psi = -\frac{\partial \Psi}{\partial r}\mathbf{i}_r - \frac{1}{r}\frac{\partial \Psi}{\partial \theta}\mathbf{i}_\theta$$

$$= \begin{cases} -A(\mathbf{i}_r\cos\theta - \mathbf{i}_\theta\sin\theta), & r < R \\ -(C - 2D/r^3)\cos\theta\,\mathbf{i}_r + (C + D/r^3)\sin\theta\,\mathbf{i}_\theta, & r > R \end{cases}$$

The magnetic field far from the sphere must approach the uniform applied field $H_0\mathbf{k}$, where $\mathbf{k}$ is the unit vector in the z direction. In spherical coordinates $\mathbf{k} = \mathbf{i}_r\cos\theta - \mathbf{i}_\theta\sin\theta$, so for $r \to \infty$ it may be seen from the solution for $\mathbf{H}$ that

$$C = -H_0$$

By the same token, the form of the solution for $r < R$ shows that $\mathbf{H} = A\mathbf{k}$, and hence the magnetic field within the sphere is uniform and z directed. The solution outside the sphere is the imposed field plus a contribution as if there were a magnetic dipole at the center of the sphere with moment $4\pi D$.

Because the tangential component of $\mathbf{H}$ and the normal component of $\mathbf{B}$ are continuous across the spherical interface, two additional

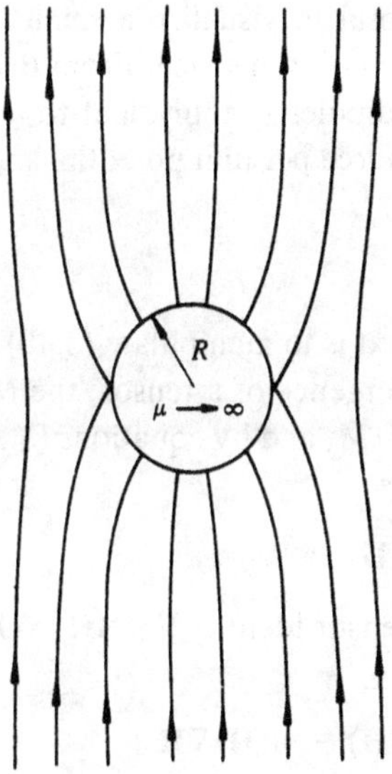

3.4 Magnetic field lines about an infinitely permeable sphere subjected to a uniform impressed magnetic field.

relationships are obtained for the unknown constants. These yield the solutions

$$A = -\frac{3\mu_1 H_0}{\mu_2 + 2\mu_1}, \qquad D = \frac{\mu_2 - \mu_1}{\mu_2 + 2\mu_1} R^3 H_0$$

where μ_1 is the permeability of the medium surrounding the sphere, and μ_2 is the permeability of the sphere. The magnetic field lines are plotted in Figure 3.4 for the case of a sphere permeability approaching infinity. In this limit, **H** within the sphere is zero, and the field lines incident on the sphere are purely radial.

3.3 Maxwell stress tensor

The physical and mathematical meaning of the stress tensor was developed in Section 1.7. As will be shown, the notion provides a powerful alternative description of magnetic force transmission to the force-at-a-distance concept inherent in Coulomb's law. A mathematical expression for the magnetic field stress tensor is developed next. It should be kept in mind that no material medium is required to transmit magnetic stress.

To begin it is helpful to visualize a small "test" cloud of magnetic dipoles surrounding a point in space. From Brown's theorem, equation (1.63), the cloud of dipoles is equivalent to a pole density distribution ρ_v. Because **H** is the force per unit pole, the apparent local force density **F** is

$$\mathbf{F} = \rho_v \mathbf{H} \tag{3.10}$$

What will be done next is to manipulate (3.10) so that the right side can be written as the divergence of a tensor, the Maxwell stress tensor. By (1.63), or $\rho_v = -\mu_0 \nabla \cdot \mathbf{M}$, and by equations (3.2) and (3.3), or $-\nabla \cdot \mathbf{M} = \nabla \cdot \mathbf{H}$, (3.10) becomes

$$\mathbf{F} = \mu_0 \mathbf{H}(\nabla \cdot \mathbf{H}) \tag{3.11}$$

With the aid of the tensor identity $\nabla \cdot (\mathbf{HH}) = \mathbf{H}(\nabla \cdot \mathbf{H}) + \mathbf{H} \cdot \nabla \mathbf{H}$, (3.11) can be written

$$\mathbf{F} = \nabla \cdot (\mu_0 \mathbf{HH}) - \mu_0 \mathbf{H} \cdot \nabla \mathbf{H} \tag{3.12}$$

However, $\mathbf{H} \cdot \nabla \mathbf{H} = \nabla(\frac{1}{2}H^2) - \mathbf{H} \times (\nabla \times \mathbf{H})$ and, with the relationship $\nabla \times \mathbf{H} = \mathbf{0}$ [equation (3.4)], (3.12) can be rearranged to become

$$\mathbf{F} = \nabla \cdot (\mu_0 \mathbf{HH}) - \nabla(\tfrac{1}{2}\mu_0 H^2) \tag{3.13}$$

It is shown presently that $\nabla \cdot (s\mathbf{I}) = \nabla s$ for any scalar function s, and thus (3.13) can be written as the divergence of a tensor in the form

$$\mathbf{F} = \nabla \cdot (\mu_0 \mathbf{HH} - \tfrac{1}{2}\mu_0 H^2 \mathbf{I}) \tag{3.14}$$

where the quantity in the parentheses is the Maxwell stress tensor **T**:

$$\mathbf{T} = \mu_0 \mathbf{HH} - \tfrac{1}{2}\mu_0 H^2 \mathbf{I} \tag{3.15}$$

or, in indicial notation,

$$T_{ij} = \mu_0 H_i H_j - \tfrac{1}{2}\mu_0 H^2 \delta_{ij} \tag{3.16}$$

Proof that $\nabla \cdot (s\mathbf{I}) = \nabla s$:

$$\nabla \cdot (s\mathbf{I}) = \mathbf{e}_k \frac{\partial}{\partial x_k} \cdot (s\delta_{ij}\mathbf{e}_i\mathbf{e}_j) \quad \text{where} \quad \delta_{ij} = \begin{cases} 0, & i \neq j \\ 1, & i = j \end{cases}$$

$$= \delta_{ij}\delta_{ik}\left(\frac{\partial s}{\partial x_k}\right)\mathbf{e}_j = \left(\frac{\partial s}{\partial x_i}\right)\mathbf{e}_i$$

$$= \left(\mathbf{e}_i \frac{\partial}{\partial x_i}\right)s = \nabla s \qquad \text{QED}$$

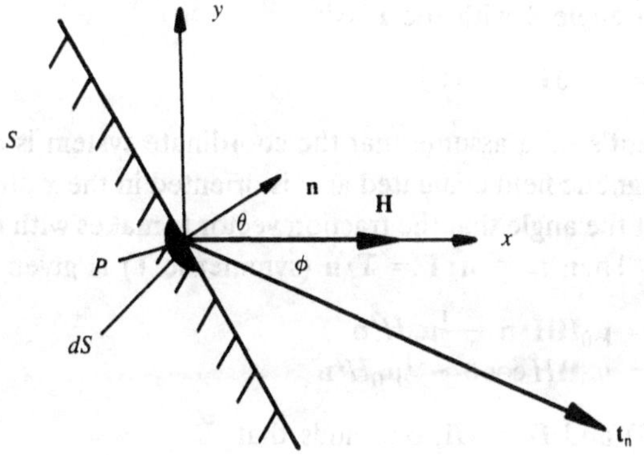

3.5 Development of equation (3.21).

From the mathematical properties of a stress tensor, the traction force on an arbitrary patch of surface dS having the outward-facing normal $\mathbf{n}$ is $\mathbf{n}\cdot\mathbf{T}\,dS$, and so the total force acting on a body occupying volume V enclosed by surface S may be written

$$\mathbf{F}_{tot} = \oint_S \mathbf{n}\cdot\mathbf{T}\,dS \tag{3.17}$$

The importance of this result is that the magnetic force acting on the contents of a "control" volume can be evaluated using only knowledge of the field existing over the surface of the volume. The shape and position of the volume may be chosen for convenience in evaluating the terms of the expression. An illustration of the utility of the concept is given in Appendix 2 as the analysis of a sheet jet.

It should be noted that although the Maxwell stress tensor was constructed on the assumption of a magnetic force exerted on a distribution of dipoles, the result is applicable in free space, i.e., where all matter is absent. That is, the moment of the dipole cloud can be assumed to vanish. Thus, in magnetostatics the integration of the Maxwell stress vector $\mathbf{n}\cdot\mathbf{T}$ over a closed surface of empty space must yield zero force.

Portrait of the Maxwell stress tensor

Consider an element dS of a surface S, shown in Figure 3.5. It is desired to compute the traction acting at point P where the unit normal

n makes an angle θ with the x axis:

$$\mathbf{n} = \cos\theta\,\mathbf{i} + \sin\theta\,\mathbf{j} \tag{3.18}$$

For argument's sake assume that the coordinate system is chosen such that the magnetic field evaluated at P is oriented in the x direction; i.e., $\mathbf{H} = H\mathbf{i}$. Let the angle that the traction vector $\mathbf{t}_n$ makes with the x axis be denoted ϕ. Then $\mathbf{t}_n \equiv \mathbf{n}\cdot\mathbf{T} = \mathbf{T}\cdot\mathbf{n}$ (symmetric $\mathbf{T}$) is given by

$$\begin{aligned}\mathbf{t}_n &= \mu_0\mathbf{H}\mathbf{H}\cdot\mathbf{n} - \tfrac{1}{2}\mu_0 H^2\mathbf{n} \\ &= \mu_0\mathbf{H}H\cos\theta - \tfrac{1}{2}\mu_0 H^2\mathbf{n}\end{aligned} \tag{3.19}$$

Using (3.18) and $\mathbf{H} = H\mathbf{i}$, one finds that

$$\begin{aligned}\mathbf{t}_n &= (\tfrac{1}{2}\mu_0 H^2\cos\theta)\mathbf{i} + (-\tfrac{1}{2}\mu_0 H^2\sin\theta)\mathbf{j} \\ &= t_x\mathbf{i} + t_y\mathbf{j}\end{aligned} \tag{3.20}$$

Thus

$$\frac{t_y}{t_x} = -\tan\theta$$

Since t_y/t_x is also equal to tan ϕ, it may be concluded that

$$|\theta| = |\phi| \tag{3.21}$$

This proves a very general result: In words, *the angle between the surface normal* **n** *and the traction vector* $\mathbf{t}_n$ *is always bisected by the magnetic field* **H**. From the foregoing analysis it is seen that the traction vector corresponds to *collinear tension* when **H** is parallel to **n** and to *transverse pressure* when **H** is perpendicular to **n**. These facts are illustrated in Figure 3.6, where $|\mathbf{t}_n| = (t_x^2 + t_y^2)^{1/2}$. Viewed in another way, the Maxwell vacuum stress tensor gives rise to tension in the direction of the field lines and to pressure in the perpendicular direction; that is, the lines of force behave like rubber bands. A careful study of Figure 3.7 should clarify some of the points made in this section.

3.4 Maxwell's equations

The historical development of electromagnetic theory followed two separate paths until the nineteenth century. One of these was the study of electric charges and their fields, and the other concerned electric currents and the magnetic fields they produce. This was the state of affairs until Faraday showed that a time-varying magnetic field can generate an electric field and Maxwell, through introduction of the dis-

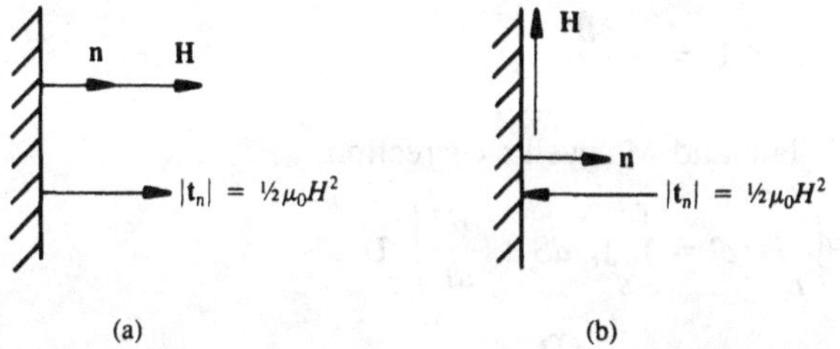

3.6 Geometrical analysis of the stress vector $\mathbf{t}_n$: (a) Collinear tension and (b) transverse pressure.

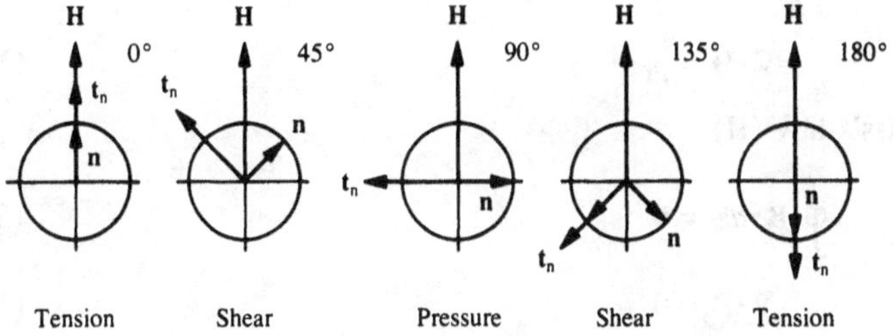

3.7 Portrait of the Maxwell vacuum stress tensor in a magnetic field **H**. As the surface normal vector **n** rotates clockwise, the stress vector $\mathbf{t}_n$ revolves counterclockwise.

placement current, showed that a time-varying electric field produces a magnetic field. The mathematical relationships governing electromagnetic phenomena are the celebrated Maxwell's equations. We cannot hope to trace the development of the full equations in a short compass, but we shall state these laws and discuss aspects of them. The predictions from these relationships have proven infallible in wonderfully diverse applications; an example is the nature of electromagnetic radiation and its propagation in vacuum at the constant speed of light. Maxwell's equations are shown here in both integral (a) and differential (b) form:

Faraday's law:

$$\oint_L \mathbf{E}' \cdot d\mathbf{l} = -\frac{d}{dt}\int_S \mathbf{B} \cdot d\mathbf{S} \tag{3.22a}$$

$$\nabla \times \mathbf{E} = -\frac{\partial \mathbf{B}}{\partial t} \tag{3.22b}$$

Ampère's law and Maxwell's correction:

$$\oint_L \mathbf{H} \cdot d\mathbf{l} = \int_S \mathbf{J}_f \cdot d\mathbf{S} + \frac{d}{dt}\int_S \mathbf{D} \cdot d\mathbf{S} \tag{3.23a}$$

$$\nabla \times \mathbf{H} = \mathbf{J}_f + \frac{\partial \mathbf{D}}{\partial t} \tag{3.23b}$$

Gauss's law (I):

$$\oint_S \mathbf{D} \cdot d\mathbf{S} = \int_V \rho_f \, dV \tag{3.24a}$$

$$\nabla \cdot \mathbf{D} = \rho_f \tag{3.24b}$$

Gauss's law (II):

$$\oint_S \mathbf{B} \cdot d\mathbf{S} = 0 \tag{3.25a}$$

$$\nabla \cdot \mathbf{B} = 0 \tag{3.25b}$$

Charge conservation and continuity of charge:

$$\oint_S \mathbf{J}_f \cdot d\mathbf{S} = -\frac{d}{dt}\int_V \rho_f \, dV \tag{3.26a}$$

$$\nabla \cdot \mathbf{J}_f = -\frac{\partial \rho_f}{\partial t} \tag{3.26b}$$

Except in this chapter, in the subsequent treatment of ferrohydrodynamics, the free charge ρ_f and electric displacement **D** are assumed absent, so that Gauss's law (I) will play no role and Ampère's law will be absent the Maxwell term. In addition, the current $\mathbf{J}_f$ is absent in nonconducting ferrofluid, making all terms in the charge-conservation equation zero.

Following standard convention, the polarization **P** and magnetization **M** are given by the defining equations

$$\mathbf{D} = \varepsilon_0 \mathbf{E} + \mathbf{P} \tag{3.27}$$

$$\mathbf{B} = \mu_0(\mathbf{H} + \mathbf{M}) \tag{3.28}$$

It has already been shown, in Chapter 1, that the definition of **M** permits the identification of $\mu_0\mathbf{M}$ with the dipolar moment density in a substance. The field variables appearing in Maxwell's equations are defined in Table 3.1. $\oint_L$ stands for a line or contour integral around a closed loop L, and $d/dt = \partial/\partial t + \mathbf{v}\cdot\nabla$ is the substantial derivative introduced in Chapter 1. The primed quantity $\mathbf{E}'$ in (3.22a) is the value measured by an observer moving with the contour L at the point in question.

Of the field variables appearing in Table 3.1, **H** and **B** have already been introduced. Magnetic fields are produced both by electric currents and magnetically polarized matter. In matter it is electron spin, which has no classical analog, that principally generates the polarization. The free current density $\mathbf{J}_f$ results from the translational motion of free electric charge having density ρ_f. The term "free" distinguishes such charge from the "bound" charge associated with electrically polarized matter. The electric field vector **E** at a point represents the force experienced by a test electric charge located at the field point. The force is operationally measurable when the field point is accessible to a probe, as in air, and within matter the field is calculable by definition from the field equations. The displacement field **D** is the electric analog of **B**, the field of magnetic induction. When the polarization **P** is known independently, **D** can be calculated from the defining equation (3.27). For example, a permanently polarized rod or electret is the analog of a permanent magnet, with **P** the analog of $\mu_0\mathbf{M}$.

Integral equations

Integral equations are especially useful in moving boundary problems and in deriving general expressions for interfacial or boundary conditions. The integral equations as written are true for stationary contours and surfaces. However, if the contours and surfaces are moving, then the field quantities must be written in the moving frame of reference (the primed frame). This in turn requires transforming the field quantities from the moving frame to the observer's frame to obtain useful working relationships. The general transformation of field variable is accomplished consistent with special relativity theory with the assertion that Maxwell's differential equations must exhibit the same form in both frames, an assertion leading to the *Lorentz transformation* (see, e.g., Jackson 1975, Chapter 11). For velocities small compared to the velocity of light (i.e., $u \ll c$), the only numerically significant transformation to be employed in the integral equations relates to the electric field **E** (see below).

The line integral in the integral form of Faraday's law is known as the

Table 3.1 *Summary of field variables and units*

	$\mathbf{E}$	$\mathbf{D}$	$\mathbf{H}$	$\mathbf{B}$	$\mathbf{J}_f$	ρ_f
Name	Electric field	Electric displacement	Magnetic field	Magnetic induction	Free current density	Free charge density
SI units	$\frac{\text{volts}}{\text{meter}}$	$\frac{\text{coulomb}}{(\text{meter})^2}$	$\frac{\text{ampere}}{\text{meter}}$	tesla	$\frac{\text{ampere}}{(\text{meter})^2}$	$\frac{\text{coulomb}}{(\text{meter})^3}$

electromotive force (EMF). Faraday's integral law states that the EMF is induced by the changing flux of **B** through the circuit L, and the flux can be changed by changing the magnetic induction or the shape or orientation or position of the circuit. The time derivative appearing on the right side of the law can be brought inside the integral with the aid of the surface integral version of the *Reynolds' transport theorem* (see Appendix 1):

$$\frac{d}{dt}\int_S \mathbf{B}\cdot d\mathbf{S} = \int_S \left\{\frac{\partial \mathbf{B}}{\partial t} + \nabla \times (\mathbf{B} \times \mathbf{v}) + \mathbf{v}(\nabla\cdot\mathbf{B})\right\}\cdot d\mathbf{S} \tag{3.29}$$

However, because $\nabla\cdot\mathbf{B} = 0$ and by Stokes's theorem (see Appendix 1), (3.29) becomes

$$\frac{d}{dt}\int_S \mathbf{B}\cdot d\mathbf{S} = \int_S \frac{\partial \mathbf{B}}{\partial t}\cdot d\mathbf{S} + \oint_L (\mathbf{B} \times \mathbf{v})\cdot d\mathbf{l} \tag{3.30}$$

Using equation (3.22a) to eliminate the integral on the left side of (3.30) gives

$$\oint_L \mathbf{E}'\cdot d\mathbf{l} = \int_S -\frac{\partial \mathbf{B}}{\partial t}\cdot d\mathbf{S} + \oint_L (\mathbf{v} \times \mathbf{B})\cdot d\mathbf{l} \tag{3.31a}$$

Combining the contour integral terms of equation (3.31a) then yields the form

$$\oint_L (\mathbf{E}' - \mathbf{v} \times \mathbf{B})\cdot d\mathbf{l} = \int_S -\frac{\partial \mathbf{B}}{\partial t}\cdot d\mathbf{S}$$

or

$$\oint_L \mathbf{E}\cdot d\mathbf{l} = \int_S -\frac{\partial \mathbf{B}}{\partial t}\cdot d\mathbf{S} \tag{3.31b}$$

where

$$\mathbf{E} = \mathbf{E}' - \mathbf{v} \times \mathbf{B} \tag{3.32}$$

This is the *Galilean transformation* relating **E** to **E**′. It is an approximation, although a very good one; through the use of Reynolds' theorem it has been tacitly assumed that $x' = x - v_x t$, $y' = y$, $z' = z$, and $t' = t$ connects coordinates of space and time in a primed system moving at speed v_x along the x axis relative to an unprimed system. The exact relationship between **E** and **E**′ is developed in many electromagnetics texts. Return now to equation (3.31b) and let Stokes's theorem be applied to transform the left side to an integral over surface area; noting the arbitrariness of the surface of integration now leads at once to the differential form of Faraday's law, equation (3.22b). This equation, and in fact the complete set of the differential Maxwell equations, are always correct as written.

Differential equations

As already noted, a magnetic field is produced not only from polarized matter but also from motion of charge, that is, from electric current. The influence of the current density $\mathbf{J}_f$ appears as the first term on the right side in the differential form of Ampère's law.

Although he was unable to prove it experimentally, Faraday believed that a time-varying electric field should also generate a magnetic field. It was left for Maxwell to show that Faraday was right and that without this amendment Ampère's law and conservation of charge are inconsistent. Thus, if the divergence of the incomplete differential form of Ampère's law is taken, it is found that

$$\nabla \cdot (\nabla \times \mathbf{H}) = 0 = \nabla \cdot \mathbf{J}_f$$

because $\nabla \cdot (\nabla \times \mathbf{A}) = 0$ for any vector field $\mathbf{A}$ with continuous partial derivatives of second order. This result, however, is in contradiction with (3.26b) if a time-varying charge is present. Maxwell realized that if the *displacement current* $\partial \mathbf{D}/\partial t$ is added to the right-hand side of Ampère's law, charge conservation is automatically satisfied.

Thus, if the divergence of (3.23b) is taken and the resulting equation is simplified with the aid of (3.24b), the charge-conservation equation, (3.26b), is obtained. Unlike in fluid mechanics, where the mass conservation equation is an independent relation, Maxwell's equations have built into them the principle of charge conservation.

Equation (3.24a), the first of Gauss's laws, is a direct consequence of Coulomb's law for point electric charges. Equation (3.25a) states that isolated magnetic poles do not exist, and to date none have been found. The differential form was derived as (3.3). The equivalence of the differential and integral forms of Gauss's laws is immediately demonstrated through use of the divergence theorem.

Note that Maxwell's equations have more dependent variables than the number of equations relating them. In working with macroscopic aspects of electromagnetism it is convenient to relate the field quantities by constitutive laws defining the functions ε, μ, and σ. For *isotropic* but *nonlinear* materials these are

$$\mathbf{D} \equiv \varepsilon(\mathbf{E})\mathbf{E} \tag{3.33}$$
$$\mathbf{B} \equiv \mu(\mathbf{H})\mathbf{H} \tag{3.34}$$
$$\mathbf{J}_f \equiv \sigma(\mathbf{E} + \mathbf{v} \times \mathbf{B}) \tag{3.35}$$

where ε is the permittivity, μ is the permeability, and σ is the electrical conductivity of the material. In free space ε and μ are constants having the values $\varepsilon_0 = 8.854 \times 10^{-12}\ \text{F}\cdot\text{m}^{-1}$ and $\mu_0 = 4\pi \times 10^{-7}\ \text{H}\cdot\text{m}^{-1}$. The speed of light in vacuum is $c = (\varepsilon_0\mu_0)^{-1} = 2.9979 \times 10^8\ \text{m}\cdot\text{s}^{-1}$. Equation (3.35) is *Ohm's law* for moving media and is correct only for systems moving at velocities much less than the speed of light.

In most work to date in FHD the medium is ferromagnetically responsive and both the free current density and Maxwell's displacement current are negligible. Hence the field equations of FHD are usually employed in the magnetostatic limit of Maxwell's equation:

$$\nabla \cdot \mathbf{B} = 0 \tag{3.36}$$
$$\nabla \times \mathbf{H} = \mathbf{0} \tag{3.37}$$

These relationships were previously developed as (3.3) and (3.4), respectively. It is not difficult to imagine circumstances in which an alternating current is present and the displacement current produces a magnetic field or in which current flows in a conducting ferrofluid, so that (3.23b) must be used in place of (3.37). Also, for example, if appreciable amounts of free charge were present, then the full description would necessarily include (3.22) and (3.24). Moreover, in dealing with the *sources* of field it is important to employ additional terms of Maxwell's equations. Illustrations of sources are provided with reference to the complete Ampère's law in the following examples.

Example – infinitely long straight conductor as illustration of Ampère's law in steady state, $\partial\mathbf{D}/\partial t = 0$ and $\mathbf{J}_f \neq 0$: A long, straight wire of radius R carries a constant current I, as shown in Figure 3.8. Determine the distribution of magnetic field $\mathbf{H}$ established by this current.

Solution: In this situation displacement current is absent because the fields are time steady. Ampère's law relates magnetic field to sources of current; from (3.23a) the appropriate form is

$$\oint_L \mathbf{H}\cdot d\mathbf{l} = \int_S \mathbf{J}_f \cdot d\mathbf{S} \tag{3.38}$$

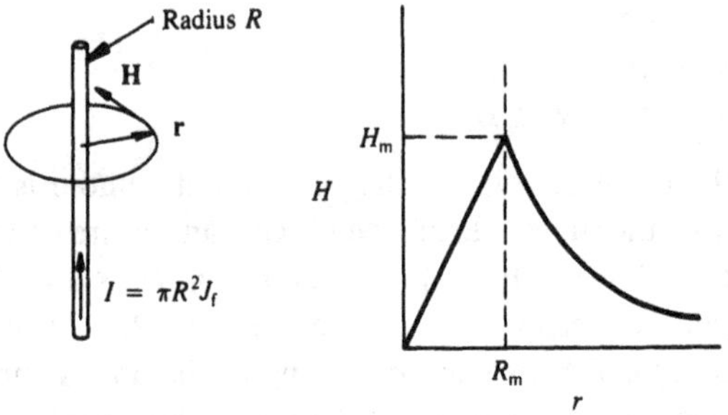

3.8 Application of Ampère's circuital law to a long current-carrying wire.

From symmetry considerations based on the circular contour indicated in Figure 3.8, the free current gives rise to an azimuthal field $H = H_\phi$ given by

$$H_\phi 2\pi r = \begin{cases} J_f \pi r^2 & r < R \\ J_f \pi R^2, & r > R \end{cases} \tag{3.39}$$

or

$$H_\phi = \begin{cases} \dfrac{J_f r}{2} = \dfrac{Ir}{2\pi R^2}, & r < R \\[2ex] \dfrac{I}{2\pi r}, & r > R \end{cases} \tag{3.40}$$

Thus, the field increases linearly with radius within the wire and dies off inversely with radius outside the wire.

Example – alternating current electrical capacitor as illustration of the unsteady state Ampère's law, $\mathbf{J}_f = \mathbf{0}$ and $\partial\mathbf{D}/\partial t \neq \mathbf{0}$: Determine the magnetic field established within the gap of an air capacitor as sketched in Figure 3.9. The metallic capacitor plates are electrically driven with an alternating source of voltage V. Neglect fringe field effects.

Solution: The alternating voltage creates an alternating current i in the conductors and an accumulation of free charge on the capacitor plates. No charge current flows in the air gap, but a uniform field $\mathbf{D} = \varepsilon_0\mathbf{E}$ is established in the gap, and its time rate of change is the dis-

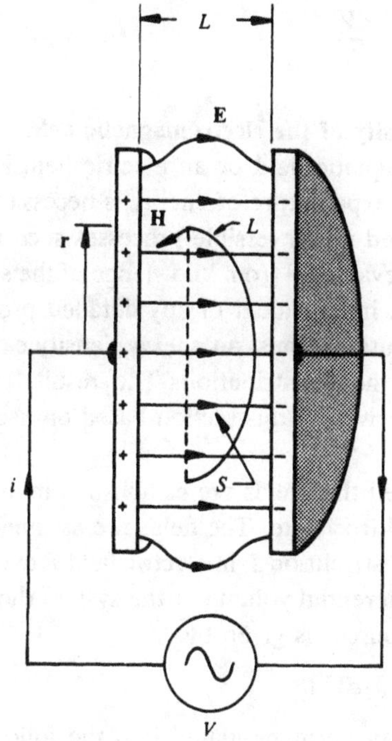

3.9 Application of Ampère's circuital law to the magnetic field produced by a displacement current.

placement current $\partial \mathbf{D}/\partial t$. From Ampère's circuital law with current $\mathbf{J}_f = \mathbf{0}$.

$$\oint_L \mathbf{H} \cdot d\mathbf{l} = \frac{d}{dt} \int_S \mathbf{D} \cdot d\mathbf{S} \tag{3.41}$$

The sketch indicates a portion of a stationary contour L of radius r enclosing an area S in the gap space. The integrals can be evaluated using symmetry considerations to give

$$H 2\pi r = \pi r^2 \frac{dD}{dt} \tag{3.42}$$

where H is constant in circular loops about the center.

Since the applied voltage $V = EL$, where L is the gap length, the field due to the displacement current is

$$H = \frac{\varepsilon_0}{2} \frac{r}{L} \frac{dV}{dt} \tag{3.43}$$

3.5 Energy density of the electromagnetic field

When a magnetic field or an electric field is established in a region of space, an expenditure of energy is necessary over and above any energy consumed in irreversible processes such as ohmic heating. This energy can be evaluated from knowledge of the spatial distribution of the field vectors, independent of any detailed process employed in reaching the distribution. Thus, an energy density can be attributed to an electromagnetic field distribution. The result, important in later developments, is derived in this section based on use of the full set of Maxwell's equations.

It will be supposed that fields are established in an arbitrary system starting from a field free state. The fields are assumed to be generated by electric current distribution $\mathbf{J}_f$ in electric field $\mathbf{E}$ over the field volume V. Thus, in any differential volume of the system the differential work done by external sources is given by

$$dW = -\mathbf{E}\cdot\mathbf{J}_f \, dV \, dt \tag{3.44}$$

The term $\mathbf{E}\cdot\mathbf{J}_f$ can be manipulated into the following form, where only electromagnetic field variables appear on the right side:

$$\mathbf{E}\cdot\mathbf{J}_f = \mathbf{E}\cdot\left(\nabla \times \mathbf{H} - \frac{\partial \mathbf{D}}{\partial t}\right) - \mathbf{H}\cdot\left(\nabla \times \mathbf{E} + \frac{\partial \mathbf{B}}{\partial t}\right) \tag{3.45}$$

$\mathbf{J}_f$ was eliminated using Ampère's law, (3.23b), in the first term on the right side, and the second term is identically zero on account of Faraday's law, equation (3.22b). If use is now made of the vector identity

$$\nabla\cdot(\mathbf{E} \times \mathbf{H}) = \mathbf{E}\cdot(\nabla \times \mathbf{H}) - \mathbf{H}\cdot(\nabla \times \mathbf{E}) \tag{3.46}$$

(3.44) can be rewritten as an integral expression for the *work done on the system by external sources to establish the field:*

$$W = \int_V \int_0^t \left(\mathbf{E}\cdot\frac{\partial \mathbf{D}}{\partial t} + \mathbf{H}\cdot\frac{\partial \mathbf{B}}{\partial t}\right) dt \, dV + \int_V \int_0^t \nabla\cdot(\mathbf{E} \times \mathbf{H}) \, dt \, dV \tag{3.47}$$

The second term on the right side can be converted to a surface integral with the aid of the divergence theorem:

$$\int_V \nabla\cdot(\mathbf{E}\times\mathbf{H})\,dV = \oint_S (\mathbf{E}\times\mathbf{H})\cdot\mathbf{n}\,dS \tag{3.48}$$

where $\mathbf{E}\times\mathbf{H}$ is called the *Poynting vector* and has units of watts per square meter. The Poynting vector formally describes the flux or flow of electromagnetic energy through space; only its integral over a closed surface has measurable significance. On the assumption that **H** and **E** go to zero on the system boundaries, (3.47) becomes

$$W = \int_V \int_0^t \left(\mathbf{E}\cdot\frac{\partial \mathbf{D}}{\partial t} + \mathbf{H}\cdot\frac{\partial \mathbf{B}}{\partial t}\right) dt\,dV \tag{3.49}$$

The volume integrations are carried out over stationary volume elements, so (3.49) becomes

$$W = \int_V \left(\int_0^D \mathbf{E}\cdot d\mathbf{D} + \int_0^B \mathbf{H}\cdot d\mathbf{B}\right) dV \tag{3.50}$$

where it is assumed that the fields are initially zero. The integrand is *energy density of the electromagnetic field* in units of newton-meters per cubic meter.

Based on the form given in (3.50), $\mathbf{H}\cdot d\mathbf{B}$ will be considered to represent the differential contribution to local energy density of the magnetic field. When **H** and **B** are collinear, as in soft magnetic materials under equilibrium conditions, the term $\mathbf{H}\cdot d\mathbf{B}$ is expressible in terms of the field magnitudes as $H\,dB$.

3.6 Transformed expression for the field energy

In the previous section (3.50) encompasses a general expression for the magnetostatic energy of an entire system, which could be composed, for example, of a volume of magnetic fluid together with regions containing no magnetic material or other regions containing a source of magnetic field – such as a current-carrying coil, permanent magnets and permeable iron structure, or other materials. However, at times it will be convenient to work with an expression that isolates just the magnetostatic energy identified with the magnetic fluid, and that is what is accomplished in this section.

Consider Figure 3.10. Let V_1 be a linear and isotropic medium of permeability μ_1, and let V_2 be the volume occupied by magnetic fluid, which will be treated as linear and isotropic with permeability μ_2. Thus, its energy density from (3.50) becomes $\frac{1}{2}\ \mathbf{H}\cdot\mathbf{B}$. The source of magnetic field is a motionless loop of wire considered perfectly conductive (i.e., of infinitely large electrical conductivity σ) and carrying an electric

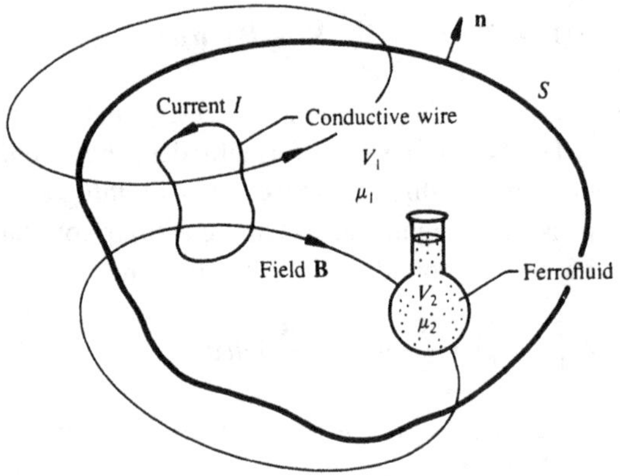

3.10 Developing an expression for magnetostatic energy of a magnetizable fluid.

current I. From Faraday's law in integral form, (3.22a), it is seen that the total flux passing through the loop $\phi = \int \mathbf{B} \cdot d\mathbf{S}$ is constant when the line integral of $\mathbf{E}'$ evaluated around the loop is zero. The loop is assumed to be perfectly conducting, so $\mathbf{E}'$ must be zero everywhere in it; otherwise, from Ohm's law, the current density $\mathbf{J}_f = \sigma \mathbf{E}'$ would become infinite. Hence the flux ϕ remains constant when the surroundings are modified as by moving in the volume of magnetic fluid. The essential point is that no additional energy storage is possible in the current loop.

A one-step process can be considered in which the magnetic fluid initially is located distantly outside the volume enclosed by surface S and then is brought into this volume, thus displacing an equal volume of the first medium. The energy change resulting from the introduction of the magnetic fluid to the region is

$$\begin{aligned}\Delta W &= \int_V \tfrac{1}{2}(\mathbf{H} \cdot \mathbf{B} - \mathbf{H}_0 \cdot \mathbf{B}_0)\, dV \\ &= \int_V \tfrac{1}{2}(\mathbf{H} + \mathbf{H}_0) \cdot (\mathbf{B} - \mathbf{B}_0)\, dV - \int_V \tfrac{1}{2}(\mathbf{H}_0 \cdot \mathbf{B} - \mathbf{H} \cdot \mathbf{B}_0)\, dV \qquad (3.51)\end{aligned}$$

where $V = V_1 + V_2$. The initial field is represented by $\mathbf{B}_0$ and $\mathbf{H}_0 = -\nabla\psi_0$, the final field by $\mathbf{B}$ and $\mathbf{H} = -\nabla\psi$, where ψ_0 and ψ are the potential functions. With the aid of the identity

$$\nabla \cdot \psi \mathbf{A} = \mathbf{A} \cdot \nabla\psi + \psi \nabla \cdot \mathbf{A} \qquad (3.52)$$

and the divergence theorem, the first integral on the right side of the final line of (3.51) may be written

$$-\int_V \tfrac{1}{2}(\nabla\psi + \nabla\psi_0)\cdot(\mathbf{B} - \mathbf{B}_0)\,dV$$

$$= \oint_S \tfrac{1}{2}(\psi + \psi_0)(\mathbf{B} - \mathbf{B}_0)\cdot\mathbf{n}\,dS$$

$$+ \int_V \tfrac{1}{2}(\psi + \psi_0)\nabla\cdot(\mathbf{B} - \mathbf{B}_0)\,dV \qquad (3.53)$$

Now the outer boundary of S may be allowed to recede to infinity so as to include all space. The integrand of the surface integral goes to zero faster than the surface area grows, so that the surface integral disappears. It is also true that $\nabla\cdot(\mathbf{B} - \mathbf{B}_0)$ is zero throughout V, and the volume integral vanishes. Thus, the right side of (3.53) is zero and hence

$$\int_V \tfrac{1}{2}(\mathbf{H} + \mathbf{H}_0)\cdot(\mathbf{B} - \mathbf{B}_0)\,dV = 0 \qquad (3.54)$$

The second integral on the right of (3.51) may written as the sum over the subvolumes V_1 and V_2:

$$-\int_{V_1} \tfrac{1}{2}(\mathbf{H}_0\cdot\mathbf{B} - \mathbf{H}\cdot\mathbf{B}_0)\,dV - \int_{V_2} \tfrac{1}{2}(\mathbf{H}_0\cdot\mathbf{B} - \mathbf{H}\cdot\mathbf{B}_0)\,dV$$

$$= -\int_{V_2} \tfrac{1}{2}(\mathbf{H}_0\cdot\mathbf{B} - \mathbf{H}\cdot\mathbf{B}_0)\,dV \qquad (3.55)$$

This is because in V_1

$$\mathbf{H}_0\cdot\mathbf{B} = \mathbf{H}_0\cdot\mu_1\mathbf{H} = \mu_1\mathbf{H}_0\cdot\mathbf{H} = \mathbf{B}_0\cdot\mathbf{H} \qquad (3.56)$$

which makes the integral over V_1 identically zero. In V_2, $\mathbf{B} = \mu_2\mathbf{H}$ and $\mathbf{B}_0 = \mu_1\mathbf{H}_0$, so that

$$-\int_{V_2} \tfrac{1}{2}(\mathbf{H}_0\cdot\mathbf{B} - \mathbf{H}\cdot\mathbf{B}_0)\,dV = -\int_{V_2} \tfrac{1}{2}(\mu_2 - \mu_1)\mathbf{H}_0\cdot\mathbf{H}\,dV$$

Therefore,

$$\Delta W = -\int_{V_2} \tfrac{1}{2}(\mu_2 - \mu_1)\mathbf{H}_0\cdot\mathbf{H}\,dV \qquad (3.57)$$

If $\mu_1 = \mu_0$, then in V_2

$$\mathbf{B} = \mu_2\mathbf{H} = \mu_0(\mathbf{H} + \mathbf{M})$$

so that

$$\mathbf{M} = \left(\frac{\mu_2}{\mu_0} - 1\right)\mathbf{H} \tag{3.58}$$

and (3.57) becomes

$$\Delta W = -\int_{V_2} \tfrac{1}{2}\mu_0 \mathbf{M}\cdot\mathbf{H}_0\, dV \tag{3.59}$$

Equation (3.59) expresses, in terms of the magnetization vector **M**, the energy effect in introducing a volume of magnetic fluid into a fixed-source static magnetic field in free space. It is remarkable in that first, the field energy that originally was expressed as an integral over all space is here expressed as an integral over just the fluid volume and, second, it is the original field $\mathbf{H}_0$ and not the final field **H** that figures into the expression. These features prove very convenient in handling certain problems that arise; an example is the problem of spacing in a magnetic fluid labyrinth, discussed in Chapter 7.

Comments and supplemental references

We cannot, of course, come even close to giving a full picture of one of the oldest and richest subjects of physics, classical electromagnetic theory. This subject, along with classical and quantum mechanics, forms the core of present-day training for physicists and is of considerable importance in electrical engineering curricula. Consequently there are numerous good books that the interested reader can turn to for further study. Excellent, modern introductions are found in the texts of

Zahn (1979)

which emphasizes problem solving, and

Reitz, Milford, and Christy (1979)

More advanced treatments are those of

Landau and Lifshitz (1960)
Panofsky and Phillips (1962)
Jackson (1975)

and the somewhat older, but still classic, texts of

Jeans (1925)
Stratton (1941)

Also of relevance and great value are books on field theory and potential theory. Two fine examples are

Moon and Spencer (1961)
Kellogg (1953)

A splendid graduate-level text that develops the concepts of electrohydrodynamics in parallel with some aspects of magnetism is that of

Melcher (1981)

For a development that is radically different from traditional treatments, see the mathematically awesome but precise article by

Truesdell and Toupin (1960)

whose principal objective is "to isolate those aspects of the theory which are independent of the assigned geometry of space-time from those whose formulation and interpretation depend or imply a particular space-time geometry."

The derivation of the macroscopic Maxwell's equations by suitably averaging over aggregates of atoms dates back to Lorentz. For a fresh look at the subject see

Russakoff (1970)

4

STRESS TENSOR AND THE EQUATION OF MOTION

In ordinary hydrodynamics, the only body force acting from the outside on the entire volume of a fluid is the force of gravity. In magnetohydrodynamics, an ionized gas or liquid metal carrying an electric current in the presence of a magnetic field is subject to the Lorentz body force, which represents the interaction of the current with the magnetic field. In electrohydrodynamics, ions or other particles that bear an electric charge are acted on by an electrodynamic force when they are exposed to an electric field. In the case of a nonconducting ferrofluid, there are no electric currents or electric charges. The body force originates from the interaction of a magnetic field with the ferromagnetic dipole moment characteristic of each colloidal particle.

The question of the spatial distribution of the force has been a subject of controversy for nearly a hundred years, as outlined by Slepian (1950), Penfield and Haus (1967), and Byrne (1977). Proposals may be found in the literature of a multiplicity of force laws with their corresponding stress tensors, all yielding different distributions of the force. In many of the older theories the elastic properties of the medium are either ignored or supposed unaffected by the fields, and a confusion arises in reconciling the various "pressures" found in the literature. A satisfactory derivation of the volume forces from an energy principle accounting for elastic deformations seems to have been proposed first by Korteweg (1880) and developed by Helmholtz (1882) with restriction to linear media.

Because colloidal ferrofluids exhibit nonlinear magnetization, the present derivation will include the nonlinear effects, an extension originally obtained by Cowley and Rosensweig (1967) and Penfield and Haus (1967). The derivation is thermodynamic and invokes the conservation of energy. A single object, the magnetic stress tensor, provides a most satisfactory tool for the quantitative description of the forces and their distribution. It will be seen to account not only for the body force, but also for the surface-force density that appears in magnetic fluids.

4.1 Thermodynamic background

The first law written for a differential change in state of a homogeneous substance of unit mass is

$$\delta Q - \delta W = dU' \tag{4.1}$$

where δQ is the heat added, δW is work done by the substance on the surroundings, and U' is internal energy per unit mass (p.u.m.). When the process is carried out reversibly, the heat added is expressible in terms of entropy change:

$$\delta Q = T\,dS'$$

where S' is the entropy per unit mass of the substance. The *Helmholtz free energy* F' is the quantity

$$F' \equiv U' - TS'$$

The differential of F' is accordingly,

$$dF' = dU' - T\,dS' - S'\,dT$$

Eliminating δQ and dU' from equation (4.1) gives

$$dF' = -\delta W - S'\,dT \tag{4.2a}$$

and for an isothermal process

$$dF'_T = -\delta W \tag{4.2b}$$

That is, the isothermal change in free energy equals the negative of the work done in a reversible process, no matter what type of work is performed.

The types of work of interest here are expansion, i.e., work involving change of volume, and magnetic work, or that involving a change in the magnetostatic field energy. Thus [refer to the discussion following equation (3.50)]

$$\delta W = p\,dv - d\left(v\int H\,dB\right)$$

$$= \left(p - \int H\,dB\right) d\upsilon - \upsilon H\,dB \tag{4.3}$$

where $\upsilon = \rho^{-1}$ is the specific volume. From (4.2) and (4.3)

$$dF' = -\left(p - \int H\,dB\right) d\upsilon - S'\,dT + \upsilon H\,dB \tag{4.4}$$

and so it is seen that a functional dependence exists that may be indicated as $F' = F'(\upsilon, T, B)$. Because B depends on υ, T, and H, the dependence can also be expressed $F' = F'(\upsilon, T, H)$.

In the absence of a magnetic field, from (4.4),

$$dF' = -p\,d\upsilon - S'\,dT \tag{4.5}$$

In the absence of fields, F' is a function of υ and T, so the following differential expansion can be written:

$$dF'(\upsilon, T, 0) = \left(\frac{dF'}{d\upsilon}\right)_T d\upsilon + \left(\frac{dF'}{dT}\right)_\upsilon dT \tag{4.6}$$

Comparing (4.5) and (4.6), one concludes that pressure and entropy are dependent only on the free energy and in the following manner:

$$\left(\frac{dF'}{d\upsilon}\right)_T = -p$$

$$\left(\frac{dF'}{dT}\right)_\upsilon = -S' \tag{4.7b}$$

when fields are absent.

At times it is more convenient to work with the free energy per unit volume, denoted F:

$$F' = \upsilon F \tag{4.8}$$

From (4.8) and (4.7a) the ordinary thermodynamic pressure p, which may be designated $p(\upsilon, T)$ or $p(\rho, T)$, is given by

$$\left[\frac{\partial(\upsilon F)}{\partial\upsilon}\right]_T = -p \tag{4.9}$$

The differential of the free energy F', from equation (4.8), is

$$dF' = \upsilon\,dF + F\,d\upsilon$$

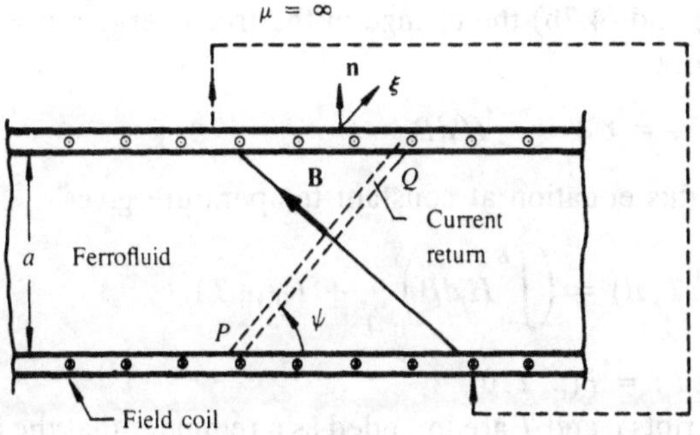

4.1 Geometry for evaluating the stress tensor in magnetizable fluid. The region exterior to the space between the two plates acts as a constant temperature reservoir, besides being a region of high permeability. The condition of high permeability implies that no field energy can be stored in the region. (*After Cowley and Rosensweig 1967.*)

Use of this expression to eliminate dF' in equation (4.4) then shows that at constant temperatures and constant volume

$$dF = H\,dB \tag{4.10}$$

This expression links the magnetostatic energy density to the thermodynamic free energy, both on a per-unit-volume (p.u.v.) basis.

4.2 Formulation of the magnetic stress tensor

Consider a homogeneous layer of fluid confined between parallel planes of separation a (see Figure 4.1). Such a layer may be considered to be representative of an elementary volume present in the fluid and small enough that the field and fluid properties are nearly uniform over the volume. A uniform magnetic field can be generated by sheet currents in the planes and returned through an external path of zero reluctance. It is convenient to think of these sheet currents as if they were made up of discrete wires or coils. The wires carrying the return current are connected at an angle ψ to the bounding planes, and hence **H** and **B** are directed at an angle $\frac{1}{2}\pi + \psi$.

Step 1 – isothermal magnetization at constant volume: Let the applied field **H** be increased while the bounding planes are held fixed (so that there is no deformation of the fluid) under isothermal conditions.

From (4.10) and (4.2b) the change in the free energy p.u.v. is just the magnetic work

$$-dF_T = \delta W = -H\,dB \tag{4.11}$$

Integrating this equation at constant temperature gives

$$F(\rho, T, B) = \left(\int_0^B H\,dB\right)_{\rho,T} + F_0(\rho, T) \tag{4.12}$$

where $F_0(\rho, T) = F(\rho, T, 0)$.

The subscripts ρ and T are intended as a reminder that the integration is carried out at constant density ρ (or specific volume υ) and constant temperature T. Equation (4.12) can be equivalently written

$$F(\rho, T, H) = F_0(\rho, T) + HB - \left(\int_0^H B\,dH\right)_{\rho,T} \tag{4.13}$$

Step 2 – isothermal deformation: Now let the fluid be deformed by an infinitesimal displacement $\boldsymbol{\xi}$ of one of the bounding planes, where $\boldsymbol{\xi}$ is not necessarily parallel to the unit normal $\mathbf{n}$ to the plane, thus increasing the gap width by $\delta a = \boldsymbol{\xi}\cdot\mathbf{n}$. During this step the temperature remains constant because of heat exchange with the thermal reservoir, and the current to the field coils is adjusted so that the flux linked by the current does not change. The latter condition implies that no electromagnetic work is delivered.

The main purpose of this analysis is to determine the form of $\mathbf{T}'_{\mathrm{m}}$, the magnetic stress tensor of the magnetic fluid. Thus, a force $-\mathbf{n}\cdot\mathbf{T}'_{\mathrm{m}}$ is exerted by the layer of fluid per unit area of the bounding plane, and the mechanical work done (per unit area) is given by

$$\delta W = -\mathbf{n}\cdot\mathbf{T}'_{\mathrm{m}}\cdot\boldsymbol{\xi} \tag{4.14}$$

From equation (4.2b) the mechanical work done in an isothermal process is equal to the decrease in the free energy, which in this case is given by

$$\delta W = -d(\mathrm{F}a)_T = -F\,da - a\,dF_T \tag{4.15}$$

where Fa is the free energy per unit surface area. Conservation of mass requires that $a\rho$, or equivalently a/υ, remain constant, and hence

$$\frac{da}{a} = \frac{d\upsilon}{\upsilon} \tag{4.16}$$

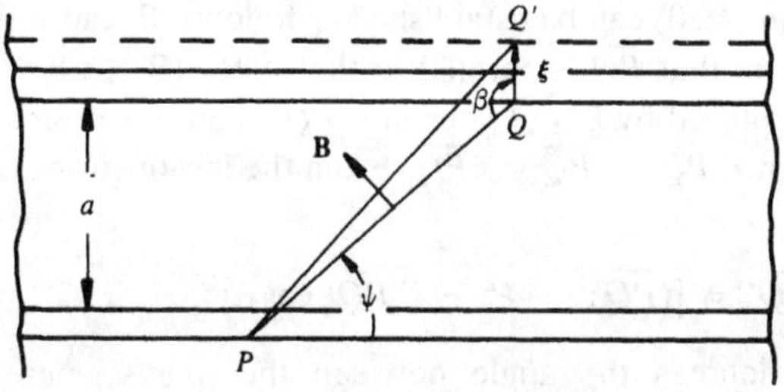

4.2 Definition of the angle β in the derivation of (4.20).

An expression for dF_T is obtained with the use of (4.13):

$$dF_T = \left(\frac{\partial F_0}{\partial \upsilon}\right)_T d\upsilon + H\,dB + B\,dH - \int_0^H \left(\frac{dB}{d\upsilon}\right)_{H,T} d\upsilon\,dH - B\,dH \qquad (4.17)$$

where $F_0 = F_0(\upsilon, T)$. Now introducing (4.12) and (4.15)–(4.17) into (4.14) gives

$$\mathbf{n}\cdot\mathbf{T}'_{\mathrm{m}}\cdot\boldsymbol{\xi} = da\left\{\left[\frac{\partial(\upsilon F_0)}{\partial \upsilon}\right]_T + \int_0^B H\,dB - \int_0^H \upsilon\left(\frac{\partial B}{\partial \upsilon}\right)_{H,T} dH\right\} + aH\,dB \qquad (4.18)$$

where υ can be brought into the integral term because the integration is done at constant mass density.

Figure 4.2 illustrates the triangular relationship of $\overline{PQ'}$ to $\overline{PQ}$ and $\boldsymbol{\xi}$. The flux ϕ of the induction field remains constant during the infinitesimal displacement; in the notation of Figure 4.2, $\phi = \text{const} = B\,\overline{PQ}$. Therefore,

$$dB/B = -d\overline{PQ}/\overline{PQ} \qquad (4.19)$$

i.e., the magnetic field decreases in proportion to the increase in distance between connected wires of the field coils. Because the line that measures this distance is parallel to $(\mathbf{B} \times \mathbf{n}) \times \mathbf{B}$, it can be shown that

$$dB/B = -[(\mathbf{B} \times \mathbf{n}) \times \mathbf{B}]\cdot\boldsymbol{\xi}/B^2 a \qquad (4.20)$$

Equation (4.20) can be established as follows. Because **B** is normal to $\overline{PQ}$ it is seen that $\overline{PQ}$ is parallel to the vector $(\mathbf{B} \times \mathbf{n}) \times \mathbf{B}$. When the plate is displaced by $\boldsymbol{\xi}$, $\overline{PQ}$ goes into $\overline{PQ}'$, and it is desired to compute the difference $\overline{PQ}' - \overline{PQ} = \delta\overline{PQ}$. From the law of cosines, $\overline{PQ}'$ is given by

$$\overline{PQ}' = [(\overline{PQ})^2 + \xi^2 + 2\,\overline{PQ}\xi\cos\beta]^{1/2} \tag{4.20a}$$

where β denotes the angle between the line segment $\overline{PQ}$ and $\boldsymbol{\xi}$. Equation (20a) can be rewritten

$$\overline{PQ}' = \overline{PQ}[1 + (\xi/\overline{PQ})^2 + 2(\xi/\overline{PQ})\cos\beta]^{1/2} \tag{4.20b}$$

Because ξ is infinitesimally small compared to $\overline{PQ}$, equation (4.20b) may be expanded with the aid of the binomial theorem $(1 + x)^n = 1 + nx + \cdots$; retaining terms of order up to $\xi/\overline{PQ}$, one has

$$\overline{PQ}' = \overline{PQ}\{1 + (\xi/\overline{PQ})\cos\beta\} + \text{higher order terms} \tag{4.20c}$$

Thus $d\overline{PQ}$ is given by

$$d\overline{PQ} = \xi\cos\beta$$

and the fractional change in $\overline{PQ}$ is

$$\begin{aligned} d\overline{PQ}/\overline{PQ} &= (\xi/\overline{PQ})\cos\beta \\ &= \frac{[(\mathbf{B} \times \mathbf{n}) \times \mathbf{B}]\cdot\boldsymbol{\xi}}{|[(\mathbf{B} \times \mathbf{n}) \times \mathbf{B}]|\,\overline{PQ}} = \frac{[(\mathbf{B} \times \mathbf{n}) \times \mathbf{B}]\cdot\boldsymbol{\xi}}{(B^2\sin\psi)\,(a/\sin\psi)} \\ &= \frac{[(\mathbf{B} \times \mathbf{n}) \times \mathbf{B}]\cdot\boldsymbol{\xi}}{B^2 a} \end{aligned} \tag{4.20d}$$

where in obtaining the second and third equalities use was made of the facts that $(\mathbf{B} \times \mathbf{n}) \times \mathbf{B}$ is parallel to $\overline{PQ}$ and of magnitude $B^2\sin\psi$ and that $a = \overline{PQ}\sin\psi$. When the last equality of (4.20d) is substituted into (4.19), (4.20) results.

Stress tensor of a magnetizable fluid

All the physical and geometric assertions needed to derive the form of the magnetic stress tensor have now been introduced. What remains is the further mathematical manipulation to put the results into a usable form.

The dot and the cross can be interchanged in a scalar triple product, so $[(\mathbf{B} \times \mathbf{n}) \times \mathbf{B}]\cdot\boldsymbol{\xi} = (\mathbf{B} \times \mathbf{n})\cdot(\mathbf{B} \times \boldsymbol{\xi})$, and (4.20) can be rearranged into the following form with the aid of the vector identity $(\mathbf{a} \times \mathbf{b})\cdot(\mathbf{c} \times \mathbf{d}) = (\mathbf{a}\cdot\mathbf{c})(\mathbf{b}\cdot\mathbf{d}) - (\mathbf{a}\cdot\mathbf{d})(\mathbf{b}\cdot\mathbf{c})$:

$$a\,dB = -B(\mathbf{n}\cdot\boldsymbol{\xi}) + (\mathbf{n}\cdot\mathbf{B})(\mathbf{B}\cdot\boldsymbol{\xi})/B \tag{4.21}$$

Substitution of (4.21) in (4.18) gives

$$\mathbf{n}\cdot\mathbf{T}'_{\mathrm{m}}\cdot\boldsymbol{\xi} = (\mathbf{n}\cdot\boldsymbol{\xi})\left\{\underset{[1]}{\left[\frac{\partial(\upsilon F_0)}{\partial\upsilon}\right]_T} + \underset{[2]}{\int_0^B H\,dB} - \underset{[3]}{\int_0^B \upsilon\left(\frac{\partial B}{\partial\upsilon}\right)_{H,T} dH}\right\}$$
$$- \underset{[4]}{HB(\mathbf{n}\cdot\boldsymbol{\xi})} + \underset{[5]}{(\mathbf{n}\cdot\mathbf{B})(\mathbf{B}\cdot\boldsymbol{\xi})(H/B)} \tag{4.22}$$

which can be easily cast into the following form:

$$\mathbf{n}\cdot\mathbf{T}'_{\mathrm{m}}\cdot\boldsymbol{\xi} = (\mathbf{n}\cdot\boldsymbol{\xi})\left\{\underset{[1]}{\left[\frac{\partial(\upsilon F_0)}{\partial\upsilon}\right]_T} - \underset{[6]}{\int_0^H\left[\frac{\partial(\upsilon B)}{\partial\upsilon}\right]_{H,T} dH}\right\}$$
$$+ \underset{[7]}{\mathbf{n}\cdot(\mathbf{BH})\cdot\boldsymbol{\xi}} \tag{4.23}$$

Equation (4.22) is simplified as follows. First, rewrite [2] as

$$(\mathbf{n}\cdot\boldsymbol{\xi})\left(HB - \int_0^H B\,dH\right) \equiv [2a] + [2b]$$

Hence [2a] and [4] cancel. Now [2b] and [3] can be combined into one term:

$$-(\mathbf{n}\cdot\boldsymbol{\xi})\int_0^H\left(\frac{\partial\upsilon B}{\partial\upsilon}\right)_{H,T} dH = [6]$$

Also because **B** and **H** are parallel and $\mathbf{B}/B$ is a unit vector, [5] becomes $(\mathbf{n}\cdot\mathbf{B})(\mathbf{H}\cdot\boldsymbol{\xi}) = \mathbf{n}\cdot(\mathbf{BH})\cdot\boldsymbol{\xi}$, or [7].

Because $\mathbf{n}\cdot\mathbf{I} = \mathbf{n}$, (4.23) can be written in the following more convenient form:

$$\mathbf{n}\cdot\mathbf{T}'_{\mathrm{m}}\cdot\boldsymbol{\xi} = \mathbf{n}\cdot\left(\left\{\left[\frac{\partial(\upsilon F_0)}{\partial\upsilon}\right]_T - \int_0^H\left[\frac{\partial(\upsilon B)}{\partial\upsilon}\right]_{H,T} dH\right\}\mathbf{I}\right)\cdot\boldsymbol{\xi}$$
$$+ \mathbf{n}\cdot(\mathbf{BH})\cdot\boldsymbol{\xi} \tag{4.24}$$

Therefore, the stress tensor $\mathbf{T}'_{\mathrm{m}}$ is given by

$$\mathbf{T}'_{\mathrm{m}} = \left\{\left[\frac{\partial(\upsilon F_0)}{\partial\upsilon}\right]_T - \int_0^H\left[\frac{\partial(\upsilon B)}{\partial\upsilon}\right]_{H,T} dH\right\}\mathbf{I} + \mathbf{BH} \tag{4.25}$$

Note that the stress tensor is symmetric, because (from $\mathbf{B} = \mu\mathbf{H}$) $\mathbf{BH} = (\mu\mathbf{H})(\mathbf{B}/\mu) = \mathbf{HB}$ and $\mathbf{I}^T = \mathbf{I}$. From (4.9) it may be recognized that the first term inside the brackets is $-p_0(\rho, T)$. The second term can be recast with the aid of the relation $B = \mu_0(H + M)$ into

$$\left[\frac{\partial(\upsilon B)}{\partial\upsilon}\right]_{H,T} = \mu_0\left[\frac{\partial(\upsilon M)}{\partial\upsilon}\right]_{H,T} + \mu_0 H \tag{4.26}$$

Using (4.26) in (4.25) and performing an easy integration then gives the key result, the *stress tensor of a magnetic fluid*:

$$\mathbf{T}'_m = -\left\{p(\rho, T) + \int_0^H \mu_0\left[\frac{\partial(\upsilon M)}{\partial\upsilon}\right]_{H,T} dH + \tfrac{1}{2}\mu_0 H^2\right\}\mathbf{I} + \mathbf{BH} \tag{4.27}$$

The stress tensor, which is expressed in (4.27) in symbolic notation, has the following form in indicial notation:

$$\mathbf{T}'_{ij} = -\left\{p(\rho, T) + \int_0^H \mu_0\left[\frac{\partial(\upsilon M)}{\partial\upsilon}\right]_{H,T} dH + \tfrac{1}{2}\mu_0 H^2\right\}\delta_{ij} + B_i H_j \tag{4.28}$$

where the subscript m is suppressed for simplicity, and where $\delta_{ij} = \mathbf{e}_i \cdot \mathbf{e}_j$ is the *Kronecker delta*; $\delta_{ij} = 1$ for $i = j$, and $\delta_{ij} = 0$ for $i \neq j$.

What is the "pressure" in a magnetized fluid?

In a nonpolar fluid at rest only normal stresses are exerted, the normal stress is independent of the direction of the normal to the surface element across which it acts, and the stress tensor has the form $\mathbf{T} = -p\mathbf{I}$, where p is the static fluid pressure and may be a function of position in the fluid. For a viscous fluid in motion this is no longer true, because tangential stresses are present: The normal component of the stress acting across a surface element then depends on the direction of the normal to the element. Thus, the simple notion of a pressure acting equally in all directions is lost even in most cases of an ordinary fluid in motion.

It has been found useful in the mechanics of ordinary fluids to define a scalar property analogous to the static fluid pressure in the sense that it is a measure of the local intensity of the compression of the fluid. The magnitude of the normal stress on the surface of a small sphere centered on a fixed point is $\mathbf{n}\cdot\mathbf{T}\cdot\mathbf{n} = \mathbf{T}:\mathbf{nn}$, where $\mathbf{n}$ is the unit outward-facing normal to the surface. When the sphere is sufficiently small, $\mathbf{T}$ approaches a constant tensor and the average of the normal component of stress can be written

$$1/4\pi \int \mathbf{n}\cdot\mathbf{T}\cdot\mathbf{n}\, d\Omega = \mathbf{T}:\int \tfrac{1}{4\pi}\mathbf{nn}\, d\Omega$$

where Ω denotes solid angle. Expressing $\mathbf{nn}$ in dyadic form and carrying out the integration term by term, using a spherical coordinate system, reduces the integral on the right side to the term $\frac{1}{3}\mathbf{I}$. (The procedure is left as an exercise for the reader.) The result may be expressed equivalently in terms of the trace of $\mathbf{T}$, $\mathrm{tr}\,\mathbf{T}$, as

$$\text{average normal stress component} = \tfrac{1}{3}\mathbf{T}:\mathbf{I} = \tfrac{1}{3}\mathrm{tr}\,\mathbf{T}$$

Because no particular orientation of axes was specified, this derivation illustrates that the trace of the tensor is invariant with respect to the orientation of the coordinate axes. In an ordinary fluid the pressure p at a point in a moving fluid is *defined* as the mean normal stress with sign reversed,

$$p = -\tfrac{1}{3}\mathrm{tr}\,\mathbf{T}$$

which for a fluid at rest reduces to the hydrostatic pressure. Correspondingly, for a magnetic fluid, which is polar, one-third the trace of the magnetic-stress tensor of (4.27) yields

$$\begin{aligned}\tfrac{1}{3}\mathrm{tr}\,\mathbf{T}'_{\mathrm{m}} &= -p(\rho, T) - \int_0^H \mu_0\left[\frac{\partial(Mv)}{\partial v}\right]_{H,T} dH - \tfrac{1}{2}\mu_0 H^2 + \tfrac{1}{3}\mathbf{B}\cdot\mathbf{H}\\ &= -p(\rho, T) - \int_0^H \mu_0\left[\frac{\partial(Mv)}{\partial v}\right]_{H,T} dH + \tfrac{1}{6}\mu_0 H^2\,(2\mu/\mu_0 - 3)\end{aligned}$$

Based on the averaging analysis that has been developed here, this result would appear to represent the mean normal component of stress averaged over a small sphere centered on a fixed point in the magnetized fluid. The expression reduces to the hydrostatic pressure when the magnetic field is absent, as it must. However, for the case of a nonmagnetic fluid subjected to a magnetic field, the expression reduces to $-p(\rho, T) - \frac{1}{6}\mu_0 H^2$, which differs from the hydrostatic pressure by one-third the trace of the Maxwell vacuum stress tensor. Evidently that amount should be amended to the expression above. It is plausible that experiments in phase-change cavitation could be used to test candidate expressions for the compression.

The thermodynamic pressure $p(\rho, T)$, which has appeared naturally as a result of the derivation, represents the pressure that would be present in the fluid if the fluid were unpolarized yet held at the same density and temperature. It can be separated from the expression for $\mathbf{T}'_m$ in order to emphasize the magnetic aspect of the result. That is, a new tensor $\mathbf{T}_m$ is defined such that $\mathbf{T}_m \equiv \mathbf{T}'_m + p(\rho, T)\mathbf{I}$:

$$\mathbf{T}_m = -\left\{\int_0^H \mu_0 \left[\frac{\partial(vM)}{\partial v}\right]_{H,T} dH + \tfrac{1}{2}\mu_0 H^2\right\}\mathbf{I} + \mathbf{BH} \tag{4.29}$$

4.3 Magnetic body-force density

The magnetic force per unit volume corresponding to a magnetic stress tensor $\mathbf{T}_m$ is

$$\mathbf{f}_m = \nabla \cdot \mathbf{T}_m \tag{4.30}$$

in accord with equation (1.57). For brevity $\mathbf{T}_m$ can be written

$$\mathbf{T}_m = -a\mathbf{I} + \mathbf{BH}$$

where

$$a = \mu_0 \int_0^H \left(\frac{\partial M v}{\partial v}\right)_{H,T} dH + \tfrac{1}{2}\mu_0 H^2$$

The divergence of $\mathbf{T}_m$ therefore has the form

$$\nabla \cdot \mathbf{T}_m = -\nabla a + \mathbf{H}(\nabla \cdot \mathbf{B}) + \mathbf{B} \cdot \nabla \mathbf{H}$$

Here it may be recognized from Maxwell's relation $\nabla \cdot \mathbf{B} = 0$ that the middle term is identically zero. Also, because the field vectors are collinear by assumption, $\mathbf{B} \cdot \nabla \mathbf{H} = (B/H)\mathbf{H} \cdot \nabla \mathbf{H}$. Using the vector identity $\mathbf{H} \cdot \nabla \mathbf{H} = \frac{1}{2}\nabla(\mathbf{H} \cdot \mathbf{H}) - \mathbf{H} \times (\nabla \times \mathbf{H})$ with the current-free magnetostatic Ampère's law $\nabla \times \mathbf{H} = \mathbf{0}$ permits one to express $\mathbf{B} \cdot \nabla \mathbf{H}$ in terms of the magnitudes of the field vectors:

$$\mathbf{B} \cdot \nabla \mathbf{H} = (B/H)\tfrac{1}{2}\nabla H^2 = B \nabla H \tag{4.31}$$

The same derivation but with $\mathbf{M}$ substituted for $\mathbf{B}$ and $\mathbf{H}_0$ for $\mathbf{H}$ transforms the force density $\mu_0 \mathbf{M} \cdot \nabla \mathbf{H}_0$ of (1.10) to the form $\mu_0 M \nabla H_0$. In the present case the force density can thus be written

$$\mathbf{f}_m = \nabla \cdot \mathbf{T}_m$$

$$= -\nabla\left[\mu_0 \int_0^H \left(\frac{\partial M\upsilon}{\partial \upsilon}\right)_{H,T} dH + \tfrac{1}{2}\mu_0 H^2\right] + B\nabla H \tag{4.32}$$

or, with use of $B = \mu_0(H + M)$,

$$\mathbf{f}_\mathrm{m} = -\nabla\left[\mu_0 \int_0^H \left(\frac{\partial M\upsilon}{\partial \upsilon}\right)_{H,T} dH\right] + \mu_0 M\nabla H \tag{4.33}$$

The term $\mu_0 M\nabla H$ is suggestive of (1.10), the Kelvin force density $\mu_0 M\nabla H_0$ on an isolated body, but here the local field H appears in place of the applied field H_0. Equation (4.33) is the force density of Cowley and Rosensweig (1967).

It may be shown, choosing a control volume enveloping a magnetized body and invoking Newton's third law, that integration of the force density $\mathbf{f}_\mathrm{m}$ of (4.33) gives the net force expressed in (1.10) when the latter is applied in the limit of soft magnetic property.

In (4.33) the product $M\upsilon$ represents the magnetic moment per unit mass of the mixture. This can be written alternatively $M\upsilon = n\upsilon\bar{m}$ because $M = n\bar{m}$, when n is number density of magnetic colloid particles and $\bar{m}$ is the average magnetic moment of the particle solid. Now, $n\upsilon$ is the number of particles per unit mass of mixture and thus remains constant with change of mixture volume. Also, in a dilute system $\bar{m}$ depends only on the external field H_0 unless compression changes the magnetization of the solid particle. In a typical hydrocarbon-base ferrofluid, the carrier is about a hundred times as compressible as magnetite, so it will be assumed that magnetostriction of the particle solid is negligible. Accordingly, $M\upsilon$ is independent of υ, and the integral term in (4.33) vanishes for a dilute colloid. What remains is the Kelvin force density with $H = H_0$. The additional terms in (4.33) when the fluid is not dilute evidently represent the influence of dipole interactions.

There is an arbitrariness of the grouping of magnetic terms in (4.27), (4.32), and (4.33), and this has led to some confusion in the literature. What follows is a development of many of the forms that have appeared from time to time in the magnetics literature.

Alternative general forms

In the sequel repeated use will be made of the following relation:

$$\nabla \int_0^H M\,dH = M\nabla H + \int_0^H \nabla M\,dH \tag{4.34}$$

which is a simple extension of the *Leibniz formula* for differentiating an integral. In (4.34) ∇M is evaluated at constant H, and thus (4.34) allows for the possibility that fluid magnetizability may vary with position. For example, let the integral in (4.33) be expanded as follows:

$$\mathbf{f}_m = -\nabla\left[\mu_0 \int_0^H \upsilon\left(\frac{\partial M}{\partial \upsilon}\right)_{H,T} dH + \mu_0 \int_0^H M\, dH\right] + \mu_0 M \nabla H$$

$$= -\nabla\left[\mu_0 \int_0^H \upsilon\left(\frac{\partial M}{\partial \upsilon}\right)_{H,T} dH\right] - \mu_0 \int_0^H \nabla M\, dH \qquad (4.35)$$

where to obtain the second line use was made of (4.34).

Certain pressure-like quantities appear repeatedly and so are now defined: The *magnetostrictive pressure* is

$$p_s \equiv \mu_0 \int_0^H \upsilon\left(\frac{\partial M}{\partial \upsilon}\right)_{H,T} dH \qquad (4.36a)$$

The *fluid-magnetic pressure*

$$p_m \equiv \mu_0 \int_0^H M\, dH = \mu_0 \bar{M} H \qquad (4.36b)$$

where the *field-averaged magnetization* as previously defined is

$$\bar{M} = \frac{1}{H}\int_0^H M\, dH \qquad (4.37)$$

Equation (4.35) for the force density is written in terms that are free of B or $\mathbf{B}$. Equation (4.29) can be put on the same basis using $\mathbf{B} = \mu\mathbf{H}$, giving the pressure-free form

$$\mathbf{T}_m = -\left[\mu_0 \int_0^H \left(\frac{\partial M\upsilon}{\partial \upsilon}\right)_{H,T} dH\right]\mathbf{I} + (\mu\mathbf{HH} - \tfrac{1}{2}\mu_0 H^2 \mathbf{I}) \qquad (4.38)$$

In terms of the definitions (4.36a) and (4.36b), $\mathbf{T}_m$ can be written

$$\mathbf{T}_m = -(p_s + p_m + \tfrac{1}{2}\mu_0 H^2)\mathbf{I} + \mu\mathbf{HH} \qquad (4.39)$$

Another, equivalent form of equation (4.38) is

$$\mathbf{T}_m = -\mathbf{I}\int_0^H \left[\mu - \rho\left(\frac{\partial \mu}{\partial \rho}\right)_{H,T}\right] H\, dH + \mu\mathbf{HH} \qquad (4.40)$$

where $\mu = \mu(H, T, \rho)$. The transformation of (4.38) to this form is left as an exercise for the reader.

Alternative reduced forms

The foregoing relationships are exact at equilibrium for *non-linear* materials. When the magnetization vanishes, the expressions for $\mathbf{T}_m$ must reduce to the free-space Maxwell stress tensor, and those for $\mathbf{f}_m$ must become zero.

Setting $M = 0$ in (4.35) indeed gives $\mathbf{f}_m = \mathbf{0}$. Likewise, if one sets $M = 0$ in (4.38), the bracket multiplying $\mathbf{I}$ disappears, and because μ becomes μ_0, equation (4.38) reduces to the expression developed previously for the Maxwell vacuum stress, (3.15). These are correspondences that necessarily must exist if the expressions for $\mathbf{T}_m$ are valid.

For nonisothermal but sensibly constant-density materials, the magnetization thus depends only on the field H and temperature T, so that $M = M(H, T)$ and thus, from (4.34),

$$\nabla \int_0^H M\,dH = M\nabla H + \int_0^H \frac{\partial M}{\partial T}\nabla T\,dH \qquad (4.41)$$

There are two important circumstances under which $\nabla \int_0^H M\,dH = M\nabla H$:

1. The flow field is isothermal, and hence $\nabla T = \mathbf{0}$.
2. The temperature range of operation is far from the Curie temperature, and thus $\partial M/\partial T = 0$ for most practical purposes. (A plot of M versus T would be rather flat from $T = 0$ K to near T_c and then very quickly fall to zero.)

Therefore, under most ordinary circumstances the last term on the right side of (4.41) can safely be neglected. When this is the case, it is seen from (4.35) and (4.36a) that only the magnetostrictive term

$$\mathbf{f}_m = -\nabla p_s \qquad (4.42)$$

remains. Strictive effects have no net influence on the flow field of incompressible fluids provided there is no physicochemical phase change, and the magnetostriction term can be omitted from the analysis with no effect on the calculated results. This behavior is illustrated in Chapter 5 in applying the ferrohydrodynamic Bernoulli equation. In problems concerned with compressibility, such as acoustic wave propagation or phase change as in cavitation, the effects of magnetostriction should not be neglected without further study.

As another point of correspondence, it will now be shown that the general force density expression (4.35) reduces to the expression of Korteweg and Helmholtz valid for *linearly magnetizable media*. To begin, the middle term on the right side of the first equality is rewritten using the condition that in a linear material μ is assumed to depend only on ρ and T and not on H:

$$\mu_0 \int_0^H M\,dH = \int_0^H (\mu - \mu_0) H\,dH = (\mu - \mu_0)H^2/2 \tag{4.43}$$

where use has been made of $B = \mu_0(H + M) = \mu H$. Now the strictive term can be manipulated by setting $\rho = 1/\upsilon$:

$$\mu_0 \int_0^H \upsilon\left(\frac{\partial M}{\partial \upsilon}\right)_{H,T} dH = -\mu_0 \int_0^H \rho\left(\frac{\partial M}{\partial \rho}\right)_{H,T} dH = -\int_0^H \rho\left(\frac{\partial \mu}{\partial \rho}\right)_T H\,dH$$
$$= -\frac{H^2}{2}\rho\left(\frac{\partial \mu}{\partial \rho}\right)_T \tag{4.44}$$

Finally, consider the last term of (4.35):

$$\mu_0 M \nabla H = \mu_0(\mu/\mu_0 - 1)H \nabla H = (\mu - \mu_0)\nabla(H^2/2) \tag{4.45}$$

Now putting (4.43)–(4.45) into (4.35) yields

$$\mathbf{f}_\mathrm{m} = \nabla\left[\frac{H^2}{2}\rho\left(\frac{\partial \mu}{\partial \rho}\right)_T\right] - \nabla\left[(\mu - \mu_0)\frac{H^2}{2}\right] + (\mu - \mu_0)\nabla\left(\frac{H^2}{2}\right) \tag{4.46}$$

Finally, collecting terms after performing a differentiation gives the force density expression due to Korteweg and Helmholtz:

$$\mathbf{f}_\mathrm{m} = \nabla\left[\left(\frac{H^2}{2}\right)\rho\left(\frac{\partial \mu}{\partial \rho}\right)_T\right] - \frac{H^2}{2}\nabla\mu \tag{4.47}$$

For fluids with constant μ this force density disappears in the fluid. However, a surface force density appears due to the step change in μ at the interface.

The form of the Korteweg–Helmholtz stress tensor corresponding to equation (4.47) can be established as follows. Substitution of (4.43) and (4.44) into (4.29) and use of the relation $\mathbf{B} = \mu\mathbf{H}$ gives

$$\mathbf{T}_\mathrm{m} = -\left[-\frac{H^2}{2}\rho\left(\frac{\partial \mu}{\partial \rho}\right)_T + (\mu - \mu_0)\frac{H^2}{2} + \tfrac{1}{2}\mu_0 H^2\right]\mathbf{I} + \mu\mathbf{H}\mathbf{H} \tag{4.48}$$

Collecting terms then shows that the stress tensor for *compressible linear media* reads

$$\mathsf{T}_m = -\left[\frac{\mu_0}{2} H^2 - \frac{H^2}{2}\rho\left(\frac{\partial\mu}{\partial\rho}\right)_T\right]\mathsf{I} + \mu\mathbf{HH} \tag{4.49}$$

In incompressible flow problems the middle term on the right side can be lumped with the thermodynamic pressure or effectively dropped with no effect on the calculated flow dynamics. The resulting stress tensor can be termed incomplete.

In vacuum $\mu = \mu_0$ and $\rho = 0$, so assuming $(\partial\mu/\partial\rho)_T$ is finite reduces (4.49) to the Maxwell surface stress tensor,

$$\mathsf{T}_m = -(\mu_0/2) H^2\mathsf{I} + \mu_0\mathbf{HH} \tag{4.50}$$

Yet another form of the force density, useful in treating *incompressible nonlinear media*, will further illustrate the diversity of forms appearing in the literature. Let the term $\mu_0 H\nabla H = \nabla(\mu_0\int_0^H H\,dH)$ be added and subtracted from the right side of (4.33), which, neglecting striction, thus becomes

$$\mathbf{f}_m = -\nabla\left(\mu_0\int_0^H H\,dH\right) - \nabla\left(\mu_0\int_0^H M\,dH\right) + \mu_0 M\nabla H + \mu_0 H\nabla H \tag{4.51}$$

or

$$\mathbf{f}_m = -\nabla\int_0^H B\,dH + B\nabla H = -\int_0^H \nabla B\,dH \tag{4.52}$$

where ∇B is to be evaluated at constant H. The second equality (4.52) produces the form desired, which is a compact one.

Exercise: Derive the expression for the stress tensor corresponding to (4.52).

Solution: From (4.31) it is seen that $B\nabla H = \mathbf{B}\cdot\nabla\mathbf{H}$. Furthermore, $\mathbf{B}\cdot\nabla\mathbf{H} = \nabla\cdot(\mathbf{BH}) - \mathbf{H}(\nabla\cdot\mathbf{B}) = \nabla\cdot(\mathbf{BH}) = \nabla\cdot(\mathbf{HB})$, where a well-known vector identity was used to obtain the second equality, $\nabla\cdot\mathbf{B} = 0$ was used to get the third equality, and the last equality was obtained by use of the symmetry of the dyadic $\mathbf{BH}$. Thus (4.52) can be written

$$\mathbf{f}_m = -\nabla\cdot\left\{\left(\int_0^H B\,dH\right)\mathsf{I}\right\} + \nabla\cdot(\mathbf{HB}) \tag{4.53}$$

Table 4.1 *Mutually consistent stress tensors and associated force*

Formulation	Stress tensor $\mathbf{T}_m$	Body force density $\mathbf{f}_m$
Compressible nonlinear Cowley and Rosensweig (1967); Penfield and Haus (1967)	$-\left[\mu_0\int_0^H\left(\frac{\partial Mv}{\partial v}\right)_{H,T}dH+\frac{\mu_0}{2}H^2\right]\mathbf{I}+\mathbf{HB}$	$-\nabla\left\{\mu_0\int_0^H\left(\frac{\partial Mv}{\partial v}\right)_{H,T}dH\right\}+\mu_0 M\nabla H$
Incomplete nonlinear	$-\left(\mu_0\int_0^H M\,dH+\frac{\mu_0}{2}H^2\right)\mathbf{I}+\mathbf{HB}$	$-\nabla\left\{\mu_0\int_0^H M\,dH\right\}+\mu_0 M\nabla H$ $=-\mu_0\int_0^H\nabla_H M\,dH$
Incomplete nonlinear Chu (1959)	$-\left(\int_0^H B\,dH\right)\mathbf{I}+\mathbf{HB}$	$-\nabla\int_0^H B\,dH+B\nabla H=-\int_0^H\nabla_H B\,dH$
Incomplete nonlinear Zelazo and Melcher (1969)	$-W'\mathbf{I}+\mu\mathbf{HH}$	$-\sum_i\frac{\partial W'}{\partial\alpha_i}\nabla\alpha_i$
Compressible linear Korteweg and Helmholtz (see Melcher 1963)	$-\left(\frac{\mu}{2}H^2-\frac{H^2}{2}\rho\frac{\partial\mu}{\partial\rho}\right)\mathbf{I}+\mu\mathbf{HH}$	$\nabla\frac{H^2}{2}\rho\frac{\partial\mu}{\partial\rho}-\frac{H^2}{2}\nabla\mu=\frac{1}{2}\rho\nabla\left(\frac{\partial\mu}{\partial\rho}H^2\right)$
Incomplete linear	$-\frac{1}{2}\mu H^2\mathbf{I}+\mu\mathbf{HH}$	$-\frac{H^2}{2}\nabla\mu$
Vacuum stresses Maxwell	$-\frac{1}{2}\mu_0 H^2\mathbf{I}+\mu_0\mathbf{HH}$	0

Because $\mathbf{f}_m = \nabla\cdot\mathbf{T}_m$ and equation (4.53) is in the form of divergence operations upon terms, an expression due to Chu (1959) can be written immediately for the stress tensor, applicable to *incompressible nonlinear media*:

$$\mathbf{T}_m = -\left(\int_0^H B\,dH\right)\mathbf{I} + \mathbf{HB} \tag{4.54}$$

Because elastic terms are omitted, this tensor should also be termed incomplete.

Formulation of the magnetic force by Zelazo and Melcher (1969) in terms of the coenergy W', defined as

$$W' \equiv \int_0^H B\,dH = \tfrac{1}{2}\int_0^{H^2} \mu(\alpha_1, \cdots, \alpha_n, H^2)\,dH^2 \tag{4.55}$$

where the αs represent intensive properties of the fluid, such as temperature and composition, yields another equivalent formalism. In the form given, it generalizes the magnetic treatment to account for spatial variation of the properties within the fluid region. However, the fluid properties are restricted to convect with material particles.

densities for isotropic magnetic fluids

Normal surface traction, t_{nn}	Surface traction difference, $-[t_{nn}]$
$-\mu_0\int_0^H\left(\frac{\partial M\upsilon}{\partial \upsilon}\right)_{H,T} dH + \frac{\mu_0}{2}(H_n^2 - H_t^2) + \mu_0 M_n H_n$	$\mu_0\int_0^H\left(\frac{\partial M\upsilon}{\partial \upsilon}\right)_{H,T} dH + \frac{\mu_0}{2}M_n^2$
$-\mu_0\int_0^H M\,dH + \frac{\mu_0}{2}(H_n^2 - H_t^2) + \mu_0 M_n H_n$	$\mu_0\int_0^H M\,dH + \frac{\mu_0}{2}M_n^2$
$-\int_0^H B\,dH + H_n B_n$	
$-W' + \mu H_n^2$	
$\frac{\mu}{2}(H_n^2 - H_t^2) + \frac{H^2}{2}\rho\frac{\partial \mu}{\partial \rho}$	$\frac{1}{2}\mu_0 MH - \frac{1}{2}H^2\rho\frac{\partial \mu}{\partial \rho} + \frac{1}{2}\mu_0 M_n^2$
$\frac{\mu}{2}(H_n^2 - H_t^2)$	$\frac{1}{2}\mu_0 MH + \frac{1}{2}\mu_0 M_n^2$
$\frac{\mu_0}{2}(H_n^2 - H_t^2)$	

Table 4.1 summarizes alternative expressions for the stress tensor and body force density. As in the case of striction, magnetization force-density terms that take the form of the gradient of a pressure have no influence on the hydrostatics or hydrodynamics of incompressible magnetic liquids. Again, such terms are frequently omitted in analyzing problems, and the omission is fully justified.

The last two columns of Table 4.1 list expressions for the normal component t_{nn} of the surface-traction vector and for the negative of its difference evaluated across an interface. The component t_{nn} is defined as $t_{nn} = \mathbf{n}\cdot\mathbf{T}_m\cdot\mathbf{n}$, and the surface traction difference is denoted $-[t_{nn}]$, where the brackets indicate the difference of the enclosed quantity across the interface between phases.

Example: Prove that the tangential component of the surface stress is continuous across an interface for any formulation of stress tensor in Table 4.1.

Solution: Each tensor listed in the table can be expressed in the form

$$\mathbf{T}_m = \gamma \mathbf{I} + \mathbf{HB}$$

where γ is a scalar function. The stress vector that results is thus of the form

$$\mathbf{t}_n = \mathbf{n}\cdot\mathbf{T}_m = \gamma\mathbf{n} + H_n\mathbf{B}$$

The stress vector may be resolved into components along the normal $\mathbf{n}$ and a tangential vector $\mathbf{t}$ located at the interface. The tangential component of $\mathbf{t}_n$ is given by

$$\mathbf{t}_n\cdot\mathbf{t} = \gamma\mathbf{n}\cdot\mathbf{t} + H_n\mathbf{B}\cdot\mathbf{t} = 0 + H_n B_t = H_n \mu H_t = H_t B_n$$

Because B_n is continuous, $[H_t B_n] = B_n[H_t]$, and because $[H_t] = 0$, it is seen that $[H_t B_n] = 0$; i.e., the tangential component of the surface stress is the same on both sides of the interface.

It has been shown, in all formulations, that with $\mathbf{t}$ a unit tangential vector, the tangential component of the surface stress vector is $\mathbf{n}\cdot\mathbf{T}_m\cdot\mathbf{t} = H_n B_t$, which is continuous across the interface. Thus, magnetic surface stress is directed normal to the magnetic fluid interface.

Remarks concerning striction in compressible media

Within the scientific literature no general consensus has existed concerning the correctness of the Korteweg–Helmholtz force density (4.47). A recent objection, arising in the ferrofluids literature, is expressed by Shaposhnikov and Shliomis (1975). The expression has had to exist side by side with rival expressions, one being the Kelvin-type expression $\mu_0(\mathbf{M}\cdot\nabla)\mathbf{H}$, which for isotropic media with $\mu = \mu(H)$ reduces to $\mu_0 M \nabla H = \frac{1}{2}\nabla(\mu_0 HM) - \frac{1}{2}H^2\nabla\mu$. What has become clear, however, is that the Korteweg–Helmholtz expression stands out for its ability to explain experimental results in a straightforward way. In an experiment by Hakim and Higham (1962), which examined the analogous force expression for a dielectric liquid, a parallel-plate condenser was completely immersed in the liquid. The analogous second term to the right in (4.47) was negligible, and the striction term contributed a significant force drawing the liquid into a field region between the plates, thus increasing the density of the liquid. The density was measured optically, and the striction term was evaluated using the Clausius–Mossotti expression of classical electromagnetism relating the dielectric constant to the mass density. Good agreement was found between the measurements of the Hakim and Higham experiment and the prediction of Korteweg–Helmholtz theory. An analysis of the experiment is presented in Section 5.6. For a recent discussion citing additional evidence in favor of equation (4.47) see Brevik (1982).

4.4 Equation of motion for a magnetic fluid

Formulating a momentum equation for a magnetic fluid is decisive to the study of ferrohydrodynamics. A momentum equation was first proposed by Neuringer and Rosensweig (1964). The equation continues to serve as the point of departure for most problems in which viscous friction or inertia along with magnetic forces play an important role. Consider the dynamic equilibrium per unit volume of an "infinitesimal" element of magnetic fluid. The element is assumed to be large enough to contain a large number of colloidal magnetic particles yet small in size compared to the dimensions of the flow field.* The time rate of change of momentum for the constant mass contained in a deformable element having volume $dx\,dy\,dz$ is

$$\frac{D}{Dt}(\rho\mathbf{v}\,dx\,dy\,dz) = \rho\,dx\,dy\,dz\frac{D\mathbf{v}}{Dt} + \mathbf{v}\frac{D\rho\,dx\,dy\,dz}{Dt}$$

Because the mass $\rho\,dx\,dy\,dz$ is constant, the last term vanishes and Newton's law normalized to unit volume can be written

$$\rho\frac{D\mathbf{v}}{Dt} = \underset{\substack{\text{Pressure}\\ \text{force}}}{\mathbf{f}_p} + \underset{\substack{\text{Viscous}\\ \text{force}}}{\mathbf{f}_v} + \underset{\substack{\text{Gravity}\\ \text{force}}}{\mathbf{f}_g} + \underset{\substack{\text{Magnetic}\\ \text{force}}}{\mathbf{f}_m} \qquad (4.56)$$

where D/Dt is the substantial derivative. In Chapter 1 the substantial derivative was defined to follow the fluid motion. To be precise, the following convention will be adopted:

$$\frac{D}{Dt} = \frac{\partial}{\partial t} + \mathbf{v}\cdot\nabla \qquad (4.57)$$

is the *substantial*, *material*, or *convective* derivative, i.e., the rate of change following the mass motion $\mathbf{v}$;

$$\frac{d}{dt} = \frac{\partial}{\partial t} + \mathbf{q}\cdot\nabla \qquad (4.58)$$

*Alternatively, the equation can be regarded as exact in representing the *expectation value* for an ensemble of systems.

is the derivative following any motion $\mathbf{q}$; and the two are identical when $\mathbf{v} = \mathbf{q}$. The distinction between (4.57) and (4.58) has importance in the study of antisymmetric stress and in the study of multiphase flow, as developed in Chapters 8 and 9. It is the form (4.57) that appears on the left side of (4.56).

The right side of (4.56) is the sum of the body forces normalized to a unit volume. The terms familiar from fluid mechanics are the (thermodynamic) pressure gradient

$$\mathbf{f}_p = -\nabla p(\rho, T) \tag{4.59}$$

the viscous force p.u.v.

$$\mathbf{f}_v = \nabla \cdot \mathbf{T}_v \tag{4.60}$$

and the gravitational force p.u.v.

$$\mathbf{f}_g = \rho \mathbf{g} \tag{4.61}$$

where $\mathbf{T}_v$ is the *viscous stress tensor* and $\mathbf{g}$ is the local acceleration due to gravity. Viscous stress is related to the velocity by the *Newtonian constitutive relation*,

$$\mathbf{T}_v = \eta[\nabla\mathbf{v} + (\nabla\mathbf{v})^T] + \lambda(\nabla \cdot \mathbf{v})\mathbf{I} \tag{4.62}$$

The viscous force density corresponding to equation (4.62) is

$$\mathbf{f}_v = \nabla \cdot \mathbf{T}_v = \eta \nabla^2 \mathbf{v} + (\eta + \lambda)\nabla(\nabla \cdot \mathbf{v}) \tag{4.63}$$

In equation (4.62) η and λ are called the coefficients of viscosity; η is often called the *ordinary* or *first coefficient of viscosity* and λ the *second coefficient of viscosity*. The second coefficient of viscosity has importance in the absorption of energy associated with the longitudinal vibrations of sound and ultrasound transmission (Karim and Rosenhead 1952). The ratio λ/η is about 2.4 for water at a frequency of 5 MHz and exceeds 100 for benzene at 2 MHz. To an excellent approximation in most flows, ordinary liquids and magnetic liquids both are sensibly incompressible, $\nabla \cdot \mathbf{v} = 0$, and thus λ does not enter into these problems.

Equation (4.63) can be established as follows:

$$\begin{aligned}\nabla \cdot \nabla\mathbf{v} &= \mathbf{e}_i \frac{\partial}{\partial x_i} \cdot \left(\mathbf{e}_j \mathbf{e}_k \frac{\partial v_k}{\partial x_j}\right) = \mathbf{e}_k \delta_{ij} \frac{\partial^2 v_k}{\partial x_i \partial x_j} \\ &= \frac{\partial^2}{\partial x_i^2}(v_k \mathbf{e}_k) = \nabla^2 \mathbf{v}\end{aligned} \tag{4.64}$$

$$\nabla\cdot(\nabla\mathbf{v})^{\mathrm{T}} = \mathbf{e}_i\frac{\partial}{\partial x_i}\cdot\left(\mathbf{e}_k\mathbf{e}_j\frac{\partial v_k}{\partial x_j}\right) = \mathbf{e}_j\,\delta_{ik}\,\frac{\partial^2 v_k}{\partial x_i\,\partial x_j}$$

$$= \frac{\partial^2 v_i}{\partial x_i\,\partial x_j}\,\mathbf{e}_j = \mathbf{e}_j\,\frac{\partial}{\partial x_j}\left(\frac{\partial v_i}{\partial x_i}\right) = \nabla(\nabla\cdot\mathbf{v}) \tag{4.65}$$

Also, from the property of the unit dyadic it follows that

$$\nabla\cdot[(\nabla\cdot\mathbf{v})\mathbf{I}] = \nabla(\nabla\cdot\mathbf{v}) \tag{4.66}$$

Substituting these expressions into the first equality of (4.63), with the use of (4.62), leads to the second equality of (4.63).

From (4.62) it follows that, for an incompressible liquid (because then $\nabla\cdot\mathbf{v} = 0$),

$$\mathbf{T}_{\mathrm{v}} = \eta[\nabla\mathbf{v} + (\nabla\mathbf{v})^{\mathrm{T}}] \tag{4.67}$$

and from (4.63) the *viscous force density for an incompressible liquid* is given by

$$\mathbf{f}_{\mathrm{v}} = \eta\,\nabla^2\mathbf{v} \tag{4.68}$$

Substituting this expression for $\mathbf{f}_{\mathrm{v}}$, the expression for $\mathbf{f}_{\mathrm{p}}$ from (4.59), and $\mathbf{f}_{\mathrm{g}}$ from (4.61) into the equation of motion (4.56) gives the form

$$\rho\frac{D\mathbf{v}}{Dt} = -\nabla p + \rho\mathbf{g} + \mathbf{f}_{\mathrm{m}} + \eta\,\nabla^2\mathbf{v} \tag{4.69}$$

Then introducing the expression for $\mathbf{f}_{\mathrm{m}}$ from (4.32) gives the following definite form of the equation of motion for a magnetic fluid:

$$\rho\frac{D\mathbf{v}}{Dt} = -\nabla p + \eta\,\nabla^2\mathbf{v} + \rho\mathbf{g}$$

$$-\nabla\left[\mu_0\int_0^H\left(\frac{\partial M\upsilon}{\partial\upsilon}\right)_{H,T}dH\right] + \mu_0 M\,\nabla H \tag{4.70}$$

Alternative forms of the equation of motion

It will prove useful to write (4.70) in various equivalent ways. Let a composite pressure p^* be defined as

$$p^* \equiv p(\rho, T) + \mu_0\int_0^H\left(\frac{\partial M\upsilon}{\partial\upsilon}\right)_{H,T}dH$$

$$= p(\rho, T) + \mu_0\int_0^H\left(\upsilon\frac{\partial M}{\partial\upsilon}\right)_{H,T}dM + \mu_0\int_0^H M\,dH$$

$$= p(\rho, T) + p_{\mathrm{s}} + p_{\mathrm{m}} \tag{4.71}$$

where p_s and p_m were defined in (4.36a) and (4.36b), respectively. With the introduction of the definition for p^*, the equation of motion (4.70) takes the form

$$\rho \frac{D\mathbf{v}}{Dt} = -\nabla p^* + \mu_0 M \nabla H + \eta \nabla^2 \mathbf{v} + \rho \mathbf{g} \tag{4.72}$$

or, equivalently,

$$\rho \frac{D\mathbf{v}}{Dt} = -\nabla(p + p_s + p_m) + \mu_0 M \nabla H + \eta \nabla^2 \mathbf{v} + \rho \mathbf{g} \tag{4.73}$$

In an isothermal flow field, (4.73) may be simplified. Refer to (4.36b) and (4.41); when ∇T is zero, $-\nabla p_m$ reduces to the term $-\mu_0 M \nabla H$, which cancels the term $\mu_0 M \nabla H$ on the right side of (4.73), leaving

$$\rho \frac{D\mathbf{v}}{Dt} = -\nabla(p + p_s + \rho g z) + \eta \nabla^2 \mathbf{v} \tag{4.74}$$

Any of the formulations leads to the same predicted flow field when used in a consistent manner with the corresponding surface stress condition. In the absence of magnetization any of the forms (4.70) or (4.72)–(4.74) reduce to the well-known Navier–Stokes equation of fluid mechanics.

Comments and supplemental references

The derivation of the expression for the magnetic stress tensor for nonlinearly magnetizable media was based on the principle of conservation of energy. This result was obtained by

Cowley and Rosensweig (1967)

and also, from a relativistic, virtual power treatment, by

Penfield and Haus (1967), (C-6) in Table C.2

Unfortunately, most electromagnetism texts either omit any discussion of stress in continua under applied electric/magnetic fields or barely scratch the surface of the subject. One notable exception is the classic treatise of

Landau and Lifshitz (1960), Sections 15–17, 34, 38, 56

where an expression for the stress tensor in both a fluid and solid continuum under an applied electric field is derived and then specialized to linear materials. A few of the standard electromagnetism texts contain some good material on the subject. Three examples, listed in the order of increased sophistication, are those by

Abraham and Becker (1937), Chapter 5

Panofsky and Phillips (1962), Chapter 6
Stratton (1941), Chapter 2

A recent treatment that develops and compares electrical and magnetic force densities in the manner of continuum electromechanics is that of

Melcher (1981), Chapter 3

The magnetostrictive contribution to the force density arises from the change of magnetization with compression or expansion of the magnetic fluid. This behavior should influence the propagation of sound waves in a ferrofluid. Ultrasonic attenuation has been treated by

Gotoh, Isler, and Chung (1980)
Tarapov and Patsegon (1980)

5

THE FERROHYDRODYNAMIC BERNOULLI EQUATION

One of the most useful relations in ordinary fluid mechanics is the equation the Swiss mathematician Daniel Bernoulli presented in his *Hydrodynamica* of 1738. The equation relates the pressure, the velocity, and the elevation of a fluid in a gravitational field. With it one can calculate such quantities as the variation of pressure with depth in a reservoir, the lift on the wing of an airplane, or the force generated by wind on a sail.

Many problems in ferrohydrodynamics can be similarly analyzed in a penetrating yet simple manner using the augmented Bernoulli equation to be developed in this chapter. The utility of this equation is great in problems of static equilibrium and steady-state flow in which viscous drag effects are absent or else negligible. In other cases it is often possible to determine the order of magnitude of an effect using the FHD Bernoulli relationship when obtaining an exact solution would require a prohibitive expense of time and resources.

The generalized Bernoulli equation was derived in Section 1.6 in an incomplete form. The topic is now revisited, with the development based on the deeper understanding of bulk and interfacial magnetic stress that has been built up along the way. Sections 5.1–5.3 develop this important relationship; its application is then illustrated in detail in a spectrum of problems.

5.1 **Derivation**

The ferrohydrodynamic Bernoulli equation will be obtained as a general integral of the equation of motion developed as (4.72):

$$\rho\left(\frac{\partial \mathbf{v}}{\partial t} + \mathbf{v}\cdot\nabla\mathbf{v}\right) = -\nabla p^* + \mu_0 M \nabla H + \eta \nabla^2 \mathbf{v} + \rho\mathbf{g} \qquad (5.1)$$

This equation represents a generalization of the Navier–Stokes equation of conventional fluid mechanics such that a magnetic body force $\mu_0 M \nabla H$ appears on the right side and the composite pressure p^* appears in place of pressure p.

From a well-known vector identity, $\nabla^2 \mathbf{v} = \nabla(\nabla \cdot \mathbf{v}) - \nabla \times (\nabla \times \mathbf{v})$ where $\nabla \times \mathbf{v} \equiv \mathbf{\Omega}$ is called the *vorticity*. For an incompressible liquid, then $\nabla^2 \mathbf{v} = -\nabla \times \mathbf{\Omega}$. If the fluid is inviscid ($\eta = 0$) or the flow is *irrotational* ($\mathbf{\Omega} = \mathbf{0}$), the viscous term in the equation of motion is identically zero so that

$$\rho(\partial \mathbf{v}/\partial t) + \rho \mathbf{v} \cdot \nabla \mathbf{v} = -\nabla p^* + \mu_0 M \nabla H + \rho \mathbf{g} \tag{5.2}$$

From another vector identity $\mathbf{v} \cdot \nabla \mathbf{v} = \nabla(\frac{1}{2}v^2) - \mathbf{v} \times (\nabla \times \mathbf{v})$, and with the use of (4.41), (5.2) can be rewritten

$$\rho \frac{\partial \mathbf{v}}{\partial t} - \rho \mathbf{v} \times \mathbf{\Omega} = -\nabla\left(p^* + \rho \frac{v^2}{2} + \rho g h - \mu_0 \int_0^H M\, dH\right) - \int_0^H \left(\frac{\partial M}{\partial T}\right)_{H,\upsilon} (\nabla T)_{H,\upsilon}\, dH \tag{5.3}$$

In going from (5.2) to (5.3) it was assumed that $g = |\mathbf{g}|$ is constant and h is the elevation in the direction opposite to gravity above some reference level. For irrotational flow, $\mathbf{\Omega} = \mathbf{0}$, so there exists a velocity potential ϕ such that $\mathbf{v} = -\nabla \phi$. Then if $\nabla T = \mathbf{0}$ or $(\partial M/\partial T)_{H,\upsilon} = 0$ [see the discussion following equation (4.41)], (5.3) can be written

$$\nabla(-\rho\, \partial\phi/\partial t + p^* + \tfrac{1}{2}\rho v^2 + \rho g h - \mu_0 \bar{M} H) = \mathbf{0}$$

where $\bar{M}$ is the field-averaged magnetization, defined in (4.37):

$$\bar{M} = \frac{1}{H} \int_0^H M\, dH \tag{5.4}$$

When the equation preceding (5.4) is integrated, the quantity in parentheses can at most equal a function of time $f(t)$:

$$-\rho\, \partial\phi/\partial t + p^* + \tfrac{1}{2}\rho v^2 + \rho g h - \mu_0 \bar{M} H = f(t) \tag{5.5}$$

Equation (5.5) is the time-dependent ferrohydrodynamic Bernoulli equation. It should be kept in mind that it has been derived under the following assumptions:

incompressible, Newtonian liquid
$\nabla \times \mathbf{H} = \mathbf{0}$
nonlinear magnetic material without hysteresis

collinear magnetization and field: $\mathbf{M}\|\mathbf{H}$
irrotational flow field; i.e., $\nabla \times \mathbf{v} = \boldsymbol{\Omega} = \mathbf{0}$
constant gravitational force
$\begin{cases}\text{isothermal flow field } \textit{or} \\ \text{temperature independent magnetization}\end{cases}$

For steady-state or time-invariant flow, $\partial\phi/\partial t = 0$ and $f(t) = \text{const}$, so the generalized Bernoulli equation reduces to

$$p^* + \tfrac{1}{2}\rho v^2 + \rho g h - \mu_0 \bar{M} H = \text{const} \tag{5.6}$$

or, equivalently, using the appropriate definitions

$$p + p_s + \tfrac{1}{2}\rho v^2 + \rho g h = \text{const} \tag{5.7}$$

The meaning of this relationship will be discussed in detail after the boundary conditions appropriate to these flows are developed. Because the boundary conditions often exert a major influence on the fluid configuration, a digression will be made to develop the boundary conditions with care.

5.2 Boundary conditions

To complete the Bernoulli description, it is necessary to obtain the surface tension and momentum boundary conditions at the interface between two immiscible fluids both of which are inviscid and magnetized. The interfacial force due to the existence of surface tension may be developed with reference to Figure 5.1, which illustrates a rectangular element of interface film bounded by lines of curvature in two planes at right angles. Surface tension acts tangential to the surface along the edges of the element, producing differential forces of magnitude $df_1 = \sigma\, ds_1$ and $df_2 = \sigma\, ds_2$, where σ is the interfacial tension. These forces can be resolved along the direction of the normal $\mathbf{n}$, giving $df_{1,n} = -\sin(d\theta_2/2)\, df_1$, and $df_{2,n} = -\sin(d\theta_1/2)\, df_2$. Because the angles are small, the sine of an angle is well approximated by the angle, so that $df_{1,n} = -d\theta_2\, df_1/2$ and $df_{2,n} = -d\theta_1\, df_2/2$. In terms of the radii of curvature of the surface, $d\theta_2 = ds_2/R_2$ and $d\theta_1 = ds_1/R_1$. The surface force density equals the sum of these forces on the four edges, normalized to the surface area $dA = ds_1\, ds_2$ and produces the force density.

$$-\sigma(1/R_1 + 1/R_2)$$

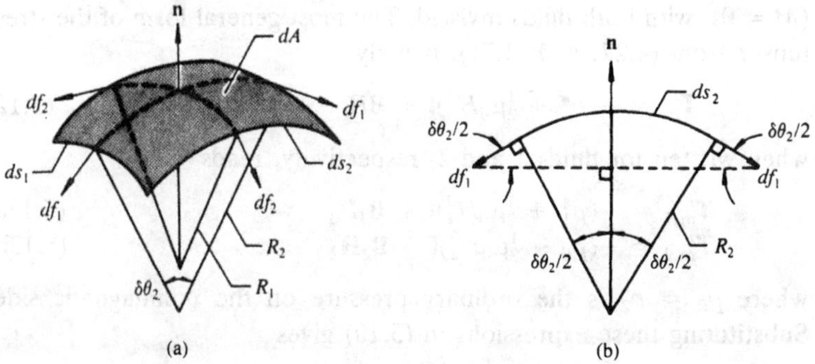

5.1 Geometry of the interfacial element of area acted on by the forces of surface tension. (a) overview; (b) arc of surface ds_2 in plane 2.

If $\mathscr{H}$ denotes the arithmetic mean curvature,

$$\mathscr{H} \equiv \tfrac{1}{2}(1/R_1 + 1/R_2) \tag{5.8}$$

the surface force density may be expressed

$$p_c \equiv 2\mathscr{H}\sigma \tag{5.9}$$

where p_c is the *capillary pressure* produced on an interface having curvature and acted upon by forces of surface or interfacial tension. For a sphere, $R_1 = R_2 = R$ and $\mathscr{H} = 1/R$, whereas for a cylinder with $R_1 = R$ and $R_2 \rightarrow \infty$, $\mathscr{H} = 1/2R$. Using (5.9), one can now write a general form of the interfacial momentum transfer equation for inviscid isothermal systems of constant composition:

$$\mathbf{n}\cdot(\mathbf{T}'_{m,2} - \mathbf{T}'_{m,1}) - \mathbf{n}2\mathscr{H}\sigma = 0 \tag{5.10}$$

where $\mathbf{T}'_{m,2}$ and $\mathbf{T}'_{m,1}$ are the magnetic stress tensors, including pressure stress, evaluated in the media adjacent to the interface. Even when σ is not zero, at a flat interface the surface tension force $2\mathscr{H}\sigma$ is zero because R_1 and R_2 are infinitely large and the stress must be continuous across the interface. Therefore

$$\mathbf{n}\cdot(\mathbf{T}'_{m,2} - \mathbf{T}'_{m,1}) = \mathbf{n}\cdot[\mathbf{T}'_{m}] = 0 \tag{5.11}$$

where the bracket notation stands for the jump in quantity as the interface is crossed.

A more definite form of the FHD boundary condition will now be derived from (5.10). Consider the interface between fluid 1, which is taken to be magnetic ($M > 0$), and fluid 2, which is nonmagnetic

($M = 0$), with both fluids inviscid. The most general form of the stress tensor from (4.27) and (4.71), namely,

$$\mathbf{T}'_{\mathrm{m}} = -(p^* + \tfrac{1}{2}\mu_0 H^2)\mathbf{I} + \mathbf{BH} \tag{5.12}$$

when written for fluids 1 and 2, respectively, reads

$$\mathbf{T}'_{\mathrm{m},1} = -(p_1^* + \tfrac{1}{2}\mu_0 H_1^2)\mathbf{I} + \mathbf{B}_1\mathbf{H}_1 \tag{5.13a}$$
$$\mathbf{T}'_{\mathrm{m},2} = -(p_0 + \tfrac{1}{2}\mu_0 H_2^2)\mathbf{I} + \mathbf{B}_2\mathbf{H}_2 \tag{5.13b}$$

where $p_0 = p_2$ is the ordinary pressure on the nonmagnetic side. Substituting these expressions in (5.10) gives

$$[(p_1^* - p_0) + \tfrac{1}{2}\mu_0(H_1^2 - H_2^2)]\mathbf{n} + B_{\mathrm{n}}(\mathbf{H}_2 - \mathbf{H}_1) - \mathbf{n}2\mathscr{H}\sigma = \mathbf{0} \tag{5.14}$$

The magnetic term outside the brackets in (5.14) is obtained as follows:

$$\mathbf{n}\cdot(\mathbf{B}_2\mathbf{H}_2 - \mathbf{B}_1\mathbf{H}_1) = (\mathbf{n}\cdot\mathbf{B}_2)\mathbf{H}_2 - (\mathbf{n}\cdot\mathbf{B}_1)\mathbf{H}_1 = B_{2\mathrm{n}}\mathbf{H}_2 - B_{1\mathrm{n}}\mathbf{H}_1 \tag{5.15}$$

Recall that it was found in Chapter 3, by application of the integral form of Gauss's law $\oint_S \mathbf{B}\cdot d\mathbf{S} = 0$ [see (3.8)], that the normal component of **B** is continuous at the interface:

$$\mathbf{n}\cdot(\mathbf{B}_2 - \mathbf{B}_1) = \mathbf{n}\cdot[\mathbf{B}] = B_{2\mathrm{n}} - B_{1\mathrm{n}} = 0 \tag{5.16}$$

Therefore $B_{2\mathrm{n}} = B_{1\mathrm{n}} = B_{\mathrm{n}}$ in (5.15), thus completing the proof. The terms $\mathbf{H}_2 - \mathbf{H}_1$ and $H_1^2 - H_2^2$ in (5.14) can also be simplified. Recall from (3.9) that

$$\mathbf{n} \times (\mathbf{H}_2 - \mathbf{H}_1) = \mathbf{n} \times [\mathbf{H}] = \mathbf{0} \tag{5.17}$$

and so the tangential component of **H** is continuous at the interface:

$$H_{2\mathrm{t}} - H_{1\mathrm{t}} = 0 \tag{5.18}$$

On account of (5.18)

$$H_1^2 - H_2^2 = H_{1\mathrm{n}}^2 - H_{2\mathrm{n}}^2 \tag{5.19}$$

and

$$\mathbf{H}_2 - \mathbf{H}_1 = \mathbf{n}(H_{2\mathrm{n}} - H_{1\mathrm{n}}) \tag{5.20}$$

Now, by the defining equation $B = \mu_0(H + M)$, $H_{2\mathrm{n}} = B_{\mathrm{n}}/\mu_0$ and $H_{1\mathrm{n}} = B_{\mathrm{n}}/\mu_0 - M_{\mathrm{n}}$; hence from (5.19) and (5.20)

$$H_{1\mathrm{n}}^2 - H_{2\mathrm{n}}^2 = (H_{1\mathrm{n}} + H_{2\mathrm{n}})(H_{1\mathrm{n}} - H_{2\mathrm{n}}) = M_{\mathrm{n}}^2 - 2B_{\mathrm{n}}M_{\mathrm{n}}/\mu_0 \tag{5.21a}$$

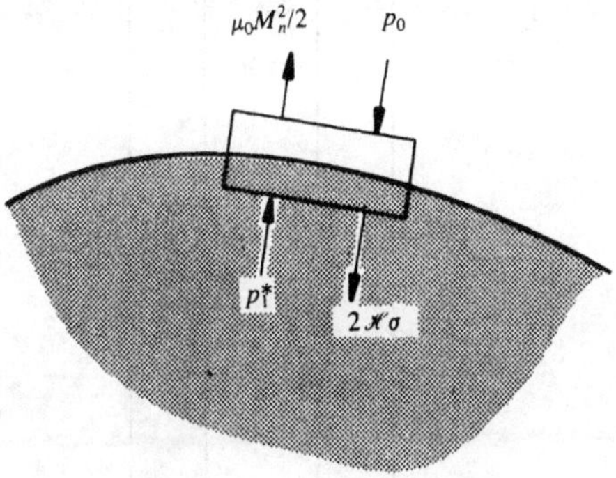

5.2 Balance of forces in the FHD boundary condition of equation (5.22) with medium 1 magnetizable and medium 2 nonmagnetizable.

and

$$H_{2n} - H_{1n} = M_n \tag{5.21b}$$

If these results are substituted into (5.19) and (5.20) and the resultant expressions are substituted into (5.14), the *FHD boundary condition in the absence of viscous forces* is obtained:

$$p_1^* + p_n = p_{0,2} + p_c \tag{5.22}$$

where p_c was introduced in (5.9) and

$$p_n \equiv \mu_0 M_n^2/2 \tag{5.23}$$

will be called the *magnetic normal traction.* It is a familiar requirement that the pressure be continuous across a plane fluid boundary for ordinary fluids, but this condition no longer holds for fluids possessing magnetization. As illustrated in Figure 5.2, the magnetic stress at an interface in this formulation produces a traction $\frac{1}{2}\mu_0 M_n^2$. In terms of the definition of p^* [see (4.71)], this boundary condition may be reexpressed

$$p_1 + p_s + p_m + p_n = p_0 + p_c \tag{5.24}$$

The ferrohydrodynamic relations governing the steady flow of inviscid, incompressible magnetic liquids are summarized in Table 5.1. Note that in the equivalent form the Bernoulli equation does not contain the fluid-magnetic pressure term p_m. However, p_m then appears explicitly in the boundary condition.

Now it can be seen that the elementary derivation of the generalized Bernoulli equation in Section 1.6 is deficient in not including the mag-

Table 5.1 *Summary of FHD relationships*

Definitions of bulk pressures			
Magnetostrictive pressure		$p_s = \mu_0 \int_0^H v\left(\frac{\partial M}{\partial v}\right)_{H,T} dH$	(4.36a)
Fluid-magnetic pressure		$p_m = \mu_0 \int_0^H M\,dH = \mu_0 \bar{M} H$	(4.36b)
Composite pressure		$p^* = p(\rho, T) + p_s + p_m$	(4.71)
Thermodynamic pressure		$p = p(\rho, T)$	
Expressions for interfacial force densities			
Magnetic normal pressure		$p_n = \frac{1}{2}\mu_0 M_n^2$	(5.23)
Capillary pressure		$p_c = 2\mathscr{H}\sigma$	(5.9)
FHD Bernoulli equation		**Boundary condition**	
Usual: $p^* + \frac{1}{2}\rho v^2 + \rho g h - p_m = \text{const}$	(5.6)	$p^* + p_n = p_0 + p_c$	(5.22)
Equivalent: $p + p_s + \frac{1}{2}\rho v^2 + \rho g h = \text{const}$	(5.7)	$p + p_s + p_m + p_n = p_0 + p_c$	(5.24)

Note: FHD Bernoulli equations apply to the steady flow of inviscid, incompressible magnetic fluid; p_0 is pressure in nonmagnetic fluid.

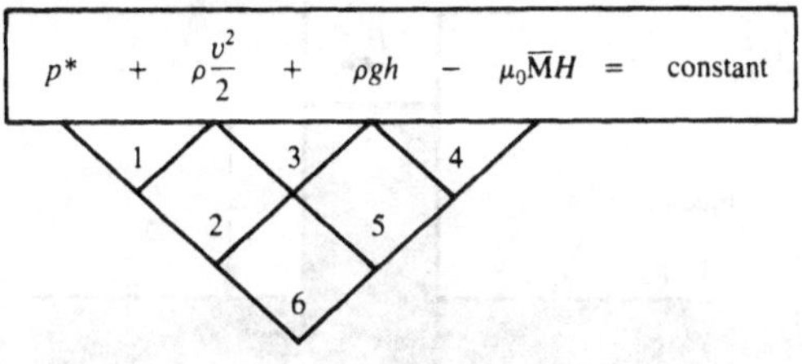

5.3 Categories of flows described by the FHD Bernoulli relationship. In reading the diagram, 3 represents those flows in which only kinetic energy and gravity terms are important, 2 represents those situations in which pressure and gravity are important, and so on. Magnetism yields the novel families 4, 5, and 6. (*After Rosensweig 1966b.*)

netic normal pressure, represented by the term p_n, or the capillary pressure p_c in a statement of boundary conditions. An experimental investigation of the elongation of a ferrofluid droplet found the droplet aspect ratio l/d_0 (l is the drop length, d_0 the diameter of the original droplet) to be a unique function of the dimensionless group $S \equiv \mu_0 d_0 M^2/\sigma$ with values of l/d_0 ranging from 1 to 5 over the range $0 < S < 150$ (Arkhipenko, Barkov, and Bashtovoi 1978). Basaran (1984) generates surface shape of polarized drops using the Galerkin finite element numerical technique.

5.3 Categories of equilibrium inviscid flows

The steady-state FHD Bernoulli equation describes several new classes of fluid flow in addition to the conventional flows, as diagrammed in Figure 5.3. In the absence of an applied field, p^* reduces to $p(\rho, T)$ and the fluid-magnetic pressure is identically zero. Then with any one term absent the remaining terms provide familiar examples from ordinary fluid mechanics:

1. describing the operation of venturi meters and Pitot tubes and giving the pressure at the edge of a boundary layer or on a bluff body;
2. giving the pressure distribution in a tank of liquid or the loading on a dam; and
3. yielding an expression for the efflux rate of material from a hole in the tank.

In the presence of an applied magnetic field, pairing the fluid-magnetic pressure with each of the remaining terms produces new interactions

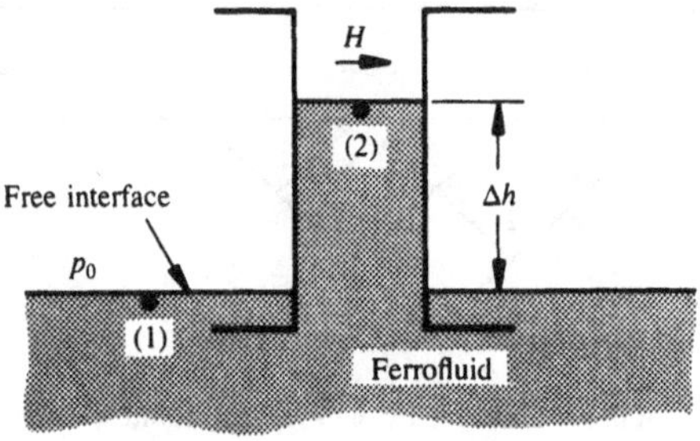

5.4 The classical Quincke experiment, showing the rise of magnetic fluid between the poles of a magnet and illustrating category 4 behavior of Figure 5.3. (*After Jones 1978.*)

and *doubles the number of classes of fluid phenomena.* Some examples within these classes are

4. describing the shape of a static meniscus;
5. governing the behavior of a jet of magnetic liquid under an applied magnetic field (one form of magnetic ink jet printing relies on this interaction); and
6. describing the operation of magnetic fluid seals, bearings, load supports, and sink–float material separation devices.

As has been indicated, many problems in FHD can be solved in a simple manner using the FHD Bernoulli equation. The utility of the equation is great in problems of steady-state flow in which viscous drag effects are negligible. Problems of ferrohydrostatics represent a subcategory in which these restrictions are met exactly. Often it is possible to determine the order of magnitude of an effect using the FHD Bernoulli relationship when obtaining an exact solution could require a prohibitive cost of time and resources.

In the remainder of this chapter a number of representative solutions are generated for the basic flows of a magnetic fluid.

5.4 Applications of the FHD Bernoulli equation

Classical Quincke problem

This problem has practical utility in the measurement of magnetization in magnetic liquids (Bates 1963). Figure 5.4 illustrates an ideal geometry consisting of two parallel magnetic poles. The height and width of the poles are much greater than the separation, and also the poles are assumed to be highly permeable, so the magnetic field between them is uniform. The poles are partially immersed in a reservoir

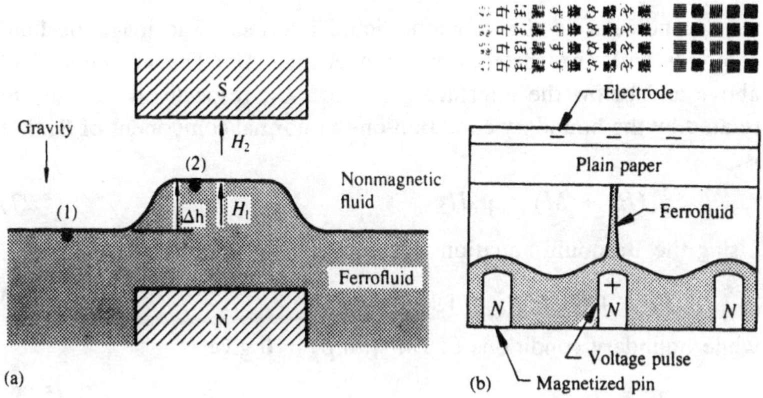

5.5 (a) Uniform magnetic field imposed normal to a free ferrofluid interface. The fluid magnetization is assumed smaller than required to produce the normal field instability. (b) Printing process utilizing a wavy magnetic-fluid surface established in a spatially periodic magnetic field. Printing occurs on command, in which an electric pulse is made to eject a droplet of the fluid, which serves as the ink, from a surface peak. The sample of printing shown results from the simultaneous ejection of droplets from selected peaks along multiple rows of droplets onto a moving page. (*After Maruno, Yubakami, and Soga 1983.*)

of magnetic liquid of density ρ. The pole spacing is sufficiently wide that capillarity is not important, so the height to which the liquid rises between the poles is a function of the applied magnetic field H. Because the density of air is very small compared to that of the magnetic liquid, the external ambient pressure may be assumed constant at p_0.

This problem is easily solved with the Bernoulli equation (5.6), considering a point 1 chosen at the free surface outside the field and a point 2 at the free surface in the field region. From (5.6),

$$p_1^* + \rho g h_1 = p_2^* + \rho g h_2 - \mu_0 \bar{M} H \tag{5.25}$$

From the boundary condition formulated as (5.22), with M_n and $\mathscr{H}$ both zero,

$$p_1^* = p_0 \qquad \text{and} \qquad p_2^* = p_0$$

Combining these relationships yields

$$\Delta h = h_2 - h_1 = \mu_0 \bar{M} H / \rho g \tag{5.26}$$

Normally the experiment is carried out with fluid in a vertical glass tube.

Surface elevation in a normal field

For the problem shown in Figure 5.5 (Jones 1978) the plane-parallel poles of a magnet produce a vertical field, so the magnetic field is perpendicular to the magnetic liquid interface. The magnetic fluid responds with the surface elevation change Δh. The magnetic field above and below the interface is H_2 and H_1, respectively. These are related by the boundary condition on the normal component of **B**; that is,

$$\mu_0(H_1 + M) = \mu_0 H_2 \tag{5.27}$$

Using the Bernoulli equation (5.6) gives

$$p_1^* + \rho g h_1 = p_2^* + \rho g h_2 - \mu_0 \bar{M} H_1 \tag{5.28}$$

while boundary conditions (5.22) with $p_c = 0$ give

$$p_1^* = p_0 \tag{5.29}$$

$$p_2^* = p_0 - (\mu_0/2)M^2 \tag{5.30}$$

Combining these gives an expression for Δh:

$$\Delta h \equiv h_2 - h_1 = \frac{1}{\rho g}\left(\mu_0 \bar{M} H_1 + \mu_0 \frac{M^2}{2}\right) \tag{5.31}$$

Compared to the Quincke result of (5.26), the surface elevation is greater in this problem by the amount $\mu_0 M^2/2$ for the same value of the field H in the fluid.

Berkovsky and Orlov (1973) have analytically investigated a number of other problems in the shape of a free surface of a magnetic fluid.

A high-speed high-resolution plain-paper recording process (see Figure 5.5b) has been developed in which spatial undulations of the magnetic fluid surface are established in the field of a comb of magnetized pins (Maruno, Yubakami, and Soga 1983). The interface shape represents a solution of the ferrohydrodynamic Bernoulli equation, similar to that illustrated in Figure 5.5a but with the difference that the south pole piece is distant.

Magnetic nozzle

The jet flow of Figure 5.6 illustrates the coupling that may occur between flow speed and applied magnetic field. The magnetic field is provided by a uniformly wound current-carrying solenoid. Attraction of the fluid by the field accelerates the fluid motion along the direction of its path which is assumed horizontal. The magnetic boundary condition on field at station 2 requires continuity of the tangential component of **H**. Thus

$$H_2 = H_2' \tag{5.32}$$

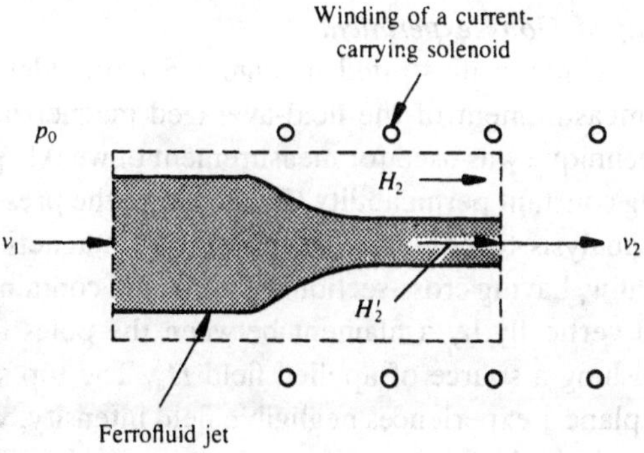

5.6 A free jet of magnetic fluid changes cross section and velocity in flowing through this magnetic nozzle, illustrating category 5 of Figure 5.3. (*After Rosensweig 1966b.*)

From the Bernoulli relationship (5.5),

$$p_1^* + \tfrac{1}{2}\rho v_1^2 = p_2^* + \tfrac{1}{2}\rho v_2^2 - \mu_0 \bar{M} H_2 \tag{5.33}$$

The boundary conditions on the fluid parameters from (5.22) with $p_n = 0$ and $p_c = 0$ give

$$p_1^* = p_0 \tag{5.34}$$
$$p_2^* = p_0 \tag{5.35}$$

Thus

$$v_2^2 - v_1^2 = 2\mu_0 \bar{M} H_2/\rho \tag{5.36}$$

The incompressible fluid velocity satisfies the continuity relationship $\nabla\cdot\mathbf{v} = 0$. This may be integrated as follows using the divergence theorem and assuming the jet possesses a round cross section everywhere:

$$\int_V \nabla\cdot\mathbf{v}\, dV = \int_S \mathbf{v}\cdot\mathbf{n}\, dS$$
$$= v_1(\pi/4)\, d_1^2 - v_2(\pi/4)\, d_2^2 = 0 \tag{5.37}$$

Combining (5.36) and (5.37) gives for the jet diameter ratio

$$\frac{d_1}{d_2} = \left(1 + \frac{2\mu_0 \bar{M} H}{\rho v_1^2}\right)^{1/4} \tag{5.38}$$

Modified Gouy experiment

The technique illustrated in Figure 5.7 provides a means for gravimetric measurement of the field-averaged magnetization $\bar{M}$. Originally the technique was used for measurement of weakly paramagnetic liquids having constant permeability (Bates 1963); the present treatment extends the analysis to nonlinear media of high magnetic moment. A tube of weight w_t having cross-sectional area a_t and containing ferrofluid is suspended vertically by a filament between the poles of an electromagnet furnishing a source of applied field H_a. The top surface of the ferrofluid at plane 1 experiences negligible field intensity, while at plane 2 the field H_2 within the fluid is uniform at its maximum value. The force F is given as the sum of pressure forces and weight of the containing tube:

$$F = (p_2^* - p_0)a_t + w_t \tag{5.39}$$

Note that p^* is regarded as capable of exerting a normal stress on a surface in the same manner as ordinary pressure. From the Bernoulli equation (5.6) applied between sections 1 and 2,

$$p_1^* + \rho g h_1 = p_2^* + \rho g h_2 - \mu_0 \bar{M} H_2 \tag{5.40}$$

From (5.22) with $p_n = 0$ and $p_c = 0$, the boundary condition at the free surface is

$$p_1^* = p_0 \tag{5.41}$$

Combining these and solving for $\bar{M}$ gives

$$\bar{M} = \frac{(F - F_0)/a_t}{\mu_0 H_2} \tag{5.42}$$

where $F_0 = w_t + \rho g(h_2 - h_1)a_t$ represents the force when the field is absent. H_2 is less than the applied field H_a owing to the influence of the sample shape. Introducing the definition of demagnetization coefficient

$$D \equiv (H_a - H)/M$$

in the present problem gives

$$H_2 = H_a - DM \tag{5.43}$$

For a circular cylinder having length that greatly exceeds its diameter, the demagnetization coefficient $D = \frac{1}{2}$.

The next section develops a problem that was important in the early

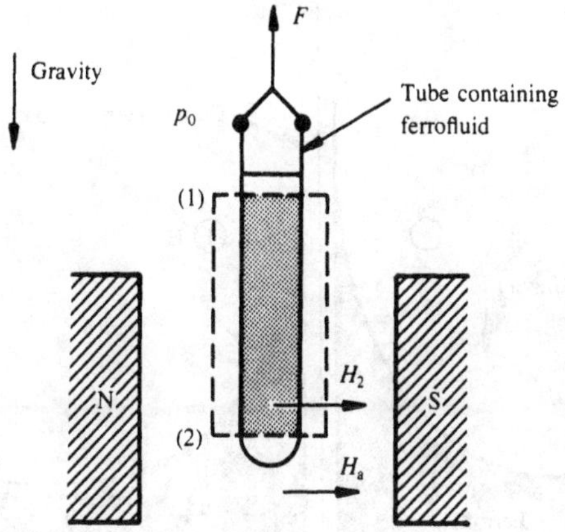

5.7 Analysis of the modified Gouy relationship illustrating category 6 of Figure 5.3. (*After Rosensweig 1979a.*)

formulation of ferrohydrodynamics principles, the phenomenon of a fluid climbing a current-carrying rod (Neuringer and Rosensweig 1964).

Conical meniscus

It is desired to find the response of an initially flat pool of magnetic liquid when a steady current is passed through a wire which emerges vertically from it. From electromagnetism [see (3.40)], the steady current I produces an azimuthal field external to the wire with magnitude

$$H = I/2\pi r \tag{5.44}$$

Because **M** is collinear with **H**, it is also azimuthal, and hence $M_n = 0$ at the free surface. Although it is necessary to solve the problem only once, it will be worthwhile at this stage to attack the problem in several different ways. The first two will employ the FHD Bernoulli equation arranged in two forms, and the third will use the form of the stress tensor due to Chu introduced in Chapter 4. A sketch of the system is given in Figure 5.8.

Method 1: Apply the FHD Bernoulli equation in the form of equation (5.6) between points 1 and 2, recognizing that the system is motionless:

$$p_1^* + \rho g h_1 - (\mu_0 \bar{M} H)_1 = p_2^* + \rho g h_2 - (\mu_0 \bar{M} H)_2 \tag{5.45}$$

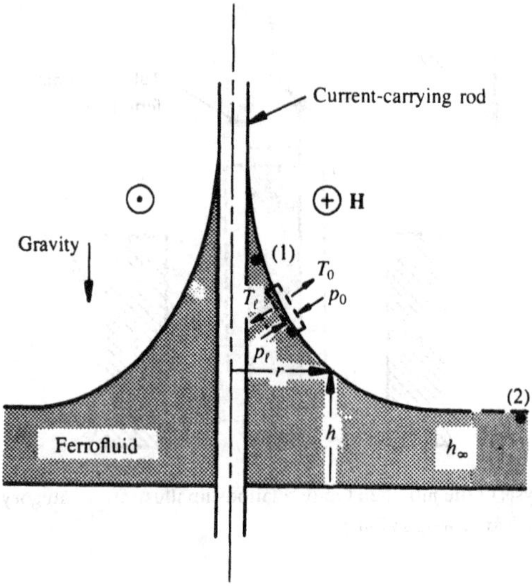

5.8 The conical meniscus in ferrohydrostatic equilibrium.

Because 2 is located far from the wire, the magnetic field approaches zero asymptotically [by (5.44)], and the last term of (5.45) goes to zero as $r \to \infty$ [or as $h_2 \to h(\infty)$]. To complete the solution of the problem, it is necessary to relate p_1^* and p_2^* to p_0, the environmental pressure, through the use of the appropriate FHD boundary condition. Thus, from (5.22) neglecting capillary-pressure effects,

$$p_1^* = p_0 - \tfrac{1}{2}\mu_0 M_n^2 = p_0 \qquad \text{(at 1)} \tag{5.46}$$

$$p_2^* = p_0 - \tfrac{1}{2}\mu_0 M_n^2 = p_0 \qquad \text{(at 2)} \tag{5.47}$$

where use was made of the facts that $M_n = 0$ at the free surface at location 1 and that $M = 0$ at location 2 because $H = 0$ there. Substituting (5.46) and (5.47) into (5.45), one finds after minor rearrangement that

$$\Delta h = h(r) - h(\infty) = \mu_0 \bar{M} H / \rho g \tag{5.48}$$

This solution shows that the height of the free surface provides a measure of the magnetic field intensity – it furnishes a kind of gaussmeter.

Method 2: The FHD Bernoulli equation in the form of equation (5.7) will now be applied between stations 1 and 2 on the magnetic fluid side of the interface.

$$(p + p_s + \rho g h)_1 = (p + p_s + \rho g h)_2 \tag{5.49}$$

Because $H \to 0$ as $r \to \infty$, $p_{s,2} = 0$. Applying (5.24), the appropriate form of the boundary condition, at 1 and 2 gives

$$p_1 + p_{s,1} + p_{m,1} = p_0 \quad \text{(at 1)} \tag{5.50}$$

$$p_2 = p_0 \quad \text{(at 2)} \tag{5.51}$$

Now (5.49) can be arranged to give

$$p_2 - p_1 = p_{s,1} + \rho g(h_1 - h_2) \tag{5.52}$$

and (5.50) and (5.51) can be combined to give

$$p_2 - p_1 = p_{s,1} + p_{m,1} \tag{5.53}$$

Eliminating common terms from (5.52) and (5.53) allows the immediate conclusion that

$$\Delta h = p_{m,1}/\rho g = \mu_0 \bar{M} H/\rho g \tag{5.54}$$

Note that striction was never evaluated in this problem. The striction terms come in and then out of the analysis, which is the rule in all incompressible flow problems. However, compressibility effects need to be considered in problems of optics, acoustics, and cavitation.

Method 3: Finally, the problem will be solved using the stress tensor of Chu, which was developed as equation (4.54):

$$\mathbf{T}_m = \left(-\int_0^H B\,dH\right)\mathbf{I} + \mathbf{HB} \tag{5.55}$$

The force density corresponding to this tensor is given by (4.52):

$$\mathbf{f}_m = -\int_0^H (\nabla B)_H\,dH \tag{5.56}$$

In this example the magnetic liquid has the same composition at all points, and thus $(\nabla B)_H = \mathbf{0}$, giving $\mathbf{f}_m = \mathbf{0}$. The equation of motion consistent with this formulation is readily found to be

$$\mathbf{0} = -\nabla p_0 + \rho \mathbf{g} = -\nabla(p_0 + \rho g h) \tag{5.57}$$

The interfacial condition to be satisfied must be worked out anew, consistent with the stress tensor, (5.55). If subscripts n and t denote, respectively, "normal" and "tangential," the magnetic stress components on the two sides of the meniscus are given by

Liquid side:

$$\mathbf{n}\cdot\mathbf{T}_{\mathrm{m}}\cdot\mathbf{t} = T_{\mathrm{nt}} = 0 \tag{5.58}$$

$$\mathbf{n}\cdot\mathbf{T}_{\mathrm{m}}\cdot\mathbf{n} = T_{\mathrm{nn}} = T_{\mathrm{l}} = -\int_0^H B_{\mathrm{l}}\,dH \tag{5.59}$$

Vapor side:

$$T_{\mathrm{nt}} = 0 \tag{5.60}$$

$$T_{\mathrm{nn}} = T_{\mathrm{e}} = -\int_0^H B_{\mathrm{e}}\,dH \tag{5.61}$$

In obtaining (5.58)–(5.61), use was made of the following relations:

$$(\mathbf{n}\cdot\mathbf{I})\cdot\mathbf{n} = \mathbf{n}\cdot\mathbf{n} = 1 \tag{5.62}$$

$$(\mathbf{n}\cdot\mathbf{I})\cdot\mathbf{t} = \mathbf{n}\cdot\mathbf{t} = 0 \tag{5.63}$$

$$\mathbf{n}\cdot\mathbf{H} = H_{\mathrm{n}} = 0 \quad \text{(along the liquid–gas interface)} \tag{5.64}$$

The subscript l denotes "liquid" and e "environment," so $B_{\mathrm{l}} = \mu_0(H + M)$ and $B_{\mathrm{e}} = \mu_0 H$ (because $M = 0$ on the vapor side). The total stress $\mathbf{T}$ can now be written

$$\underset{\substack{\text{total}\\ \text{stress}}}{\mathbf{T}} = \underset{\substack{\text{thermodynamic}\\ \text{pressure part}}}{-p_0(\rho, T)\mathbf{I}} + \underset{\substack{\text{viscous}\\ \text{stress}}}{\mathbf{T}_{\mathrm{v}}} + \underset{\substack{\text{magnetic}\\ \text{stress}}}{\mathbf{T}_{\mathrm{m}}} \tag{5.65}$$

In this problem $\mathbf{T}_{\mathrm{v}} = \mathbf{0}$ and the general interfacial condition $\mathbf{n}\cdot[\mathbf{T}] = \mathbf{0}$ applied to (5.65) gives

$$[T_{\mathrm{nn}}] = p_{\mathrm{e}} - p_{\mathrm{l}} \tag{5.66}$$

where the subscript 0 in the pressure terms is suppressed for simplicity. Substituting for T_{nn} from (5.59) and (5.61) and using condition (5.64) gives

$$\mu_0 \int_0^H M\,dH = \mu_0 \bar{M} H = p_{\mathrm{e}} - p_{\mathrm{l}} \tag{5.67}$$

The thermodynamic pressure in each phase is governed by (5.57), so, neglecting the density of the vapor and taking $h = h(\infty)$ as the datum level for pressure in the liquid, one finds that the pressure in each phase is given by

$$p_e = \text{const} \tag{5.68}$$
$$p_l - p_e = \rho g[h(\infty) - h] \tag{5.69}$$

Substituting (5.69) into (5.67) to eliminate $p_l - p_e$ then recovers equation (5.48), which is the required result.

The initial portion of the magnetization versus field curve is linear. Thus for small applied fields $M = \chi_i H$, where χ_i is the initial susceptibility [see (2.28)], and

$$\bar{M} = \frac{1}{H}\int_0^H M\,dH = \tfrac{1}{2}\chi H = \tfrac{1}{2}M \tag{5.70}$$

Substituting the field solution (5.44) and (5.70) into (5.48), one finds for *small applied fields* that

$$\Delta h = \frac{\mu_0}{\rho g}\,\frac{\chi_i I^2}{8\pi^2 r^2} \tag{5.71}$$

For *large applied fields*, $M = M_s$ and $\bar{M} = M_s$, and the elevation of the meniscus is given by

$$\Delta h = \frac{\mu_0}{\rho g}\,\frac{M_s I}{2\pi r} \tag{5.72}$$

Origin of the radial force

It is worthwhile to pause and consider the origin of the magnetic force in this problem; recall that the force and the field at any spatial point are oriented at right angles to each other, and at first consideration it may seem that no force should be present. Figure 5.9 depicts the top view of the system and shows a fluid element at radius **r** and azimuthal angle ϕ having length $\delta\mathbf{l}$ and magnetization **M** oriented along the local field **H**, which has azimuthal direction. Suppose the element possesses poles of opposite sign at its ends; it can then be realized that the local field at the two ends produces forces $\mathbf{f}_+$ and $\mathbf{f}_-$ of equal magnitude but differentially different orientation. The forces $\mathbf{f}_+$ and $\mathbf{f}_-$ may be resolved into their azimuthal and radial components with the evident result that the azimuthal components oppose each other and cancel while the radial components together produce a force that is directed radially inward toward the origin, i.e., toward the center of the wire. This attraction of fluid toward the wire results in the fluid climbing to higher elevations at

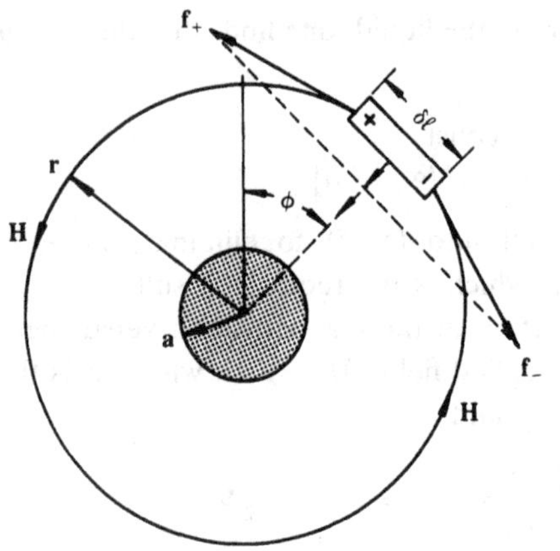

5.9 Physical origin of the radial inward force in the conical meniscus problem.

positions closer to the wire. Posing this problem and determining its resolution were crucial to the development of the FHD Bernoulli equation. Fundamentally, the magnetic force is vectorial in character, although the FHD Bernoulli equation succeeds in reducing the problem to a description in terms of field magnitudes only. A photograph of the conical meniscus phenomenon is shown in Figure 5.10.

The remote generation of forces in magnetic fluids has led to the conception of numerous devices and the successful development of a number of technological applications, a trend that is continuing. Often a small amount of magnetic fluid plays a critical role and makes entire devices possible. An excellent example is provided by magnetic-fluid shaft seals. These are devices that permit rotary motion of a cylindrical shaft relative to a stationary housing, with the magnetic liquid forming a perfect barrier in the gap between these parts, held in position by magnetic force.

Magnetic-fluid rotary-shaft seals

Sealing regions of differing pressure is now the most well developed of all proposed FHD applications. Figure 5.11a illustrates schematically how a ferromagnetic liquid is used as a leak-proof dynamic seal between a rotating shaft and stationary surroundings. A pair of bearings, not shown, is required to centrally position and support the shaft; otherwise the shaft would rub against the pole pieces, for it is

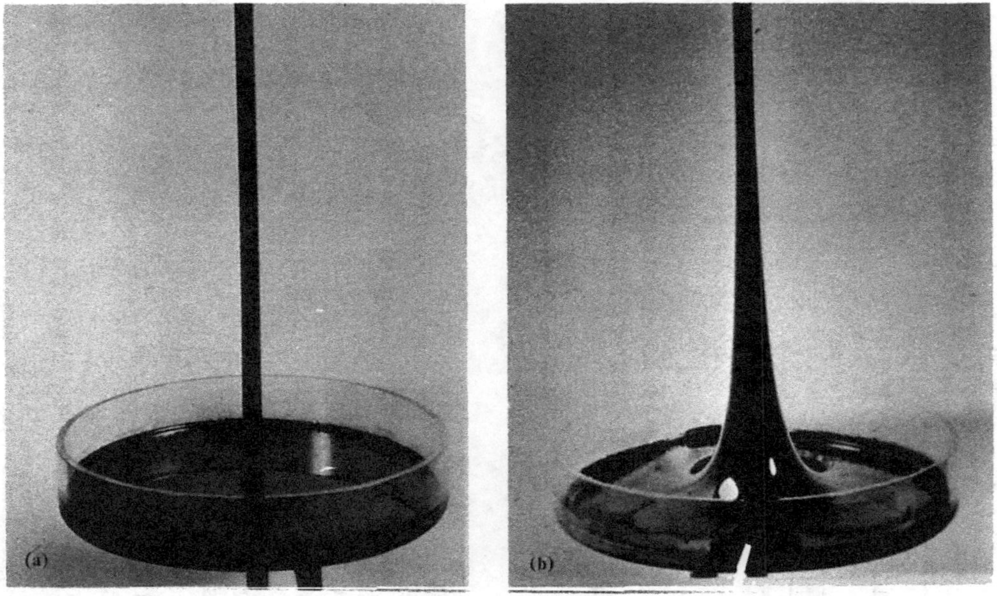

5.10 Balance of magnetic and gravitational energies at any point on the free liquid surface is demonstrated when an electric current is passed through a vertical rod running through a pool of ferrofluid. (a) The experimental arrangement in the absence of current. (b) When the current is turned on, the fluid leaps upward and assumes the shape shown. At each point in the bulk ferrofluid, the sum of the magnetic, gravitational, and pressure terms in the FHD Bernoulli equation equals the same constant. (*Photos by the author; originally appeared in International Science and Technology, July 1966.*)

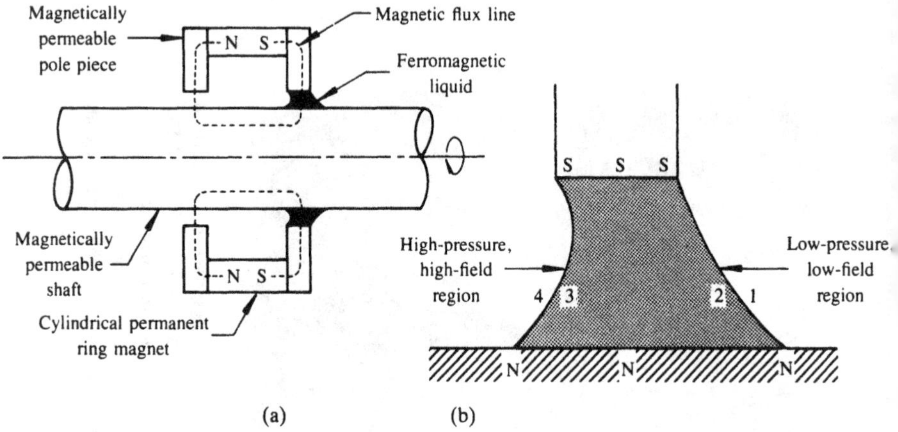

5.11 (a) Components of a magnetic fluid rotary shaft seal. (*From Rosensweig 1979a.*) (b) Sketch to analyze the pressure supported across one stage of the seal.

magnetically attracted to them. The seals are in widespread commercial use as pressure-, vacuum-, and exclusion-sealing devices. A magnetically permeable shaft is used as part of a low-reluctance magnetic circuit containing an axially magnetized ring magnet mounted between stationary pole blocks. In this configuration the magnetic field is directed radially across the gap. This arrangement permits pressure differences of up to $\sim 10^5$ pascals (Pa) to be supported across a single-stage seal and, furthermore, is particularly adaptable to staging, whereby much larger pressure differences can be sustained.

To analyze the magnetic liquid seal shown in Figure 5.11a, consider its idealization, shown in Figure 5.11b. The following simplifying assumptions are made:

The magnetic field is uniform and tangential to the meniscus at interfaces 4–3 and 2–1.
The gravitational force is negligible.
The analysis is restricted to a nonrotating shaft.

From the Bernoulli equation (5.6) applied between points 3 and 2,

$$p_3^* - p_2^* = \mu_0[(\bar{M}H)_3 - (\bar{M}H)_2] \tag{5.73}$$

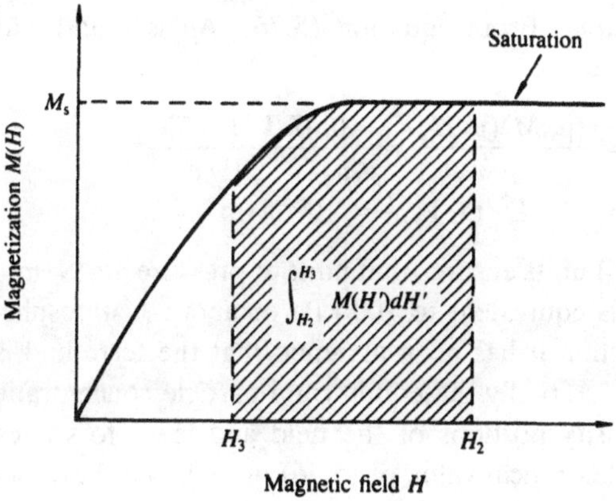

5.12 The shaded area under the magnetization curve relates to pressure difference across a single stage of a magnetic fluid seal. (*After Perry and Jones 1976.*)

The normal component M_n of the magnetization is zero at both interfaces, so the FHD boundary condition (5.22) gives

$$p_3^* = p_4 \quad \text{and} \quad p_2^* = p_1 \tag{5.74}$$

Substituting (5.74) into (5.73) gives the following expression for the burst pressure of a static seal:

$$\Delta p = p_4 - p_1 = \mu_0 \int_{H_2}^{H_3} M \, dH \tag{5.75}$$

The meaning of this integral is illustrated in Figure 5.12 as an area under the magnetization curve.

In a well-designed seal the field H_2 is negligible compared to H_3, so that the burst pressure of a single-stage static seal is given to a good approximation (Rosensweig 1971) by

$$\Delta p = \mu_0 \bar{M} H \tag{5.76}$$

where $\bar{M}H$ is evaluated at the peak value of the field.

Example: Calculate the burst pressure of a single-stage ferrofluid seal for which the maximum value of $\mu_0 H$ is 1.8 T (18,000 G) and $\mu_0 M$ = 0.07 T (700 G).

Solution: From equation (5.76), Δp is found with minor rearrangement to be

$$\Delta p = \frac{(\mu_0 M)(\mu_0 H)}{\mu_0} = \frac{0.07\ \mathrm{T}\cdot 1.8\ \mathrm{T}}{4\pi\cdot 10^{-7}\ \mathrm{H}\cdot\mathrm{m}^{-1}}$$
$$= 10^5\ \mathrm{T}^2\cdot\mathrm{m}\cdot\mathrm{H}^{-1} = 10^5\ \mathrm{N}\cdot\mathrm{m}^{-2}$$

Working in SI units ensures that units of pressure are $\mathrm{N}\cdot\mathrm{m}^{-2}$. The value $10^5\ \mathrm{N}\cdot\mathrm{m}^{-2}$ is equivalent to 100 kPa or about 1 atmosphere.

In this section it has been assumed that the ferrofluid is uniform in composition. Actually the equilibrium particle concentration is greater in high-intensity portions of the field and leads to values of Δp that exceed the theoretical value of (5.76) in seals that have been idle for a period of time. Because viscosity increases rapidly as particle concentration approaches its maximum value, such a seal is stiff to rotate initially. Rotary motion of the seal shaft stirs the fluid, tending to equalize the particle concentration, reduce the torque, and return the operating pressure capability to the predicted value.

The relationship of (5.76) was tested quantitatively by Perry and Jones (1976) using a ferrofluid plug held magnetically by external pole pieces and contained in a vertical glass tube (see Figure 5.13). Static loading of this seal was accomplished with a leg of immiscible liquid overlaying the plug. From Figure 5.14 it is seen that calculated and measured static-burst pressures are in very good agreement for various ferrofluid seals operated over a range of magnetic inductions.

The fluid–field configuration illustrated in Figure 5.11a is also that existing in the stator gap of moving-coil loudspeakers. The ferrofluid can be positioned without a container to function as a thermal contact. In this manner heat is removed from the coil, permitting the use of greater audio power (Hathaway 1979; Melillo and Raj 1981).

Another class of problems that can be treated advantageously concerns the response of objects immersed in a magnetic fluid. These Archimedean flows permit qualitative understanding and quantitative analysis of the mysterious levitational phenomena in ferrohydrodynamics.

5.5 Earnshaw's theorem and magnetic levitation

In 1839 Samuel Earnshaw put forth the theorem that *a charged body placed in an electrostatic field cannot be maintained in stable equilibrium under the influence of electric forces alone* (Stratton 1941, p. 116). From the considerations of Chapter 1, a magnetized body is

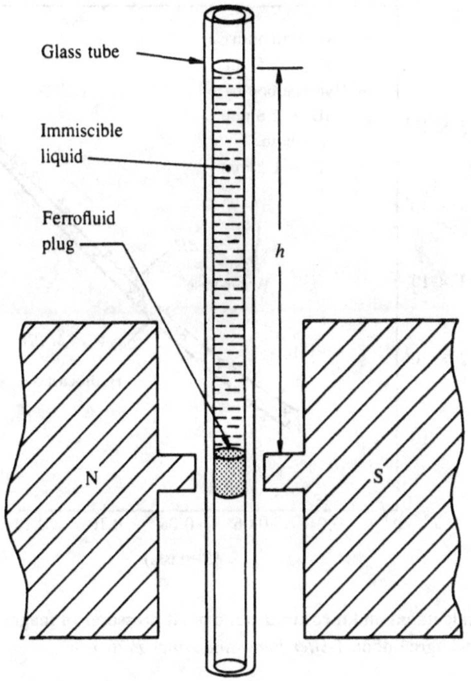

5.13 Experiment to determine the static loading of a one-stage magnetic fluid plug seal. (*After Perry and Jones 1976.*)

effectively a collection of "charges" and hence, by *Earnshaw's theorem*, it cannot be held in equilibrium by a static magnetic field.

The next paragraphs sketch a proof of Earnshaw's theorem for a point pole (charge), leaving the proof for an extended body containing a collection of poles (charges) to the reader as an exercise. In a region of space containing no volume distribution of poles (charges), the following Maxwell relations hold at any point:

$$\nabla \cdot \mathbf{H} = 0 \qquad (\nabla \cdot \mathbf{E} = 0) \tag{5.77}$$

$$\nabla \times \mathbf{H} = \mathbf{0} \qquad (\nabla \times \mathbf{E} = \mathbf{0}) \tag{5.78}$$

As discussed in Chapter 3, (5.78) implies that the magnetic (electric) field can be written as the gradient of a scalar function $\psi(x, y, z)$, called the *potential* of the field

$$\mathbf{H} = -\nabla\psi \qquad (\mathbf{E} = -\nabla\psi) \tag{5.79}$$

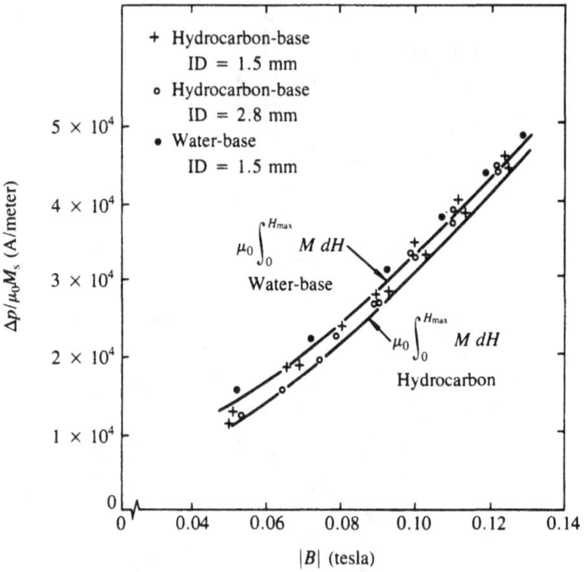

5.14 The calculated and measured static burst pressures of magnetic fluid seals are in good agreement. (*After Perry and Jones 1976.*)

Equations (5.77) and (5.78) then show that ψ satisfies Laplace's equation,

$$\nabla^2 \psi = 0 \tag{5.80}$$

The next step is to show that *the function ψ cannot be distributed with an absolute maximum or minimum inside this region*. To show this, use will be made of a proof due to Landau and Lifshitz (1960 p. 3.) It will be supposed that the potential has a maximum at some point P not on the boundary of the region. Then this point P can be surrounded by a small closed surface on which the normal derivative is everywhere negative; $\partial\psi/\partial n < 0$. Hence the integral over this surface $\oint_S (\partial\psi/\partial n)\, dS < 0$. However,

$$\oint_S \frac{\partial \psi}{\partial n}\, dS = \oint_S \mathbf{n} \cdot \nabla \psi\, dS = \int_V \nabla^2 \psi\, dV = 0 \tag{5.81}$$

the last result following from (5.80). Clearly, $\oint_S (\partial\psi/\partial n)\, dS$ cannot equal zero and be less (or greater) than zero at the same time, and hence the claim is proved by contradiction.

Previously it was seen that **H** (**E**) can be regarded as the force per unit pole (charge). Therefore, a correct expression for potential energy along an arbitrary path L within the field region, written on a per-unit-pole (charge) basis, is

Potential energy per unit pole (charge)

$$= -\int_L \mathbf{H}\cdot d\mathbf{r} \quad \left(= -\int_L \mathbf{E}\cdot d\mathbf{r}\right) \tag{5.82}$$

$$= \int_L \nabla\psi\cdot d\mathbf{r} = \psi \tag{5.83}$$

where the constant of integration was dropped in (5.83) for convenience. With the last identification it follows that *a test pole (charge) introduced into the field cannot be in stable equilibrium, for there is no point at which its potential energy would have a minimum.* (QED)

Simplified treatment of the levitation of a nonmagnetic body

Earnshaw's theorem appears to preclude the possibility of levitating a body with the aid of a magnetostatic field, but, as the author discovered (Rosensweig 1966a,b), *stable levitation is possible if a nonmagnetic object is immersed in a magnetic liquid and is brought into the presence of a suitable nonuniform magnetic field.* A heuristic argument is given first, followed by a rigorous account.

Consider a container of magnetic fluid placed in a zero-gravity environment. Assume that initially the magnetic field is absent so that the pressure is uniform at a constant value within the magnetic fluid region. Two opposed sources of magnetic field having equal strength are next brought up to the vicinity of the container, as shown in Figure 5.15a, creating a field distribution that is zero midway between the sources and increases in magnitude in every direction away from the midpoint. Because the magnetic fluid is attracted to regions of higher field magnitude, the fluid is attracted away from the midpoint. However, the magnetic fluid is incompressible and fills the container, so the response it provides is an effective increase in pressure as one moves away from the center.

Now consider the response of a nonmagnetic object introduced into this environment. If the object is placed at the center point, as shown in Figure 5.15b, it will remain there in the absence of other forces, as pressure forces are symmetrically distributed over its surface. If it is displaced from the equilibrium position, the object experiences un-

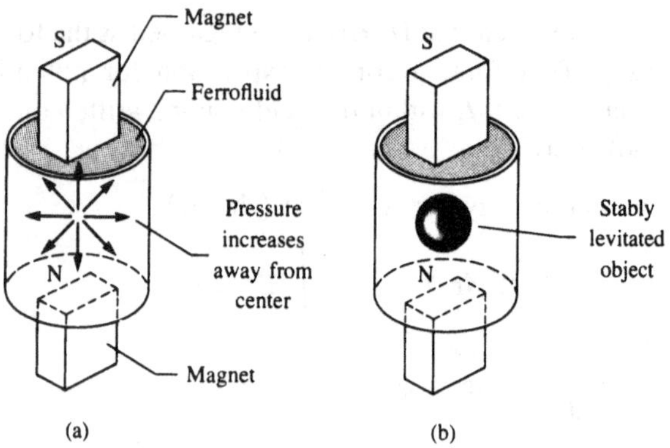

5.15 Levitation of a nonmagnetic body in a magnetized fluid is another demonstration of the expanded Bernoulli relation. (a) Pressure in the fluid is lowest at the center and increases with distance from it. (b) When a nonmagnetic object such as a glass marble is placed in the container, it moves to the center and remains there in equilibrium. (*After Rosensweig 1982a.*)

balanced pressure stresses that establish a restoring force. This is the phenomenon of *passive levitation of a nonmagnetic body*. No expenditure of energy is required to sustain the levitation of the object.

Phenomenon of self-levitation

A related phenomenon is that of self-levitation, which occurs when a magnet is immersed in magnetic fluid, as shown in Figure 5.16. The field surrounding the magnet is symmetrically disposed, so the pressure in the magnetic liquid is symmetrically distributed, and the magnet experiences an equilibrium of magnetic forces when the magnet is centered. If the magnet should be displaced from the center, the symmetry of the field is broken; consideration of the permeable path encountered by the magnetic flux leads to the conclusion that the field is then greater over that magnet surface located furthest away from the center. Employing once more the notion that the effective pressure is greatest where magnetic field is greatest, one can conclude that the

5.16 Self-levitation of a magnetic object in magnetic fluid. The x-ray images show a disk magnet stably suspending itself in a beaker of ferrofluid. The magnet, which is nearly four times as dense as the fluid, is seen hovering above the bottom of the beaker in the side view (b). As the plan view (a) illustrates, the magnet is repulsed from the surrounding beaker wall. (*From Rosensweig 1966c.*)

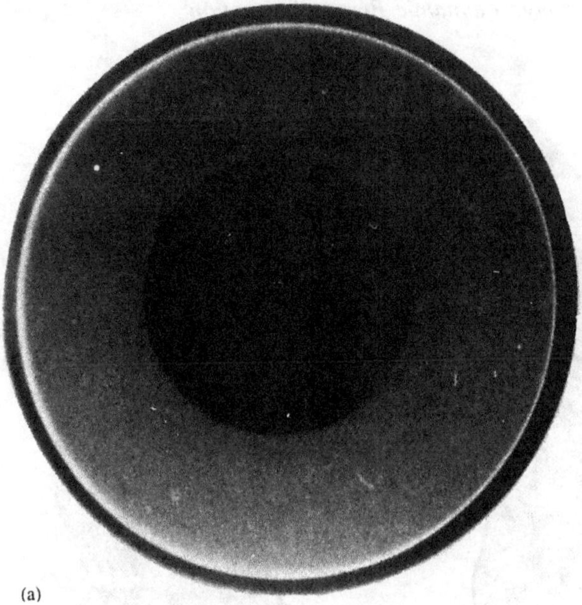

(a)

(b)

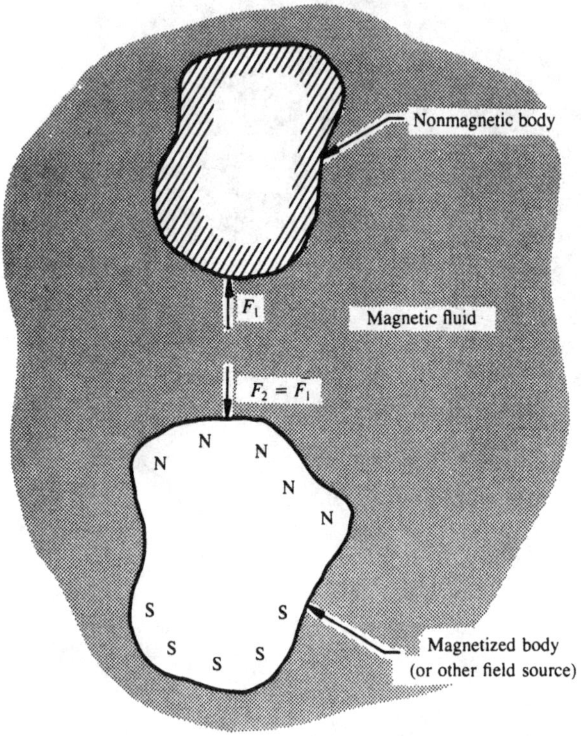

5.17 Mutual repulsion of a magnet and a nonmagnetic object when both are immersed in a ferrofluid. (*After Rosensweig 1966b.*)

magnet is subject to a restoring force tending to return it to the center. This is the phenomenon of *self-levitation in a magnetic fluid* (Rosensweig 1966c). The principle has been employed to advantage in accelerometers, motion dampeners, passive levitational bearings, and inclinometers (see, e.g., Rosensweig 1979a; Schaufeld 1982; Bailey 1983).

As another variation on the theme, consider Figure 5.17, which shows the mutual repulsion of a magnet and nonmagnetic object when both are immersed in a magnetic fluid in a region removed from any fluid boundary. Note that this interaction is without analog in ordinary magnetostatics, in which both objects must possess magnetic moments if there is to be a static interaction between them. Of course, the mutual force shown in Figure 5.17 is not due to a direct interaction between the

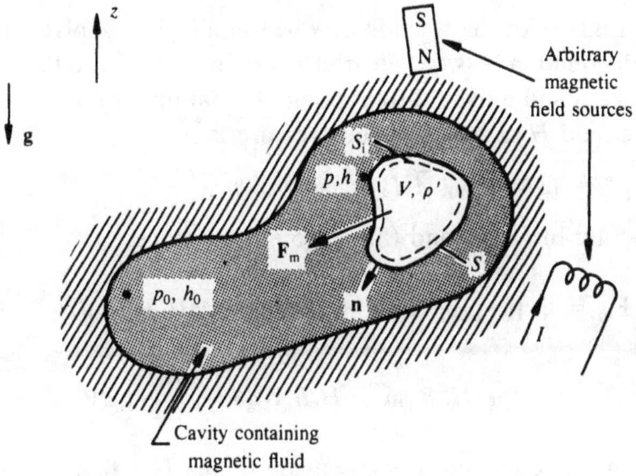

5.18 Development of the generalized expression for the force acting on an arbitrary magnetized or unmagnetized body immersed in a magnetic fluid and subjected to an arbitrary magnetic field.

two bodies; the force arises because of the magnetic fluid attracted into the space between them.

A general, quantitative analysis incorporating these effects is developed next, leading to (5.99) below.

Analysis of forces on an immersed body

Consider a body, either magnetic or nonmagnetic, immersed in a magnetic fluid in the presence or absence of arbitrary sources of magnetic field, as shown in Figure 5.18. The net force acting on the body is given by

$$\mathbf{F}'_{\mathrm{m}} = \oint \mathbf{n}\cdot\mathbf{T}'_{\mathrm{m}}\,dS + \int \rho'\mathbf{g}\,dV = \oint \mathbf{t}_{\mathrm{n}}\,dS + \int \rho'\mathbf{g}\,dV \tag{5.84}$$

where [see (4.27) and (4.71)]

$$\mathbf{T}'_{\mathrm{m}} = -(p^* + \tfrac{1}{2}\mu_0 H^2)\mathbf{I} + \mathbf{BH} \tag{5.85}$$

The stress vector $\mathbf{t}_{\mathrm{n}}$ corresponding to the stress tensor $\mathbf{T}'_{\mathrm{m}}$ of equation (5.85) is found from $\mathbf{t}_{\mathrm{n}} = \mathbf{n}\cdot\mathbf{T}'_{\mathrm{m}}$, which yields

$$\mathbf{t}_{\mathrm{n}} = -(p^* + \tfrac{1}{2}\mu_0 H^2)\mathbf{n} + \mathbf{H}B_{\mathrm{n}} \tag{5.86}$$

This expression for the traction may be simplified by applying the FHD Bernoulli equation between an arbitrary point adjacent to the outside of the surface S and a fixed distant point at a datum level h_0 at which the magnetic field H is zero and the pressure is p_0:

$$p^* + \rho g h - \mu_0(\bar{M}H) = p_0 + \rho g h_0 \tag{5.87}$$

With the aid of (5.86) and (5.87), (5.84) becomes

$$\mathbf{F}'_{\mathrm{m}} = \oint_S [(-p_0 - \rho g h_0 + \rho g h - \mu_0 \bar{M}H - \tfrac{1}{2}\mu_0 H^2 + H_{\mathrm{n}}B_{\mathrm{n}})\mathbf{n} + H_{\mathrm{t}}B_{\mathrm{n}}\mathbf{t}]\, dS + \int_V \rho' \mathbf{g}\, dV \tag{5.88}$$

In writing this equation use was made of the fact that $\mathbf{H}$ can be decomposed into a component normal to S and another lying in the tangent plane at the point in question:

$$\mathbf{H} = H_{\mathrm{n}}\mathbf{n} + H_{\mathrm{t}}\mathbf{t} \tag{5.89}$$

According to the divergence theorem, for any scalar quantity A

$$\oint_S A\mathbf{n}\, dS = \int_V \nabla A\, dV \tag{5.90}$$

$$\oint_S -(p_0 + \rho g h_0)\mathbf{n}\, dS = \int_V -\nabla(p_0 + \rho g h_0)\, dV = 0 \tag{5.91}$$

and using this result with (5.88) yields

$$\mathbf{F}'_{\mathrm{m}} = \oint_S [(-\mu_0 \bar{M}H - \tfrac{1}{2}\mu_0 H^2 + H_{\mathrm{n}}B_{\mathrm{n}})\mathbf{n} + H_{\mathrm{t}}B_{\mathrm{n}}\mathbf{t}]\, dS + \int_V (\rho - \rho')g\mathbf{k}\, dV \tag{5.92}$$

where $-g\mathbf{k} = \mathbf{g}$, $\mathbf{k}$ denoting a unit vertical vector. Now, if the magnetic field is everywhere zero, then (5.92) reduces to the usual *Archimedes' law of buoyancy*:

$$\mathbf{F}'_{\mathrm{m}}(H = 0) = \int_V (\rho - \rho')g\mathbf{k}\, dV = (\rho - \rho')g\mathbf{k}\, V \tag{5.93}$$

for any volume V, as it must. The last equality, however, is true only if the densities as well as the gravitational acceleration are uniform over the volume.

Return now to the more general case of (5.92) with $H \neq 0$, drop the ordinary buoyancy term, use the definition of $\bar{M}$ and the defining relation $B = \mu_0(H + M)$; the magnetic force $\mathbf{F}'_m$ acting on the body can then be written

$$\mathbf{F}'_m = \oint_S \left[\left(H_n B_n - \int_0^H B\, dH \right) \mathbf{n} + H_t B_n \mathbf{t} \right] dS \tag{5.94}$$

Equation (5.94) allows the total magnetically induced force on an arbitrary magnetic or nonmagnetic body immersed in ferrofluid to be computed from the field solutions.

If the immersed body is nonmagnetic, the integral in (5.94) evaluated over the surface S_i just inside the body disappears because there is no magnetic force on the matter within the surface:

$$\mathbf{0} = \oint_{S_i} \left[\left(H_n B_n - \int_0^H B\, dH \right) \mathbf{n} + H_t B_n \mathbf{t} \right] dS \tag{5.95}$$

Subtracting (5.95) from (5.94) and using the boundary conditions $[B_n] = 0$ and $[H_t] = 0$ shows that the tangential force is zero and that the normal force integrated over the body is given by

$$\mathbf{F}'_m = \oint_S \left(B_n[H_n] - \left[\int_0^H B\, dH \right] \right) \mathbf{n}\, dS \tag{5.96}$$

where, for example, $[H_n] = H_n^+ - H_n^-$, and $+$ and $-$ refer to just outside and inside the surface S, respectively. To simplify (5.96) use will be made of equations (5.21a) and (5.21b), which lead to

$$[H_n] = -M_n \tag{5.97}$$

$$\begin{aligned} \left[\int_0^H B\, dH \right] &= \left[\tfrac{1}{2}\mu_0 H^2 + \mu_0 \int_0^H M\, dH \right] \\ &= [\tfrac{1}{2}\mu_0 H_n^2] + \mu_0 \int_0^H M^+ dH \\ &= \tfrac{1}{2}\mu_0 M_n^2 - B_n M_n + \mu_0 \int_0^H M\, dH \end{aligned} \tag{5.98}$$

where $[M] = M^+ - M^- = M^+ = M$ because the body is nonmagnetic. With the aid of (5.97) and (5.98), (5.96) can be written in an equally general form that emphasizes the role of the magnetization:

$$\mathbf{F}'_m = -\oint_S \left(\tfrac{1}{2}\mu_0 M_n^2 + \mu_0 \int_0^H M\,dH \right) \mathbf{n}\,dS \tag{5.99}$$

The conditions for stable levitation of a magnetic or nonmagnetic body at an interior point of the fluid space are the following:

1. $\mathbf{F} = (\rho - \rho')g\mathbf{k}V + \mathbf{F}'_m = \mathbf{0}$ (5.100)
2. A positive restoring force accompanies any small displacement away from the equilibrium point.

The following illustrates a simple but important consequence of these levitational principles as applied to a small, immersed, nonmagnetic body. Because $\frac{1}{2}M_n^2 + \bar{M}H$, which is the integrand of (5.99), increases monotonically with H, it follows from the form of (5.99) that, to levitate, a magnetic field must possess a local minimum of field magnitude; i.e., for any displacement from such an interior point (cf. Figure 5.15)

$$\delta H > 0 \tag{5.101}$$

In the limit of intense applied field, $\frac{1}{2}M_n^2/\bar{M}H \ll 1$, and $\mathbf{F}'_m$ of (5.99) can be simplified as follows:

$$\begin{aligned}\mathbf{F}'_m &= -\oint_S p_m \mathbf{n}\,dS = -\int_V \nabla p_m\,dV \approx -\nabla p_m V \\ &= -V\mu_0 M \nabla H\end{aligned} \tag{5.102}$$

where p_m is introduced from (4.36b), and it is assumed that M and ∇H are sensibly constant across V. Owing to the minus sign in (5.102), the force is equal and opposite to the magnetic force on an equivalent volume of fluid. *Repulsion and stable levitation owe their existence to the presence of this minus sign.*

Magnetic buoyancy is illustrated in the photograph of Figure 5.19. Various mineral grains are seen to float at different elevations in a paramagnetic salt solution placed in the impressed, spatially varying gradient field of the electromagnetic pole pieces. It is easy to realize that as the fluid is attracted downward the grains are forced upward. The levitation forces are stronger in colloidal ferrofluid, and it can be calculated equating normal buoyant weight to the magnetic buoyant force of equation (5.102) that any nonmagnetic substance can be floated in this manner; substantiating data are given by Kaiser and Miskolczy

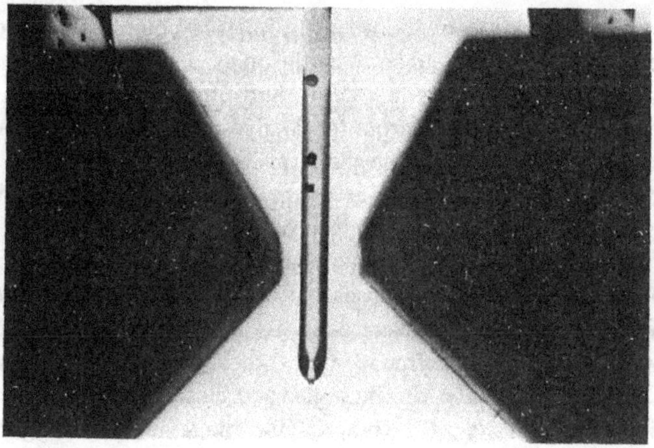

5.19 The artificial specific gravity induced in a paramagnetic salt solution separates a mixture of mineral grains. From the top down the materials are glass (s.g. 2.75), pyrite (s.g. 5.00), and galena (s.g. 7.5). (*Photograph courtesy Y. Zimmels, Technion.*)

(1970b) for diamond (specific gravity, 3.5) and other materials up to tungsten (specific gravity, 19.3). Processes for continuous separation of materials have been devised on this basis. (See, e.g., Rosensweig 1969; Khalafalla and Reimers 1973a; Shimoiizaka et al. 1980.)

Viewing a collection of small nonmagnetic spheres in a thin layer of magnetized fluid has permitted direct observation of the crystallization of magnetic "holes," forming a variety of different lattices (Skjeltorp 1983a).

5.6 Striction effect

Consider a system in which two bar magnets are immersed in a bath of magnetic fluid. The bar magnets have opposite poles facing one another and producing an intense field in the small gap between them. The gap is sufficiently far below the fluid surface that the magnetic field at the surface due to fringing from the gap is negligible. The boundary condition (5.22) applied at the flat surface gives $p_0 = p_1^*$, where the subscript 0 denotes the environmental condition and the subscript 1 denotes a position in the fluid at the surface. Applying the FHD

Bernoulli equation, (5.6), between the surface point 1 and a point 2 located in the region of field between the poles, and neglecting the effect of gravity, gives $p_1^* = p_2^* - \mu_0(\bar{M}H)_2$. Substituting the definition $p_2^* = p_2(\rho, T) + p_{s,2} + \mu_0(\bar{M}H)_2$ from (4.71) gives $p_0 = p(\rho, T) + p_s$, where the subscript 2 is dropped from the terms on the right side. Thus $\Delta p \equiv p(\rho, T) - p_0$, representing the excess pressure of the magnetized fluid relative to the environment, is given by $-p_s$.

As discussed in Section 4.3, Hakim and Higham (1962) performed the analogous experiment in a transparent dielectric fluid subjected to the applied field of immersed electrodes. By relating the pressure to the density, and thus to the refractive index, they predicted deflections of a light ray passed through the fluid gap and closely confirmed that the strictive force density of a compressible linear medium is given by equation (4.49).

Comments and supplemental references

The initial formulation of the ferrohydrodynamic Bernoulli relationship may be found in

Neuringer and Rosensweig (1964)

and a portion of the material in this chapter is drawn from the presentation of

Rosensweig (1979a)

An educational motion picture is available illustrating magnetic-fluid behavior:

Martinet (1982)

The reader desiring an introduction to the concept of vorticity and its relationship to fluid rotation will find a concise discussion by

Tritton (1977)

and a scholarly tour de force by

Truesdell (1954)

Soon after ferrofluids became available, magnetic fluid shaft seals were developed by

Rosensweig, Miskolczy, and Ezekiel (1968)

Commercialization of the concept was dependent on the discovery of a practical means for staging, whereby larger pressure differences are supported. A number of pole teeth, typically ten per centimeter, are provided on the shaft with magnetic flux supplied by one permanent magnet. The flux concentrates in the narrow gaps between the pole teeth and the pole block, gathering the ferrofluid into discrete sealed rings with tiny interstage zones between them. When the seal assembly is subjected to a large pressure difference, the seal rings bubble gas through to the interstage zones, then reseal. In this manner the pressure that can be supported with no leaking can exceed 100 kPa per centimeter of shaft length.

Rosensweig (1971)

A state of the art review of commercialized magnetic fluid seal technology is given by

Moskowitz (1975)

and uses in lubrication and damping are discussed by

Ezekiel (1975)

Further details concerning the fascinating principles and applications of the passive ferrofluidic bearings and related phenomenology may be found in

Rosensweig (1978a)

The principle is employed in the flotation of an inertial mass floated magnetically in a rotating chamber filled with magnetic fluid; the device viscously damps unwanted vibrations such as occur in numerically controlled machine tools and graphics plotters.

The total force on a homogeneous magnetically soft body has been discussed by

Byrne (1977)

In electrohydrodynamics (EHD), polarizable particles in insulating dielectric liquids experience the *dielectrophoretic* force. Uncharged droplets and bubbles can be levitated dielectrophoretically in liquids by the use of strong nonuniform electric fields. The concept, which is the analog of FHD levitation, has applications to laser-target fabrication, zero-gravity space processing, and liquid–gas separation processes. See

Jones and Bliss (1977)

Many interesting and important problems in the fluid mechanics of EHD and in FHD arise because of the presence of an interface. Thus an accurate mathematical description of the dynamics of an interface is essential for the quantitative understanding of these phenomena. For a completely general formulation, including viscous normal stress, of the dynamics of the nonmagnetic Newtonian fluid interface, see

Scriven (1960)

Higgins and Scriven (1979)

Although no examples are considered in this chapter in which capillary pressure plays a role, its influence in numerous problems of interfacial stability is major, as will be seen in Chapter 7.

6

MAGNETOCALORIC ENERGY CONVERSION

Certain magnetic materials heat up when they are placed in a magnetic field and cool down when they are removed. The phenomenon is known as the *magnetocaloric effect*. For a given material, the larger the change in magnetic field, the larger the magnetocaloric effect.

This principle was first explained in 1918 by Pierre Weiss and Auguste Picard. Nevertheless, more than twenty-five years before, Thomas Edison (1887) and Nicolai Tesla (1890) appreciated a possible application of the effect and independently, but unsuccessfully, attempted to run heat engines to produce power. In 1926 Peter Debye and William Giauque, again independently, suggested how one-step magnetic refrigeration cycles could be used to reduce the temperatures of research samples from 1 K to a small fraction of a kelvin. Giauque and Duncan MacDougall experimentally verified the method in 1933.

More recently a new idea was introduced in which the magnetic conversion of heat to work (or the reverse, refrigeration) could be made efficient at normal or elevated temperatures using regenerative thermal exchange as one step of the process (Resler and Rosensweig 1964, 1967). Direct and efficient conversion of heat to work with no mechanical moving parts was shown to be possible. A magnetic material was postulated that could flow through tubes and around bends and change its shape. Thus began a quest for a magnetic fluid. Much remains to be done using magnetic fluid for converting energy, in part because of the need for improved material properties. Nonetheless, more modest uses of the principles to enhance natural convective cooling and to achieve nonmechanical pumping of a fluid are feasible at present.

Section 6.1 develops the thermodynamics of magnetizable substances in a manner that stresses material parameters. With this information it is possible to devise power cycles and to evaluate quantitatively the thermodynamic efficiency, a program that is carried out in the remainder of the chapter.

6.1 Thermodynamics of magnetic materials

Expressions for heat addition and the work performed with magnetic materials will be needed for the analysis of power cycles, analogous to the treatment of thermodynamic cycles utilizing ordinary gases and vapors. If W denotes the work done by a unit volume of the magnetic substance, the work differential δW is given by the term $-H\,dB$, as shown in the discussion following equation (3.50). The substance will be considered incompressible. The field energy and the thermodynamic state per unit volume of the working substance change cyclically with time as the substance recirculates through the system, and it is this Lagrangian history that is analyzed. With $B = \mu_0(H + M)$, δW can be expressed $-\mu_0 H\,dH - \mu_0 H\,dM$, and over a complete cycle the first term vanishes; i.e., $\oint \mu_0 H\,dH = 0$. Thus, for cycle analysis it is permissible to write

$$\delta W = -\mu_0 H\,dM \tag{6.1}$$

The state variables will be chosen to be H, M, and the temperature T. These three variables are related by an equation of state of the form

$$M = M(H, T) \tag{6.2}$$

where H and T have been chosen as the independent variables. With the aid of (6.2), (6.1) can be rewritten

$$\delta W = -\mu_0 H\left[\left(\frac{\partial M}{\partial H}\right)_T dH + \left(\frac{\partial M}{\partial T}\right)_H dT\right] \tag{6.3}$$

The heat added to the material in terms of differential increments of T and H can be written

$$\delta Q = c(H, T)\,dT + g(H, T)\,dH \tag{6.4}$$

$c(H, T)$ is the specific heat of the magnetic material at constant H and in general is a function of H and T, as is the function g, whose form remains to be determined.

If U is the internal energy of the magnetic substance, these relationships may be augmented with a statement of the first law of thermodynamics, which can be written

$$\begin{aligned} dU &= \delta Q - \delta W \\ &= \left[c + \mu_0 H\left(\frac{\partial M}{\partial T}\right)_H\right] dT + \left[g + \mu_0 H\left(\frac{\partial M}{\partial H}\right)_T\right] dH \end{aligned} \tag{6.5}$$

Because dU is a perfect differential and can be expanded as a function of T and H, from (6.5) it is immediately evident that

$$\left(\frac{\partial U}{\partial T}\right)_H = c + \mu_0 H\left(\frac{\partial M}{\partial T}\right)_H \tag{6.6}$$

and

$$\left(\frac{\partial U}{\partial H}\right)_T = g + \mu_0 H\left(\frac{\partial M}{\partial H}\right)_T \tag{6.7}$$

The crossed partial derivatives

$$\left[\frac{\partial}{\partial H}\left(\frac{\partial U}{\partial T}\right)_H\right]_T \quad \text{and} \quad \left[\frac{\partial}{\partial T}\left(\frac{\partial U}{\partial H}\right)_T\right]_H$$

are equal, and hence with (6.6) and (6.7) it follows that

$$\left(\frac{\partial c}{\partial H}\right)_T = \left(\frac{\partial g}{\partial T}\right)_H - \mu_0\left(\frac{\partial M}{\partial T}\right)_H \tag{6.8}$$

The second law of thermodynamics states $1/T$ must be an integrating factor for the differential δQ in any reversible process, and the entropy S is thus defined by the expression

$$dS = \frac{\delta Q}{T} = \frac{c}{T}\,dT + \frac{g}{T}\,dH \tag{6.9}$$

where

$$\frac{c}{T} = \left(\frac{\partial S}{\partial T}\right)_H, \quad \frac{g}{T} = \left(\frac{\partial S}{\partial H}\right)_T \tag{6.10}$$

The entropy differential is also exact, so

$$\left(\frac{\partial c}{\partial H}\right)_T = \left(\frac{\partial g}{\partial T}\right)_H - \frac{g}{T} \tag{6.11}$$

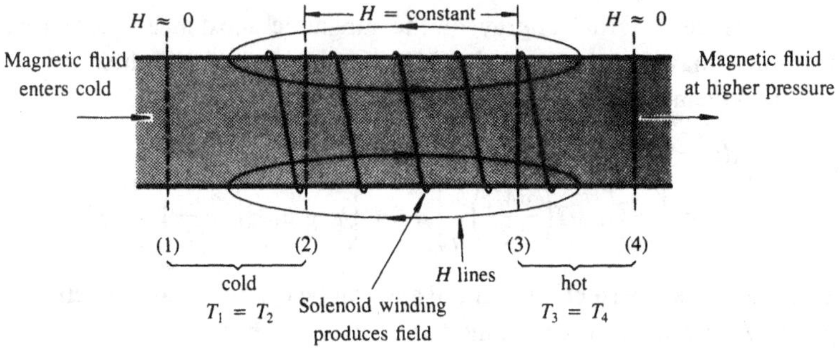

6.1 The magnetocaloric pump. An externally supplied source of heat warms the fluid between stations 2 and 3. Magnetic force drives the fluid into motion while elevating the fluid pressure. There are no moving mechanical parts.

Combining (6.8) and (6.11), one finds that

$$g = \mu_0 T \left(\frac{\partial M}{\partial T} \right)_H \tag{6.12}$$

$$\left(\frac{\partial c}{\partial H} \right)_T = \mu_0 T \left(\frac{\partial^2 M}{\partial T^2} \right)_H \tag{6.13}$$

6.2 Mechanism for power generation

Consider the situation depicted in Figure 6.1. A cold magnetic fluid is attracted into the interior of the solenoid and thus is set into motion. Inside the solenoid the material is heated; as it is warmed up, it approaches the Curie temperature θ and its magnetization decreases. The behavior is illustrated for iron, nickel, and cobalt, materials that follow a universal curve representing the Curie–Weiss law, as shown in Figure 6.2 and Table 6.1 (Bozorth 1951).

Applying the FHD Bernoulli equation between stations 1 and 2 in Figure 6.1, neglecting any change in the gravitational potential energy, and realizing that kinetic energy is constant in a tube of uniform cross section, one sees that

$$p_1^* = p_2^* - \mu_0 (\bar{M}H)_2 \tag{6.14}$$

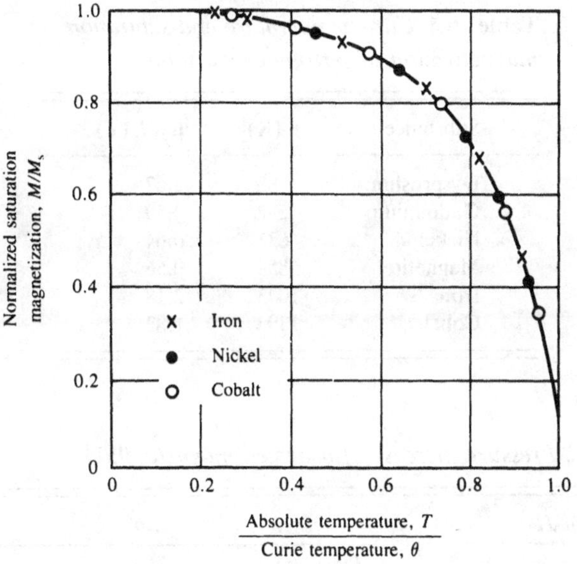

6.2 Domain magnetization as a function of temperature. Not all materials follow a universal curve. (*After Heck 1974.*)

as station 1 is in a field-free region of space. Similarly, application of the FHD Bernoulli equation between stations 3 and 4 gives

$$p_3^* - \mu_0(\bar{M}H)_3 = p_4^* \tag{6.15}$$

Inside the solenoid the FHD Bernoulli equation is not applicable, because the assumption of an isothermal flow field inherent in its derivation does not hold. Instead, going back to the equation of motion (4.72) and neglecting acceleration of the fluid, friction, and gravity, one sees that the following governing relationship holds inside the solenoid:

$$\mathbf{0} = -\nabla p^* + \mu_0 M \nabla H \tag{6.16}$$

Furthermore, because the field in the interior of the solenoid is uniform, $\nabla H = \mathbf{0}$, and consequently $p_2^* = p_3^*$. The change in pressure between stations 4 and 1 can now be computed:

$$\Delta p = p_4 - p_1 = \mu_0 H[\bar{M}(T_1) - \bar{M}(T_4)] = \mu_0 H \Delta \bar{M} \tag{6.17}$$

The component depicted in Figure 6.1 is the analog of the combustor and compressor of a conventional turbine power system. With the aid of

Table 6.1 *Curie temperature and saturation magnetization of ferromagnetic solids*

Substance	θ (K)	$\mu_0 M_s(T)$
Dysprosium	88	3.67
Gadolinium	292	2.59
Nickel	631	0.64
Magnetite	858	0.56
Iron	1043	2.18
Cobalt	1393	1.82

Table 6.2 *Pressure increase of a heated magnetic fluid*

$\mu_0 H$		Δp		
(T)	(G)	($N \cdot m^{-2}$)	(atm)	(m H_2O)
2	20,000	10^5	1	10
20	200,000	10^6	10	104
50	500,000	2.5×10^6	25	259

this simple device, the capability of converting heat into mechanical energy has now been demonstrated.

To get a feeling for the magnitude of the Δp predicted by (6.17), consider Table 6.2, which is computed for a value of $\mu_0 \Delta \bar{M} = 0.0637$ T (637 G) and some reference values for the applied field. A field of 2 T is typical of laboratory iron-yoke magnets, 20 T of advanced superconductive magnets, both operated in steady state. However, 50 T is more in the league of so-called pulsed magnets. As a measure of the magnitude of the values in the last column, the height of Niagara Falls is 51 m

6.3 Linear equation of state

Near the Curie temperature θ, it is sufficient to take a linear equation of state as an approximation to the curve of Figure 6.2 and to represent the magnetization as

$$K = -\frac{\partial M}{\partial T} \quad \text{or} \quad M = K(\theta - T) \tag{6.18}$$

The magnitude K is called the *pyromagnetic coefficient*. Because M is a linear function of T, from (6.13) it is seen that $(\partial c/\partial H)_T = 0$. Thus, for the linear equation of state

$$c = c(T) \equiv c_0 \tag{6.19}$$

where c_0 is the ordinary specific heat of the material. Also, from (6.12) it is readily seen that the function g takes the particular form

$$g = -\mu_0 KT \tag{6.20}$$

Correspondingly, the entropy S is then given by

$$S = c_0 \ln T - \mu_0 KH + \text{const} \tag{6.21}$$

The first term in (6.21) is familiar from ordinary thermodynamics and shows the increase in entropy with temperature. An interpretation of the minus sign in front of the second term is that the magnetic field lines up the dipoles and hence decreases disorder, of which entropy is a measure.

To gain some insight into the temperature changes involved, consider the process of cooling on removing a magnetic substance from a magnetic field H. For an adiabatic change of state

$$c_0 \ln T_i - \mu_0 KH_i = c_0 \ln T_f - \mu_0 KH_f \tag{6.22}$$

where i and f indicate the initial and final states, respectively. Setting $T_i = \theta$, $H_i = H$ and using the notation $T_f = T$, $H_f = 0$ gives

$$T = \theta e^{-\mu_0 KH/c_0} \tag{6.23}$$

The change in temperature

$$\Delta T_{ad} = \theta - T = \theta(1 - e^{-\mu_0 KH/c_0}) \tag{6.24}$$

is called the *adiabatic temperature change*. If the material involved is iron, for which $\theta = 1043$ K, $c_0 = 6.7 \times 10^6\ \text{J} \cdot \text{m}^{-3} \cdot \text{K}^{-1}$, and $K = 17{,}500$ $\text{A} \cdot \text{m}^{-1} \cdot \text{K}^{-1}$ ($220\ \text{G} \cdot \text{K}^{-1}$), and if the magnetic field is produced, say, by a superconducting magnet with $\mu_0 H = 20$ T (200,000 G), then the adiabatic temperature drop is

$$\Delta T_{ad} = 53\ \text{K} \tag{6.25}$$

This represents a substantial change in temperature and motivates the further development of the concept that follows.

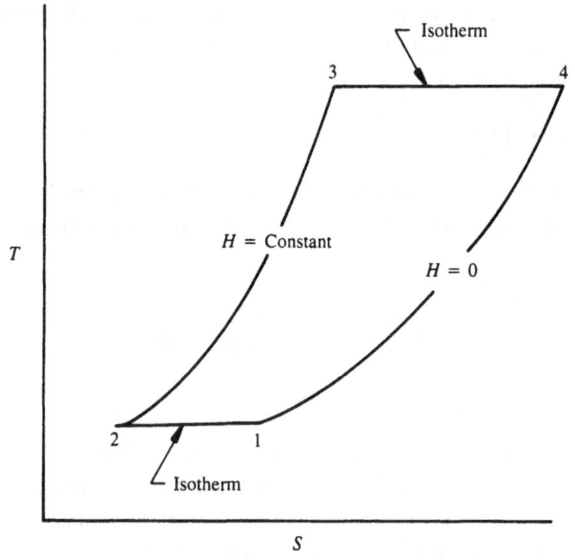

6.3 Thermodynamic cycle for a magnetocaloric heat engine. (*After Resler and Rosensweig 1967.*)

6.4 **General cycle analysis**

Consider a power cycle made up of two isothermal paths connected by constant-H-field processes, as shown in Figure 6.3. The sequence of processes experienced by a volume of fluid undergoing a cycle starting from point 1 is as follows:

1 → 2 heat rejection at constant temperature while the fluid is entering the magnetic field;

2 → 3 heat addition at constant magnetic field intensity;

3 → 4 heat addition at a constant elevated temperature while the fluid is leaving the field; and

4 → 1 heat rejection in the absence of a magnetic field.

It is now desired to derive an expression for the cyclic work. Integrating expression (6.1) for the differential work gives

$$W_{\text{cyc}} = -\oint \mu_0 H \, dM \tag{6.26}$$

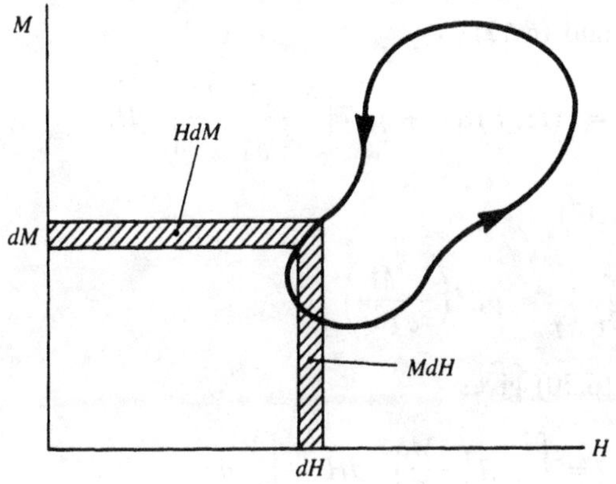

6.4 Development of the expression for cyclic-magnetic-work effect.

With the aid of the relation $d(MH) = M\,dH + H\,dM$ or Figure 6.4, one can rewrite (6.26) as

$$W_{\text{cyc}} = \oint \mu_0 M\,dH \tag{6.27}$$

because the cyclic integral of an exact differential is zero. Now summing up the contributions to the integral in (6.27) and noting that H is constant on the two paths $2 \to 3$ and $4 \to 1$ gives

$$\begin{aligned} W_{\text{cyc}} &= \int_1^2 \mu_0 M\,dH + \int_3^4 \mu_0 M\,dH \\ &= \mu_0 \int_0^H [M(H, T_1) - M(H, T_4)]\,dH \end{aligned} \tag{6.28}$$

It is reassuring that this result is equivalent to that of (6.17), obtained by use of the equation of motion.

The expression for the cyclic work could equally well be obtained by summing the heat effects. This will now be demonstrated because it proves to be an instructive exercise. From the first line of (6.5), and because U is a point function, it is seen that

$$\oint dU = 0 = \oint \delta Q - \oint \delta W \quad \text{or} \quad \oint dW = \oint dQ \tag{6.29}$$

From (6.4) and (6.12)

$$\delta Q = c(H,T)\,dT + \mu_0 T\left[\frac{\partial M(H,T)}{\partial T}\right]_H dH \tag{6.30}$$

and from (6.13)

$$\left(\frac{\partial c}{\partial H}\right)_T = \mu_0 T\left(\frac{\partial^2 M}{\partial T^2}\right)_H \tag{6.31}$$

Integrating (6.30) gives

$$\begin{aligned}\oint \delta Q &= \int_1^2 \mu_0 T\left(\frac{\partial M}{\partial T}\right)_H dH + \int_2^3 c\,dT \\ &\quad + \int_3^4 \mu_0 T\left(\frac{\partial M}{\partial T}\right)_H dH + \int_4^1 c\,dT \end{aligned} \tag{6.32}$$

$$\begin{aligned} &= \int_0^H \left[\mu_0 T_1 \frac{\partial M(H,T_1)}{\partial T} - \mu_0 T_4 \frac{\partial M(H,T_4)}{\partial T}\right] dH \\ &\quad + \int_{T_1}^{T_4} [c(H,T) - c(0,T)]\,dT \end{aligned} \tag{6.33}$$

From (6.31) it is seen that

$$c(H,T) = c(0,T) + \int_0^H \mu_0 T\left(\frac{\partial^2 M}{\partial T^2}\right)_H dH \tag{6.34}$$

and if the latter is used in (6.33), the expression for the cyclic work becomes

$$\begin{aligned}\oint \delta Q = \mu_0 \int_0^H \Bigg[& T_1 \frac{\partial M(H,T_1)}{\partial T} - T_4 \frac{\partial M(H,T_4)}{\partial T} \\ & + \int_{T_1}^{T_4} T \frac{\partial^2 M(H,T)}{\partial T^2}\,dT\Bigg] dH \end{aligned} \tag{6.35}$$

However, considerable simplification is achieved if it is noted that

$$\begin{aligned}\left(T\frac{\partial M}{\partial T}\right)_4 &= \left(T\frac{\partial M}{\partial T}\right)_1 + \int_{T_1}^{T_4} \frac{\partial}{\partial T}\left(T\frac{\partial M}{\partial T}\right) dT \\ &= \left(T\frac{\partial M}{\partial T}\right)_1 + \int_{T_1}^{T_4} T\frac{\partial^2 M}{\partial T^2}\,dT + \int_{T_1}^{T_4} \frac{\partial M}{\partial T}\,dT \end{aligned} \tag{6.36}$$

Therefore,

$$\oint \delta Q = \mu_0 \int_0^H \left(-\int_{T_1}^{T_4} \frac{\partial M}{\partial T} dT \right) dH = -\mu_0 \int_0^H \left(\int_{T_1}^{T_4} dM \right) dH$$

$$= \mu_0 \int_0^H [M(H, T_1) - M(H, T_4)] \, dH \tag{6.37}$$

which confirms (6.28).

The cycle efficiency η (not to be confused with the viscosity) can now be calculated from its definition, which is, following standard convention,

$$\eta \equiv \frac{W_{\text{cyc}}}{Q_{\text{add}}} \tag{6.38}$$

From (6.37) and the integral of (6.30), and because $T_1 = T_2$ and $T_4 = T_3$, the cycle efficiency in Figure 6.3 is easily shown to be given by the general form

$$\eta = \left\{ \mu_0 \int_0^H [M(H, T_1) - M(H, T_4)] \, dH \right\} \Big/ \left[\int_{T_1}^{T_4} c(H, T) \, dT - \mu_0 T_4 \int_0^H \left(\frac{\partial M}{\partial T} \right)_H dH \right] \tag{6.39}$$

6.5 Cycle efficiency for a linear material

If the equation of state is (6.18), then $c = c(T) = c_0$. With the linear state equation and for constant c_0, the cycle efficiency of (6.39) becomes

$$\eta = \frac{\mu_0 K H (T_4 - T_1)}{c_0 (T_4 - T_1) + \mu_0 K H T_4} \tag{6.40}$$

$$= \frac{\eta_c}{1 + \eta_c (c_0 / \mu_0 K H)} \tag{6.41}$$

where $\eta_c \equiv (T_4 - T_1)/T_4$ is the *Carnot efficiency*. Thus, it is seen that the simple cycle is less efficient than a Carnot cycle operated over the same temperature extremes.

With iron as the working substance, a field of $\mu_0 H = 20$ T, and a Carnot efficiency $\eta_c = 0.25$, (6.41) gives a cycle efficiency of 4.3%. This value is rather below the range of what is desired (but see Section 6.6).

6.6 Cycle efficiency with regeneration

The key to achieving high thermodynamic efficiency in a magnetocaloric system is the use of a regenerative cycle. The concept will be analyzed, continuing to use the approximation of a constant value of the pyromagnetic coefficient. Thus, if (6.30) is specialized for a linear material, the expression for heat flow is

$$\delta Q = c_0(T)\,dT - \mu_0 KT\,dH \tag{6.42}$$

where the first term is the differential amount of heat added in a constant-H process and the second term the differential amount of heat added during an isothermal process. Then the heat effect along each leg of the process shown in Figure 6.3 may be written as follows:

$$Q_{12} = -\mu_0 KT_1 H \quad \text{(isotherm; heat rejected)} \tag{6.43}$$
$$Q_{23} = c_0(T_4 - T_1) \quad \text{(constant field; heat added)} \tag{6.44}$$
$$Q_{34} = \mu_0 KT_4 H \quad \text{(isotherm; heat added)} \tag{6.45}$$
$$Q_{41} = -c_0(T_4 - T_1) \quad \text{(constant field; heat rejected)} \tag{6.46}$$

where c_0 is taken to be constant. Note that the heat added between the isotherms under high applied field (Q_{23}) just equals the heat rejected between the same isotherms at zero field (Q_{41}). Hence, if a regenerative cycle is used so that heat is shifted internally, that is, if the rejected heat (Q_{41}) is used to provide the heat added (Q_{23}), then the cyclic work remains unchanged, but the overall heat that needs to be supplied from an *external* source is reduced:

$$Q_{\text{add}} = \mu_0 KT_4 H \tag{6.47}$$
$$W_{\text{cyc}} = \mu_0 KH(T_4 - T_1) \tag{6.48}$$

and the cycle efficiency with perfect regeneration η_R becomes

$$\eta_R = W_{\text{cyc}}/Q_{\text{add}} = (T_4 - T_1)/T_4 \equiv \eta_c \tag{6.49}$$

Thus, with perfect regeneration, Carnot efficiency is possible in principle; no heat engine can be more efficient.

6.7 Implementing the cycle

Figure 6.5 illustrates an arrangement of functional steps serving to achieve the regenerative magnetocaloric power cycle of Figure 6.3. Magnetic fluid circulates through the solenoid, where overall it is heated from T_1 to T_4; flows to the regenerator, where it is cooled from T_4 to T_1; and then passes through a load (analogous to a turbine in a conventional power system), where it performs useful work on having its pressure

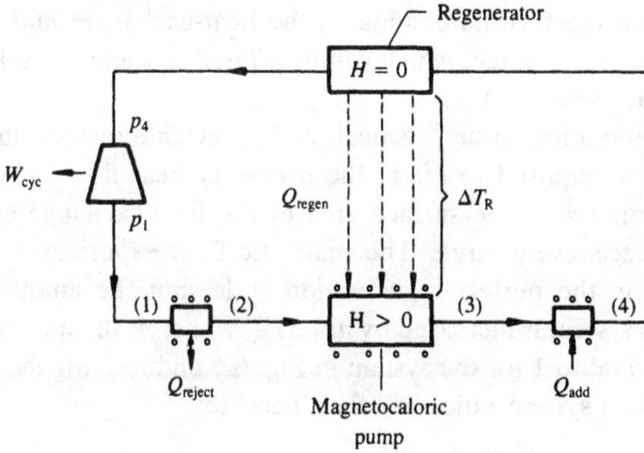

6.5 A one-fluid loop to carry out the regenerative magnetocaloric power cycle.

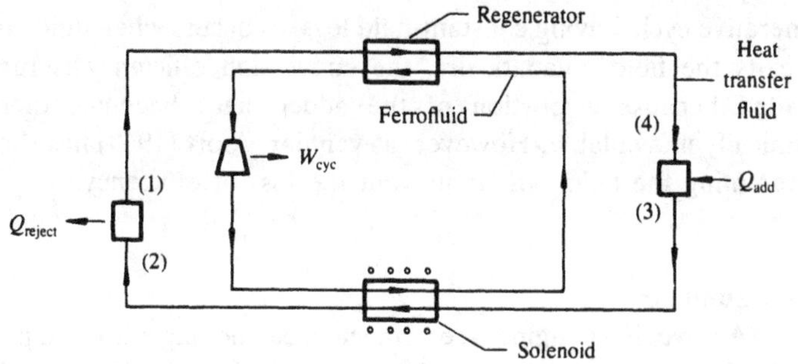

6.6 A two-fluid loop to carry out the regenerative magnetocaloric power cycle.

reduced from p_4 to p_1. Heat is transferred under ΔT_R driving force from the magnetic fluid in the regenerator, where a magnetic field is absent, to the magnetic fluid in the magnetocaloric pump, where the field is present; for perfect regeneration ΔT_R must approach zero. The need to transfer heat through the field surface presents a difficulty in some systems.

As an alternative, a second fluid acting as a heat-exchange medium circulates in a contrary direction to the flow in the magnetic fluid loop, as shown in Figure 6.6. The heat-exchange fluid circulates through the solenoid, where ideally it is cooled from T_4 to T_1, then to the heat sink, and then to the regenerator, where it is heated from T_1 to T_4 by thermal contact with the ferrofluid. Finally, the heat-exchange fluid circulates through the heat source, which can be a fossil, nuclear, or solar source or any other source of heat.

For regeneration to be practical, an appreciable temperature difference ΔT_R is required to effect the necessary heat flow between fluid streams; otherwise, the surface area of the heat-exchange equipment becomes excessively large. The magnetic fluid experiences the same history as in the perfect regeneration cycle, but the amount of heat added to a system is increased by $nc_0 \, \Delta T_R$, where n, the number of fluid loops, is equal to 1 for the system in Fig. 6.5 and to 2 for the system in Fig. 6.6. The system efficiency then becomes

$$\eta = \frac{\mu_0 KH(T_4 - T_1)}{\mu_0 KT_4 H + nc_0 \Delta T_R} = \frac{\eta_c}{1 + (nc_0 \Delta T_R)/(\mu_0 KHT_4)} \tag{6.50}$$

Because heat is transferred in two steps in the two-fluid system, the efficiency in (6.50) is lower than for the one-fluid system. For a regenerative cycle having constant-field legs, as occurs when fluid enters and exits the field adiabatically, the conversion efficiency is further degraded because a portion of the added heat becomes thermodynamically unavailable. However, as van der Voort (1969) has shown, programming the field can circumvent the loss of efficiency.

6.8 Summary

A novel heat engine free of mechanical moving parts and potentially having high reliability has been described. This magnetocaloric power system exploits the temperature dependence of magnetization to convert heat to useful work. A theoretical analysis predicting the efficiency of a practical cycle was performed for a material obeying a simple linear equation of state. If the cycle is reversed, refrigeration or heat pumping is obtained.

Magnetocaloric power devices exploit the temperature dependence of the magnetic moment to convert heat into useful work. In an analogous manner, ferroelectric devices can utilize changes of electric moment to perform the same function. Extension of the analysis to ferroelectric materials is straightforward in the absence of space-charge effects (change H to E, M to P/ε_0, and μ_0 to ε_0). To eliminate space-charge effects, alternating electric field must be applied at a frequency ω greater than the reciprocal dielectric relaxation time σ/ϵ of the fluid.

Comments and supplemental references

The development given in this chapter followed those given by

Resler and Rosensweig (1964, 1967)

The second paper also derives relationships for a material obeying a nonlinear (Weiss) law of magnetization. It is interesting to note that at the time the early paper was published, stable concentrated colloidal ferrofluids had not yet been synthesized. Resler and Rosensweig postulate a working substance consisting of such a colloidal magnetic liquid to accommodate the heat transfer requirements of a regenerative cycle.

An experimental nonregenerative convertor was operated in which an organic base ferrofluid was heated in a magnetic field of 0.8 T, generating steady flow with a thermally induced magnetization change of 0.00137 T; see

Rosensweig, Nestor, and Timmins (1965)

Japanese engineers have developed a magnetocaloric heat pipe for cooling electronic equipment; see

Matsuki, Yamasawa, and Murakami (1977)

Researchers in the United States have studied the application to a solar-collector pump.

A conceptual design for a nuclear space power generator is given by

Van der Voort (1969)

The magnetocaloric pump has also been proposed in nuclear-fusion reactors to utilize the existing high-intensity magnetic field and remove reactor heat while possibly generating electrical power; see

Roth, Rayk, and Reinmann (1970)

Making ferromagnetic particles stable in liquid metals, so that the small magnetic particles remain evenly dispersed, would be an important step toward deriving electrical energy from heat without an intermediate mechanical stage. Efforts toward developing magnetic liquids based on a liquid-metal carrier to increase the heat-transfer rate and reduce the apparatus size are reported by

Popplewell, Charles, and Chantrell (1977)

Related developments have been spurred by the advent of high-field magnets (typically as high as 10 T) and pure rare-earth elements such as gadolinium. One device is based on a wheel design in which the rim of a wheel composed of porous gadolinium rotates through a magnetic field to execute a magnetic Brayton cycle while ferrofluid circulates as a heat-exchange fluid to avoid flow control problems. The accomplishments of the Los Alamos research group are presented by

Barclay (1982)

Thermomagnetophoresis, the migration of particles of a ferrofluid in a temperature gradient within a uniform magnetic field, is treated theoretically by

Blums (1980)

7

FERROHYDRODYNAMIC INSTABILITIES

The effects considered up to this point concern the direct influence of magnetic fields on fluids. Now cases are examined in which the magnetic fluids do not merely respond to the applied magnetic field, but feed back and modify the field in a manner that alters the flow field drastically. It will be found as a common feature that, as a control parameter is varied in a continuous manner, a critical point is reached at which the fluid configuration is modified abruptly. The transition is known as a *bifurcation* and results in the establishment of a new static state, a new equilibrium flow field, a time-periodic flow, or, in some cases, an aperiodic flow.

An understanding of these instability phenomena can offer insight into the causes of *symmetry breaking* in physical systems in general and can suggest innovative means of extending the operating range of the equilibrium magnetic flow field. In other contexts, instabilities may be deliberately induced when desired.

In the absence of a magnetic field certain of these instabilities are well known as *Rayleigh–Taylor* and *Kelvin–Helmholtz* problems. The former treats the stability of a dense fluid accelerated toward a less-dense fluid, and the latter treats the behavior of a plane interface between sheared fluid layers. Other examples will be met also. The plan of this chapter is to analyze problems simple enough to permit an analytical solution yet sufficiently rich in physical content to be instructive. The particular problem considered first has much the same status in FHD that the problem of buoyancy-driven instability holds in the study of convective heat transfer. The buoyancy-driven problem is commonly termed *Bénard instability*, although what Bénard observed was surface-tension-gradient driven.

7.1 Normal-field instability

In normal-field instability, a perpendicular, uniform magnetic field applied to a pool of magnetic fluid produces the spontaneous generation of an ordered pattern of surface protuberances when the field exceeds a critical value. The appearance of the surface in reflected light will be shown in Figure 7.7.

Figure 7.1 illustrates an initially flat layer of magnetic fluid that occupies the half-space $z < 0$. Consider the carrier liquid to be an oil with such low vapor pressure that the nonmagnetic phase occupying the upper half-space ($z > 0$) can be adequately modeled as a vacuum. The analysis will investigate the stability of the equilibrium in response to a gravitational force, interfacial tension, and a magnetizing field.

The stability analysis to be carried out is called a *linear* or *small-disturbance analysis* because only small deflections of the interface will be considered. The deflection of the interface is represented as a superposition of harmonic terms, a procedure that can be broadly applied and is not dependent on the assumption of linearity. However, linearization of the equation set allows each harmonic to be considered separately, and attention can then be focused on the evolution of the fastest-growing disturbance, which will dominate the dynamics.

The analysis will be carried out in detail to exemplify general procedures employed in analyzing a variety of stability problems.

Equation set

Equations governing the motion of the magnetic liquid, in a form that is convenient for use in the present problem, were developed previously as (1.27) and (4.74), and the results are repeated here for convenience:

Continuity (incompressible):

$$\nabla \cdot \mathbf{v} = 0 \tag{7.1}$$

Equation of motion:

$$\rho\left(\frac{\partial \mathbf{v}}{\partial t} + \mathbf{v}\cdot\nabla\mathbf{v}\right) = -\nabla(p_0 + p_s + \rho g z) \tag{7.2}$$

where (4.57) was introduced into (7.2) and it is assumed that the flow is inviscid. Other governing relationships are the magnetostatic field equations developed as (3.36) and (3.37):

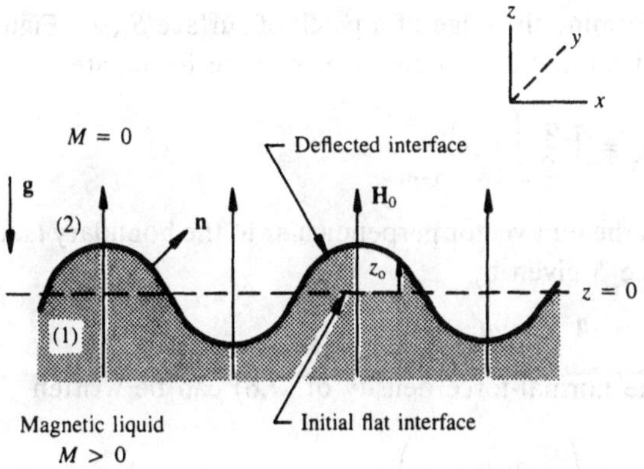

7.1 Analysis of interfacial stability of magnetic liquid in normal, uniform, applied field. Interface stays flat until field intensity exceeds a critical value.

Gauss's law:

$$\nabla \cdot \mathbf{B} = 0 \tag{7.3}$$

Ampère's law (no currents):

$$\nabla \times \mathbf{H} = \mathbf{0} \tag{7.4}$$

Interfacial force balance

Because the nonmagnetic phase is a vacuum $p_0 = 0$ and so (5.24) may be written

$$p + p_s + p_m + \tfrac{1}{2}\mu_0 M_n^2 - 2\mathscr{H}\sigma = 0 \tag{7.5}$$

which is the normal-stress boundary condition for this problem. A subscript 1 has been dropped from the first term as only one fluid phase is present and all terms are evaluated in that phase. $\mathscr{H}$ in this equation is the mean curvature, which, by definition, has a positive value when **n** projects outward from the convex side of a curved interface.

What is required next is an expression for $\mathscr{H}$ in terms of the interfacial geometry in a more general form than was derived as equation (5.8). Developing that relationship requires a digression, but one that is worthwhile as the result will have broad utility.

Consider the integrated force of surface tension along the closed contour forming the edge of a patch of surface S (see Figure 7.2). The interfacial normal-force density $\mathbf{p}_c$ may be formulated as

$$\mathbf{p}_c = \left(\frac{\sigma}{S}\oint \mathbf{t}\,dl\right)_{S\to 0} \tag{7.6}$$

where $\mathbf{t}$ is the unit vector perpendicular to the boundary l and tangent to the surface S given by

$$\mathbf{t} = d\mathbf{l} \times \mathbf{n}/dl \tag{7.7}$$

and so the normal-force density of (7.6) can be written

$$\mathbf{p}_c = \left(\frac{\sigma}{S}\oint d\mathbf{l} \times \mathbf{n}\right)_{S\to 0} \tag{7.8}$$

A generalized version of Stokes's theorem (Gibbs 1906) states that

$$\oint d\mathbf{l} \times \mathbf{A} = \int_S (\mathbf{n} \times \nabla) \times \mathbf{A}\,dS$$

$$= -\int_S \mathbf{n}\nabla\cdot\mathbf{A}\,dS + \int_S (\nabla\mathbf{A})\cdot\mathbf{n}\,dS \tag{7.9}$$

In this theorem $\mathbf{A}$ represents any vector. Letting $\mathbf{A} = \mathbf{n}$, recognizing that $(\nabla\mathbf{n})\cdot\mathbf{n} = \nabla(\mathbf{n}\cdot\mathbf{n}) = \mathbf{0}$, and allowing S to approach zero, one sees that $\mathbf{p}_c$ is

$$\mathbf{p}_c = -\sigma\mathbf{n}\,\nabla\cdot\mathbf{n} \tag{7.10}$$

showing that $\mathbf{p}_c$ is oriented normal to the interface. Comparing the expression for p_c of equation (5.9) with the present result shows that the mean curvature $\mathcal{H}$ can be written

$$2\mathcal{H} = \nabla\cdot\mathbf{n} \tag{7.11}$$

This result is valid for small and large perturbations of the interface and has broad utility in evaluating the surface force density in various coordinate geometries.

One convenient way to describe the location of the free surface is to use a relationship that gives the value of one of the coordinates in terms of the other two; e.g., for a rectangular coordinate system, $z = z_0(x, y)$. This is known as a *Monge representation*. Accordingly, the expression $z - z_0(x, y) = \text{const}$ represents contours having the shape of the interface yet displaced from it when the constant is other than zero. Thus,

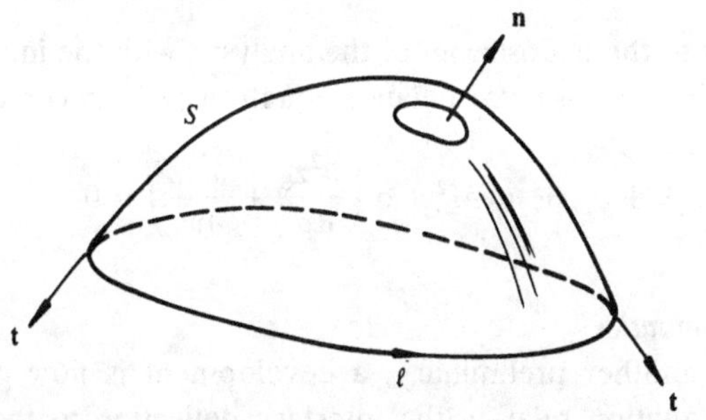

7.2 Derivation of the capillary pressure expression, equation (7.10).

$\nabla[z - z_0(x,y)]$ yields a vector oriented normal to the interface, and so the field of unit normals to the interface is given by

$$\mathbf{n} = \frac{\nabla[z - z_0(x,y)]}{|\nabla[z - z_0(x,y)]|} = \frac{\mathbf{k} - (\partial z_0/\partial x)\,\mathbf{i} - (\partial z_0/\partial y)\,\mathbf{j}}{[1 + (\partial z_0/\partial x)^2 + (\partial z_0/\partial y)^2]^{1/2}} \tag{7.12}$$

Consequently, owing to (7.11), the equation governing twice the local mean curvature is

$$-2\mathscr{H} = \left\{\left[1 + \left(\frac{\partial z_0}{\partial y}\right)^2\right]\frac{\partial^2 z_0}{\partial x^2} - 2\frac{\partial z_0}{\partial x}\frac{\partial z_0}{\partial y}\frac{\partial^2 z_0}{\partial x \partial y} + \left[1 + \left(\frac{\partial z_0}{\partial x}\right)^2\right]\frac{\partial^2 z_0}{\partial y^2}\right\} \times \left[1 + \left(\frac{\partial z_0}{\partial x}\right)^2 + \left(\frac{\partial z_0}{\partial y}\right)^2\right]^{-3/2} \tag{7.13}$$

Fortunately, only the linearized form of these relationships is needed here:

$$\mathbf{n} = \mathbf{k} - \frac{\partial z_0}{\partial x}\,\mathbf{i} - \frac{\partial z_0}{\partial y}\,\mathbf{j} \tag{7.14}$$

$$2\mathscr{H} = -\left(\frac{\partial^2 z_0}{\partial x^2} + \frac{\partial^2 z_0}{\partial y^2}\right) \tag{7.15}$$

Return now to the mainstream of the analysis; with the introduction of (7.15), the interfacial force balance equation (7.5) becomes

$$p_0 + p_s + p_m + \tfrac{1}{2}\mu_0 M_n^2 + \sigma\left(\frac{\partial^2 z_0}{\partial x^2} + \frac{\partial^2 z_0}{\partial y^2}\right) = 0 \tag{7.16}$$

Kinematics

As another preliminary, a development is now given of the kinematic equation relating the interface deflection to the motion of the adjacent fluid. When the surface position changes with time, the location of the interface may be represented as

$$z = z_0(x, y; t) \tag{7.17}$$

If a Monge function is defined by $F(x, y, z; t) \equiv z - z_0(x, y; t)$, then $DF/Dt = 0$ for a surface point; hence the *kinematic condition* at the interface is

$$\frac{DF}{Dt} = \frac{\partial F}{\partial t} + \mathbf{v}_1 \cdot \nabla F = 0 = \frac{\partial F}{\partial t} + \mathbf{v}_2 \cdot \nabla F \tag{7.18}$$

where $\mathbf{v}_1$ and $\mathbf{v}_2$ are the fluid velocities in the two phases. Because $\nabla F/|\nabla F|$ is the normal to the interface, (7.18) states that the normal components of the velocity are continuous at the interface. Equation (7.18) permits the possibility of a slip at the interface; if a viscous flow problem were considered, it would be necessary to impose the additional condition that the tangential component of the velocity be continuous at the free surface.

From the definition of F, equation (7.18) gives

$$-\frac{\partial z_0}{\partial t} + v_x\left(-\frac{\partial z_0}{\partial x}\right) + v_y\left(-\frac{\partial z_0}{\partial y}\right) + v_z(1) = 0 \tag{7.19}$$

This may be rearranged to

$$v_z = \frac{\partial z_0}{\partial t} + v_x \frac{\partial z_0}{\partial x} + v_y \frac{\partial z_0}{\partial y} \tag{7.20}$$

where $\mathbf{v} = v_x\mathbf{i} + v_y\mathbf{j} + v_z\mathbf{k}$. The linearized form of (7.20) will be used here:

$$v_z = \partial z_0/\partial t \tag{7.21}$$

Flow-field analysis

Equation (7.2) governing the motion of the magnetic liquid may be written

$$\rho(\partial \mathbf{v}/\partial t + \mathbf{v}\cdot\nabla\mathbf{v}) = -\nabla\Pi \tag{7.22}$$

where $\Pi \equiv p + p_s + \rho g z$. Following the well-established point of departure in linear stability analysis, it is useful to write a quantity as the sum of the value it had in the steady state and a small perturbation. Thus, the velocity $\mathbf{v}$ and the augmented pressure Π can be represented by the following equations:

$$\mathbf{v} \equiv \mathbf{v}_0 + \mathbf{v}_1 \quad \text{and} \quad \Pi \equiv \Pi_0 + \Pi_1 \tag{7.23}$$

where 0 indicates the steady-state value and 1 the perturbation. Thus the equation of motion, together with the fact that the initial state is quiescent ($\mathbf{v}_0 = \mathbf{0}$), becomes

$$\rho(\partial \mathbf{v}_1/\partial t + \mathbf{v}_1\cdot\nabla\mathbf{v}_1) = -\nabla\Pi_0 - \nabla\Pi_1 \tag{7.24}$$

Examining (7.22) reveals that at steady state $\nabla\Pi_0 = \mathbf{0}$, so that Π_0 is constant. The constant nature of Π_0 and the result that $\mathbf{v}_1\cdot\nabla\mathbf{v}_1$ is of second order compared to $\partial\mathbf{v}_1/\partial t$ permits rewriting (7.24) as

$$\rho\partial\mathbf{v}_1/\partial t = -\nabla\Pi_1 \tag{7.25}$$

Taking the divergence of (7.25) and using the continuity relationship $\nabla\cdot\mathbf{v}_1 = 0$, one finds that Π_1 obeys Laplace's equation:

$$\nabla^2\Pi_1 = 0 \tag{7.26}$$

Next it is assumed that the augmented pressure pertubation may be expanded in terms of its normal modes as

$$\Pi_1 = \hat{\Pi}_1(z)E \tag{7.27a}$$

where

$$E = \operatorname{Re}\exp[i(\omega t - k_x x - k_y y)] \tag{7.27b}$$

Re denotes the real part of the complex function that follows it; complex values $\omega \equiv \gamma - i\nu$ are admitted, where γ and ν are real numbers; and $\hat{\Pi}_1(z)$ is an amplitude function dependent on z. (The Addendum of this chapter is devoted to the representation of disturbance waves; the reader unfamiliar with the concept is urged to pause to develop a crystal-clear understanding of the topic before proceeding further.)

Equation (7.27) must satisfy (7.26); it follows that the amplitude function satisfies the total differential equation

$$d^2\hat{\Pi}_1/dz^2 - k^2\hat{\Pi}_1 = 0 \tag{7.28}$$

where $k^2 = k_x^2 + k_y^2$. Equation (7.28) admits the solution

$$\hat{\Pi}_1(z) = A_1 e^{kz} + A_2 e^{-kz} \tag{7.29}$$

where A_1 and A_2 are constants and k is positive. To rule out an infinite value of Π_1 as z approaches large negative values, the constant A_2 is set equal to zero. The pressure perturbation Π_1 becomes

$$\Pi_1 = A_1 e^{kz} E \tag{7.30}$$

Next consider the z component of the momentum equation:

$$\rho\frac{\partial v_z}{\partial t} = -\frac{\partial \Pi_1}{\partial z} = -A_1 k e^{kz} E \tag{7.31}$$

Integrating (7.31) gives

$$v_z = -\frac{A_1 k e^{kz}}{\rho}\int E\,dt = -\frac{A_1 k e^{kz}}{\rho i\omega} E + f(x, y, z) \tag{7.32}$$

The function $f(x, y, z)$ is set equal to zero in order to restrict the analysis to simple, harmonic disturbances.

The free-surface location may now be related to the z component of the perturbation velocity through the kinematic equation (7.21); to first order,

$$\partial z_0/\partial t = (v_z)_{z=0} = -(A_1 k/\rho i\omega)E \tag{7.33}$$

and integrating gives

$$z_0 = (A_1 k/\rho\omega^2)E \tag{7.34}$$

This is the equation governing the deflected surface; it may be rewritten in terms of the amplitude $\hat{z}_0$ of the deflection as

$$z_0 = \hat{z}_0 E \tag{7.35}$$

where $\hat{z}_0 \equiv A_1 k/\rho\omega^2$. Substituting for A_1 in terms of z_0 in (7.30) gives the solution for Π_1 in the form

$$\Pi_1 = (\rho\omega^2/k)\hat{z}_0 e^{kz} E \tag{7.36}$$

Perturbed magnetic field

To linearize the magnetic-field problem, first write

$$\mathbf{B} = \mathbf{B}_0 + \mathbf{b} \tag{7.37}$$

$$\mathbf{H} = \mathbf{H}_0 + \mathbf{h} \tag{7.38}$$

where the magnitudes of **b** and **h** are assumed to be small in comparison to B_0 and H_0. The initial magnetic field is uniform and in the z direction:

$$\mathbf{H}_0 = H_0\mathbf{k} \tag{7.39}$$

From the magnetostatic form of Ampère's law

$$\nabla \times \mathbf{H} = \nabla \times \mathbf{H}_0 = \nabla \times \mathbf{h} = \mathbf{0} \tag{7.40}$$

and hence **h** can be expressed in terms of a magnetic perturbation potential:

$$\mathbf{h} = -\nabla\phi \tag{7.41}$$

Equation (7.40) holds for the magnetic medium as well as for the vacuum region that lies above. At this point the analysis will be restricted to a linear material for which $\mathbf{B} = \mu\mathbf{H}$ with μ a constant. With the Maxwell relation $\nabla\cdot\mathbf{B} = 0$, this gives

$$\nabla\cdot\mathbf{H} = \nabla\cdot\mathbf{H}_0 = \nabla\cdot\mathbf{h} = 0 \tag{7.42}$$

Then, combining the latter result with (7.41), one finds that the magnetic perturbation potential also obeys Laplace's equation:

$$\nabla^2\phi = 0 \tag{7.43}$$

Because $\mu = \mu_0$ in the vacuum above the interface, (7.43) applies in both regions; thus if ϕ_1 is the potential in the magnetic phase and ϕ_2 the potential in the vacuum region, the following solutions may be found similarly to (7.36):

$$\phi_1 = \hat{\phi}_1 z_0 e^{kz} \tag{7.44a}$$

$$\phi_2 = \hat{\phi}_2 z_0 e^{-kz} \tag{7.44b}$$

where $\hat{\phi}_1$ and $\hat{\phi}_2$ are independent of spatial position, and the constraints were applied that ϕ_1 and ϕ_2 are bounded far from the interface.

Boundary conditions on the field vectors

Consider first the boundary condition (3.9), which states that the tangential component of the magnetic field **H** obeys $[H_t] = 0$ or

$\mathbf{n} \times [\mathbf{H}] = \mathbf{n} \times (\mathbf{H}_2 - \mathbf{H}_1) = \mathbf{0}$. From (7.14) $\mathbf{n} = (n_x, n_y, n_z) = (-\partial z_0/\partial x, -\partial z_0/\partial y, 1)$ and $[\mathbf{H}] = \mathbf{H}_2 - \mathbf{H}_1 = ([H_x], [H_y], [H_z])$; thus

$$\mathbf{i}: \qquad -\frac{\partial z_0}{\partial y}[H_z] - [H_y] = 0 \tag{7.45}$$

$$\mathbf{j}: \qquad [H_x] + \frac{\partial z_0}{\partial x}[H_z] = 0 \tag{7.46}$$

$$\mathbf{k}: \quad -\frac{\partial z_0}{\partial x}[H_y] + \frac{\partial z_0}{\partial y}[H_x] = 0 \tag{7.47}$$

Only two of these three equations are independent.

Now the initial field is in the z direction, and so

$$[H_z] = [H_0] \tag{7.48}$$

$$[H_y] = [h_y] \tag{7.49}$$

$$[H_x] = [h_x] \tag{7.50}$$

to highest order. Hence (7.45) and (7.46) can be replaced by

$$\left[h_y + H_0\frac{\partial z_0}{\partial y}\right] = 0 \tag{7.45$'$}$$

$$\left[h_x + H_0\frac{\partial z_0}{\partial x}\right] = 0 \tag{7.46$'$}$$

Next consider the boundary condition (3.8) imposed on the magnetic induction field $\mathbf{B}$: $[B_n] = \mathbf{n}\cdot[\mathbf{B}] = 0$. Now, $[\mathbf{B}] = \mathbf{B}_2 - \mathbf{B}_1 = (\mathbf{B}_{2,0} - \mathbf{B}_{1,0}) + (\mathbf{b}_2 - \mathbf{b}_1)$, so

$$\mathbf{n}\cdot[\mathbf{B}] = \mathbf{n}\cdot[\mathbf{b}] = 0 \tag{7.51}$$

With (7.14), (7.51) becomes

$$-\frac{\partial z_0}{\partial x}[b_x] - \frac{\partial z_0}{\partial y}[b_y] + [b_z] = 0 \tag{7.52}$$

The first two terms of (7.52) are second order, and therefore

$$[b_z] = 0 \tag{7.53}$$

to first order.

Applying the magnetic boundary conditions
Use of (7.41) in (7.45′) gives

$$\frac{\partial \phi_1}{\partial y} + H_{0,1}\frac{\partial z_0}{\partial y} = \frac{\partial \phi_2}{\partial y} + H_{0,2}\frac{\partial z_0}{\partial y} \tag{7.54}$$

With equations (7.44) and (7.35), this can be rewritten

$$(\phi_1 - \phi_2)/z_0 = H_{0,2} - H_{0,1} \tag{7.55}$$

for $z = 0$. However, $H_{0,2} = B_{0,2}/\mu_0 - M_{0,2}$, $H_{0,1} = B_{0,1}/\mu_0 - M_{0,1}$; $B_{0,2} = B_{0,1}$; and $M_{0,2} = 0$, $M_{0,1} \equiv M_0$; hence

$$\phi_1 - \phi_2 = z_0 M_0 \qquad (\text{at } z = 0) \tag{7.56}$$

Equation (7.53), with (7.41), becomes

$$\mu\frac{\partial \phi_1}{\partial z} = \mu_0\frac{\partial \phi_2}{\partial z} \qquad (\text{at } z = 0) \tag{7.57}$$

From solutions (7.44) for ϕ_1 and ϕ_2, (7.56) and (7.57) give

$$\hat{\phi}_1 - \hat{\phi}_2 = M_0 \tag{7.58}$$

$$-\mu/\mu_0\hat{\phi}_1 = \hat{\phi}_2 \tag{7.59}$$

Solving for $\hat{\phi}_1$ it is found that

$$\hat{\phi}_1 = \frac{M_0}{1 + \mu/\mu_0} \tag{7.60}$$

and also

$$\phi_1 = \frac{M_0 z_0}{1 + \mu/\mu_0}\, e^{kz} \tag{7.61}$$

$$\phi_2 = \frac{-(\mu/\mu_0)M_0 z_0}{1 + \mu/\mu_0}\, e^{-kz} \tag{7.62}$$

A physical appreciation of the influence of the surface deflection on the field distribution can be obtained by considering either (7.61) or (7.62) with (7.53). Thus, the z component of the perturbation induction field is continuous at the interface and is given by

$$(b_{z,1})_0 = (b_{z,2})_0 = (\mu_0 h_{z,2})_0$$

$$= \mu_0\left(\frac{\partial \phi_2}{\partial z}\right)_0 = \frac{\mu_0 k M_0 z_0}{1 + \mu_0/\mu} \tag{7.63}$$

Equation (7.63) implies there is a concentration of flux at the peaks of disturbance with b_z a maximum where z_0 is a maximum, as illustrated in Figure 7.3.

Theoretical predictions

All the expressions that are needed are now available to substitute into the boundary condition governing the normal-field instability. The equation, (7.5), is repeated here for convenience:

$$p_0 + p_s + p_m + \tfrac{1}{2}\mu_0 M_n^2 - 2\mathscr{H}\sigma = 0 \tag{7.64}$$

or, from the definitions of Π and p_m,

$$\Pi - \rho g z_0 + \mu_0 \bar{M} H + \tfrac{1}{2}\mu_0 M_n^2 - 2\mathscr{H}\sigma = 0 \tag{7.65}$$

It is now necessary to simplify the terms $\mu_0 \bar{M} H$ and $\tfrac{1}{2}\mu_0 M_n^2$.

Consider first the term $\mu_0 \bar{M} H$. Now, $\mathbf{H} = H_0\mathbf{k} + h_x\mathbf{i} + h_y\mathbf{j} + h_z\mathbf{k}$, so

$$\begin{aligned} H = (\mathbf{H}\cdot\mathbf{H})^{1/2} &= (H_0^2 + 2H_0 h_z)^{1/2} + \text{higher-order terms} \\ &\simeq H_0(1 + 2h_z/H_0)^{1/2} \simeq H_0 + h_z \end{aligned} \tag{7.66}$$

with the last result following from the binomial theorem. Then

$$\begin{aligned} \mu_0 \bar{M} H &= \mu_0 \int_0^{H_0+h_z} M'\,dH' \\ &= \mu_0 \int_0^{H_0} M'\,dH' + \mu_0 \int_{H_0}^{H_0+h_z} M'\,dH' \end{aligned} \tag{7.67}$$

where M' and H' are the dummy variables of integration; integrating gives

$$\mu_0 \bar{M} H = \mu_0 (\bar{M} H)_0 + \mu_0 M h_z \tag{7.68}$$

but $M = M_0 + m$; if the second-order term $\mu_0 m h_z$ is neglected, (7.68) becomes

$$\mu_0 \bar{M} H = \mu_0 (\bar{M} H)_0 + \mu_0 M_0 h_z \tag{7.69}$$

Now examine $\tfrac{1}{2}\mu_0 M_n^2$. Because M and H are parallel

$$\mathbf{M} = M_0\mathbf{k} + m_x\mathbf{i} + m_y\mathbf{j} + m_z\mathbf{k} \tag{7.70}$$

and, with the aid of (7.14) M_n may be written

$$M_n = \mathbf{M}\cdot\mathbf{n} = M_0 + m_z + \text{higher-order terms} \tag{7.71}$$

Therefore

$$\tfrac{1}{2}\mu_0 M_n^2 = \tfrac{1}{2}\mu_0 M_0^2 + \mu_0 M_0 m_z + \text{higher-order terms} \tag{7.72}$$

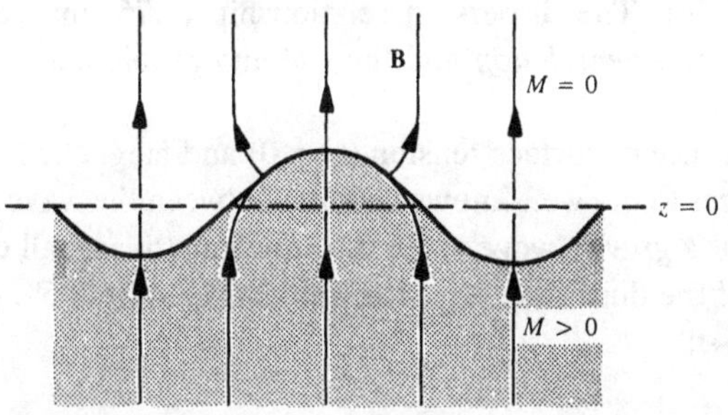

7.3 The perturbed magnetic field is focused along the wave crests.

Substituting (7.69) and (7.72) into (7.65) gives

$$[\Pi_0 + \mu_0(\bar{M}H)_0 + \tfrac{1}{2}\mu_0 M_0^2] + (\Pi_1 - \rho g z_0 + \mu_0 M_0 h_z + \mu_0 M_0 m_z - 2\mathscr{H}\sigma) = 0 \quad (7.73)$$

The term inside the square brackets is the FHD boundary condition at the original flat interface and hence vanishes. With this result and because $b_z = \mu_0(h_z + m_z)$, the traction boundary condition reduces to

$$\Pi_1 - \rho g z_0 + M_0 b_z - 2\mathscr{H}\sigma = 0 \quad (7.74)$$

Now, from (7.36), $(\Pi_1)_{z=0} = (\rho\omega^2/k)\hat{z}_0 E$; from (7.35), $z_0 = \hat{z}_0 E$; from (7.63) and (7.35),

$$(b_z)_{z=0} = \frac{k\mu_0 M_0}{1 + \mu_0/\mu}\hat{z}_0 E$$

and from (7.15) and (7.35), $2\mathscr{H} = k^2\hat{z}_0 E$. Thus

$$\frac{\rho\omega^2}{k} - \rho g + \frac{k\mu_0 M_0^2}{1 + \mu_0/\mu} - k^2\sigma = 0 \quad (7.75)$$

A slight rearrangement of (7.75) now yields a relation for the dependence of frequency on wave number:

$$\rho\omega^2 = \rho g k + \sigma k^3 - \frac{k^2\mu_0 M_0^2}{1 + \mu_0/\mu} \quad (7.76)$$

Such an equation is termed generally a *dispersion relationship*. Equations (7.28) and (7.43) are homogeneous and possess the mathematically trivial solutions $\Pi_1 = 0$ and $\phi = 0$ yet admit nontrivial solutions when the corresponding values of k and ω are interrelated through equation (7.76). The dispersion relationship (7.76) may therefore be regarded as an *eigenrelation* and the stability problem as an *eigenvalue problem*.

In the absence of surface tension ($\sigma = 0$) and magnetic field ($M = 0$), (7.76) reduces to the well-known relation between the wave number and frequency of a *gravity wave* when the wavelength is small compared to the depth of the fluid (see, e.g., Landau and Lifshitz 1959, p. 39; Lamb 1945, p. 364):

$$\omega^2 = gk \tag{7.77}$$

In terms of the classification of Table 7.1 (see the Addendum to this chapter), $\nu = 0$ and $\gamma \neq 0$, so the system propagates surface waves with no change of amplitude. The gravity-wave term dominates the right side of (7.76) when k is small or, correspondingly (because the wavelength $\lambda = 2\pi/k$), when the wavelength is long.

If, on the other hand, capillarity is the dominant effect in (7.76), then

$$\omega^2 = \sigma k^3/\rho \tag{7.78}$$

and such waves are called *capillary waves* (Lamb 1945, p. 457). The capillary wave term dominates the right side of (7.76) when disturbances of short wavelength are considered. Again, the system propagates the surface waves without change of amplitude.

Because ω appears as a squared term in (7.76) and the right side is always real, ω can only be *real* or *pure imaginary*. The important roots are.

$$\omega = \begin{cases} \gamma & (\gamma \text{ real}) \\ -i\nu & (\nu \text{ real}) \end{cases} \tag{7.79}$$

Recall from (7.35) that $z_0 = \hat{z}_0 \operatorname{Re} \exp[i(\omega t - k_x x - k_y y)]$. When ω is imaginary, therefore, an arbitrary disturbance to the interface grows with time as $e^{\nu t}$. This behavior is illustrated by a plot of the dispersion relation in the ω^2–k plane (see Figure 7.4). For small values of k the graph is linear, even if a magnetic field is present. At high values of k the capillary term dominates and all the curves are cubic in nature. Of course, if the magnetic field is zero, the system is stable to all small amplitude disturbances. However, as the magnetic field is slowly increased, the quadratic term dominates the behavior at intermediate values of k. It is evident from Figure 7.4 that the transition from a stable to an unstable state occurs if both the following conditions are met:

$$\omega^2 = 0 \tag{7.80}$$

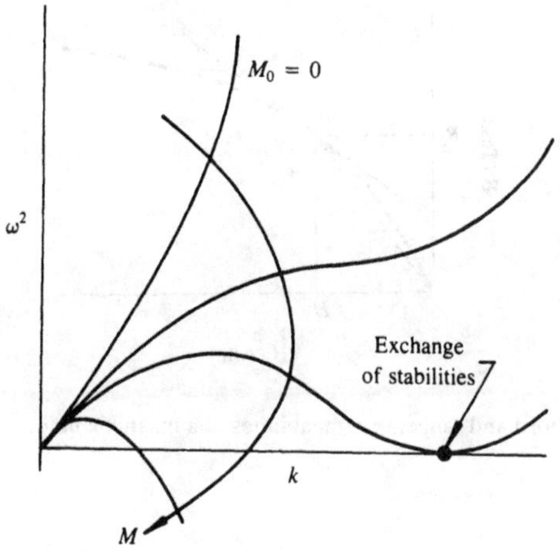

7.4 Dispersion in the ω^2–k plane. The upper quadrant corresponds to $\nu = 0$ and $\pm\gamma$ real; in the lower quadrant $\gamma = 0$ and $\pm\nu$ real.

and

$$\partial\omega^2/\partial k = 0 \tag{7.81}$$

At this transition, as the magnetization increases, ω changes from real ($\nu = 0$) to pure imaginary ($\gamma = 0$). Equations (7.80) and (7.81) are the conditions for an *exchange of stability*, a term introduced by Poincaré in the study of the shapes and stability of rotating drops. Applying these conditions gives

$$k_c = (\rho g/\sigma)^{1/2} \tag{7.82}$$

$$M_c^2 = \frac{2}{\mu_0}\left(1 + \frac{\mu_0}{\mu}\right)(\rho g \sigma)^{1/2} \tag{7.83}$$

The critical wave number k_c for the normal-field instability corresponds to the critical wave number in the Rayleigh–Taylor problem, which will be treated in Section 7.3. Equation (7.83) gives the critical value of the magnetization, i.e., the lowest value of the magnetization at which the phenomenon can be observed. Indeed, it is found experimentally that when the magnetization is increased from zero by increasing the applied magnetic field, the fluid interface is perfectly flat over a range of applied field intensities up to the point when transition suddenly occurs, an excellent test of the theory. Conversely, no increase in the applied field

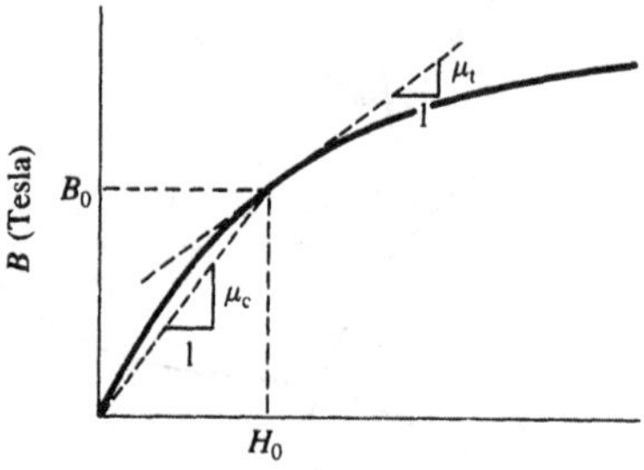

7.5 The chord and tangent permeabilities of a magnetic fluid.

H, no matter how large, can cause the interface to be unstable if the saturation magnetization of the fluid is less than the value of M_c given in (7.83).

If instead of a vacuum the superposing phase in Figure 7.1 is a nonmagnetic fluid (denoted by the subscript 2), the dispersion relation becomes

$$\omega^2(\rho_1 + \rho_2) = gk\,\Delta\rho + \sigma k^3 - \frac{k^2\mu_0 M_0^2}{1 + \mu_0/\mu} \tag{7.84}$$

where $\Delta\rho \equiv \rho_1 - \rho_2$. The appropriate expressions for the critical wave number and the critical magnetization are obtained simply by replacing ρ in (7.82) and (7.83) by $\Delta\rho$.

Although the expression for the critical wave number is the same for linear and nonlinear media, the expressions for the critical magnetization are different. The corresponding version of (7.83) appropriate for nonlinear media is

$$M_c^2 = \frac{2}{\mu_0}\left(1 + \frac{1}{r_0}\right)(g\,\Delta\rho\sigma)^{1/2} \tag{7.85}$$

where

$$r_0 \equiv (\mu_c\mu_t/\mu_0^2)^{1/2} \tag{7.86}$$

$\mu_c \equiv B_0/H_0$ is the chord permeability, and $\mu_t \equiv (\partial B/\partial H)_0$ is the tangent permeability, as defined in Figure 7.5. Thus, the relative permeability μ/μ_0 in the linear-medium relationship gets replaced by r_0, the geometric mean of the chord and tangent permeabilities.

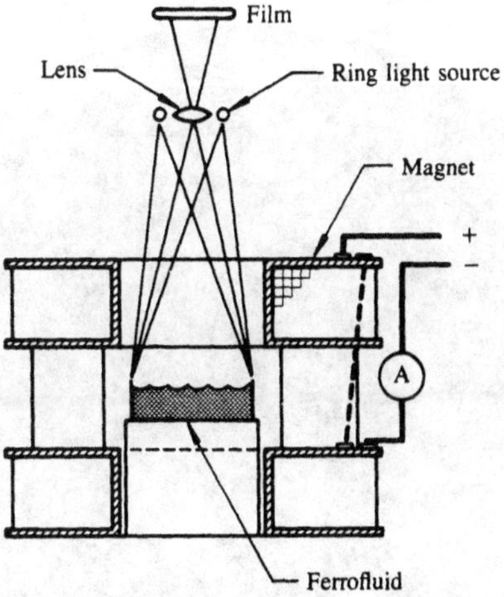

7.6 Experimental arrangement for photographing the glint pattern of Figure 7.7. (*After Rosensweig 1982b.*)

Normal-field instability experiments

Cowley and Rosensweig (1967) also made a careful experimental study of the conditions for onset of the normal-field instability phenomenon and critically compared the results with the theoretical predictions. The experimental apparatus, as sketched in Figure 7.6, consists of a pool of ferrofluid subjected to the uniform, vertically oriented magnetic field produced by a pair of electromagnets, together with a means for detecting reflected light from the free surface. The ring light source surrounding the lens acts effectively as a point source within the lens. Any local flat area on the perturbed surface – a maximum, minimum, or saddle point – reflects and produces a glint pattern, such as illustrated in Figure 7.7. The hexagonal pattern in this photograph is reminiscent of the pattern of thermal convection cells (see Figure 1 in Chandrasekhar 1961); however, *the magnetic fluid is static and free of any convective motion.* At higher than critical intensity of the applied field, the peaks rise well above the surface. The magnitude of the perturbation forces established in this system in air is such that the peak height can exceed the spacing, giving the surface a markedly setaceous appearance.

Figure 7.8 compares the predicted critical magnetization with experiment for a magnetic fluid with air and water interfaces. The subscript 0

7.7 Hexagonal cells in the normal-field instability of a magnetic fluid. (*After Cowley and Rosensweig 1967.*)

denotes the kerosene carrier liquid. The density ρ of the magnetic fluid was varied by changing the particle concentration in the kerosene carrier. Experimental points are shown with error bars representing the estimated uncertainty in observing the critical magnetization value.

The agreement between the predicted values from equation (7.85) and the experimental values of M_c is quite good. For the liquid–liquid interface between a hydrocarbon magnetic fluid and water, the critical magnetization falls to zero for zero density difference, this point occurring when densities of the top and bottom layers are equal. Under the conditions of the experiment, ρ_0 was 792 $kg \cdot m^{-3}$; because the density of water is approximately 1000 $kg \cdot m^{-3}$, the cusp point in Figure 7.8 corresponds to a relative density of 1.26.

Figure 7.9 compares the experimental values of peak spacing to the prediction of (7.82) (solid lines), and again it may be seen that the

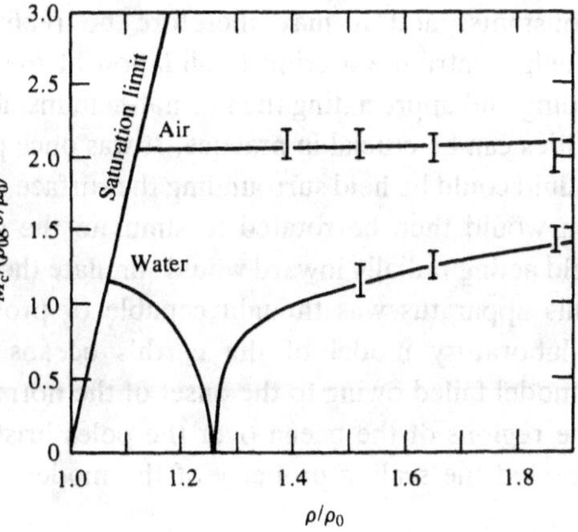

7.8 Comparison of the experimental critical magnetization with theory for the normal-field instability. (*After Cowley and Rosensweig 1967.*)

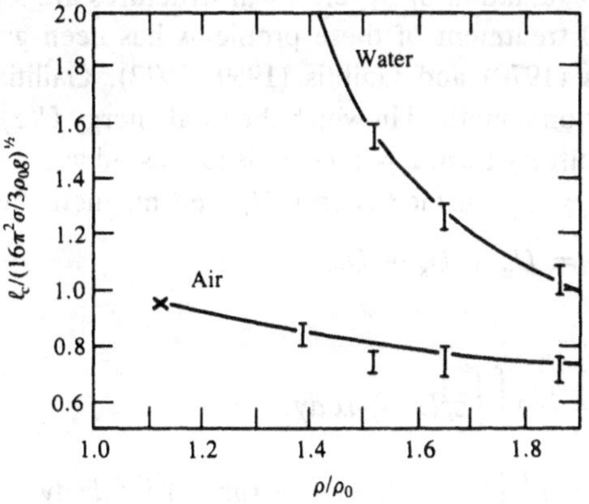

7.9 Comparison of the experimental critical spacing with theory for the normal-field instability. (*After Cowley and Rosensweig 1967.*)

theory is in very good accord with experiment. Critical magnetization for the magnetic-fluid–air interface is nearly constant because an increase in the effective permeability is offset by the buoyancy change. The theoretical curves in Figure 7.8 and 7.9 are computed with no adjustable constants, and it may therefore be realized that these experiments help contribute a critical validation of the theory.

Understanding and appreciating the normal-field instability and other FHD instabilities can be crucial in practice. It was once proposed that a layer of ferrofluid could be held surrounding the surface of a magnetized sphere, which would then be rotated to simulate the spinning of the earth. The field acting radially inward would simulate the pull of gravity, and hence this apparatus was thought capable of providing a three-dimensional laboratory model of the earth's oceans. When implemented, the model failed owing to the onset of the normal-field surface instability; the regions of the ocean over the poles bristled with liquid spikes that spoiled the scaling property of the model.

Nonlinear analysis

The linear analysis is unable to select between instability patterns resulting from the superposition of disturbances of different orientations; nor is it able to predict the amplitude of surface deflections. In addition, questions connected with the character of the transition and the evaluation of hypercritical structures are left unanswered. A theoretical treatment of these problems has been given by Zaitsev and Shliomis (1970) and Gailitis (1969, 1977). Gailitis employed an energy-variational method in which the total energy $U(z)$ of a perturbed surface of arbitrary form $z = z_0(x, y)$ is expressed as a sum of the gravitational energy U_g, surface energy U_s, and magnetic-field energy U_m:

$$U(z) = U_g + U_s + U_m \tag{7.87}$$

where

$$U_g = \tfrac{1}{2}\rho g \iint z^2(x, y)\, dx\, dy \tag{7.88}$$

$$U_s = \sigma \iint [1 + (\partial z/\partial x)^2 + (\partial z/\partial y)^2]^{1/2}\, dx\, dy \tag{7.89}$$

$$U_m = \iiiint H\, dB\, dx\, dy\, dz \tag{7.90}$$

When the magnetic susceptibility is constant, the magnetostatic energy can be transformed, as was shown in Chapter 3, to the more informative form

$$U_m = -\tfrac{1}{2}\iiint \mu_0 M H_0\, dx\, dy\, dz + \tfrac{1}{2}\iiint \mu_0 H_0^2\, dx\, dy\, dz \tag{7.91}$$

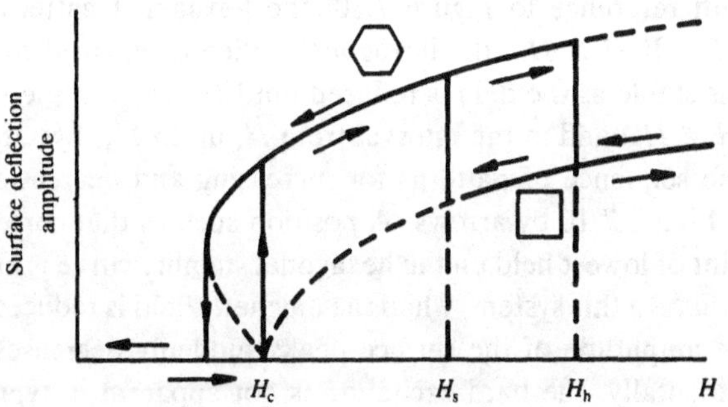

7.10 Bifurcation and hysteresis features in the nonlinear analysis of the normal-field instability. (*After Gailitis 1977.*)

The surface $z_0(x, y)$ was represented as a superposition of N one-dimensional waves of different orientations:

$$z_0(x, y) = \sum_{i=1}^{N} a_i \cos(\mathbf{k}_i \cdot \mathbf{r} + \delta_i) \tag{7.92}$$

where $\mathbf{k}_i$ and δ_i are the wave vector and the phase angle, respectively, associated with the ith mode.

A power-series expansion for $U(z)$ was used with any extremum of the series corresponding to an equilibrium form of the surface. Maxima and saddle points correspond to unstable equilibriums, whereas minima correspond to stable equilibria. In finding the stable equilibriums, consideration was given to surfaces formed by one, two, and three wave modes. The analysis showed that the surface has three possible configurations of stable equilibrium: a flat surface, an array of hexagonal waves, and an array of square waves. Figure 7.10 shows, that at the critical field H_c a "hard" excitation to a static hexagonal wave pattern arises, $z_0(H_c) \neq 0$. The hexagonal array may be of different types; the upper half plane represents an array with one crest, two troughs, and three saddle points in each elementary hexagonal cell (six troughs shared among three cells each and six saddle points shared between two cells each). The lower half plane (not shown) represents another type, with two crests, one trough, and three saddle points in each elementary cell. In stable equilibrium the theory predicts only the first type, in agreement with the observations of Cowley and Rosensweig (1967).

Again with reference to Figure 7.10, the hexagonal pattern is stable for $H < H_h$. If $H > H_h$, the hexagonal lattice is replaced by a square one that is stable as the field is reduced until $H < H_s$. In the subcritical region ($H < H_c$) and in the interval from H_s up to H_h, hysteresis takes place. The sequence of patterns for increasing and decreasing field is shown in Figure 7.10 by arrows. A position such as that corresponding to the point of lowest field on the hexagonal stability curve is known as a *turning point*. In this system, when the magnetic field is reduced past this point, the amplitude of the surface peaks suddenly decreases to zero.

Experimentally, the hard excitation is not apparent in typical ferrofluids, for which the permeability ranges up to a value of ~5. However, the hard excitation and the subcritical hysteresis features, including the existence of a turning point, have been verified by Bacri and Salin (1984) using a ferrofluid having a relative permeability of 40, a remarkably high value. The ferrofluid is obtained as a phase-separation product from a charge-stabilized colloidal dispersion. A related phenomenon is the instability of ferrofluid drops, for which Bacri and Salin (1983) observed jumps in the elongated drop shape in both increasing and decreasing applied magnetic fields.

The analysis of Gailitis (1969, 1977) in the linear limit recovers the onset relations of Cowley and Rosensweig (1967), (7.82) and (7.87). The one-dimensional nonlinear result of Zaitsev and Shliomis (1970) is also recovered by Gailitis but corresponds to a saddle point of energy within the wider class of two-wave disturbances and hence is unstable. Brancher's (1978) analysis of inviscid oscillations determined phase-plane portraits of the dynamic periodic process, but again the least-stable mode was not analyzed. Other analyses are that of Brancher (1980) accounting for the influence of viscous effects, the mathematical bifurcation analysis of Twombly and Thomas (1980), and a treatment by Malik and Singh (1983) based on a method of multiple scales that indicates that the nature of the transition is influenced critically by the value of the permeability.

Berkovsky and Bashtovoi (1980) have described the topological instability that can occur when the horizontal magnetic fluid is very thin. In this case the hexagonal pattern first develops and then, with further increase in the field, a *rupture of continuity* occurs, with individual droplets formed that preserve the hexagonal geometry. The effect is nicely produced by spreading a film of organic-base ferrofluid onto the surface of water in an open beaker and then applying the field perpendicular to the surface (Kendra 1982). This lubricates the rupture

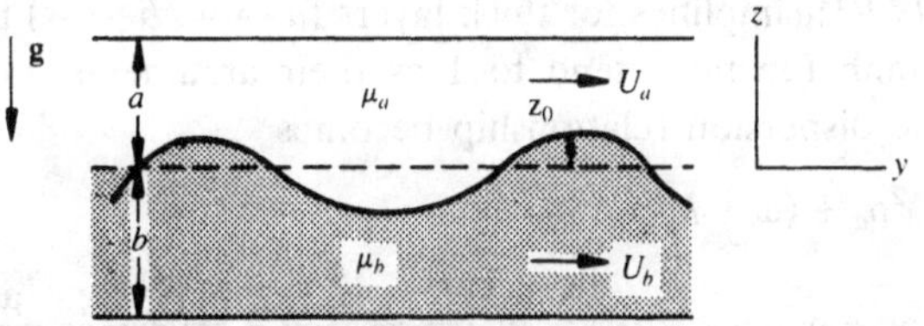

7.11 The nomenclature for the general dispersion relation. Initially the two uniform layers a and b are separated by an equilibrium plane interface of infinite extent.

process, and, as a further consequence, one witnesses a lively repulsion and escape of isolated droplets to the beaker circumference. Skjeltorp (1983b) presents a remarkable analogy in which the ferrofluid surface peaks represent a two-dimensional "crystal" exhibiting dislocations, disinclinations, and melting. The magnetic-field intensity plays the role of the temperature.

7.2 General dispersion relation for moving media with oblique magnetic field

Now a more general relationship is given for subsequent use in more complex problems. It is assumed that gravity is oriented normal to the original flat interface, the fluid layers move at speeds U_a and U_b above and below the interface, respectively, and the applied magnetic field H lies in the y–z plane (see Figure 7.11).

The deflection of the interface will be represented by a disturbance propagating along the y direction:

$$z = \hat{z}_0 \operatorname{Re} \exp[i(\omega t - k_y y)]$$

where, as in (7.27b) for E, complex values of ω are admitted with the same definition, $\omega = \gamma - i\nu$, in which γ and ν are both real numbers. The general dispersion relation in this case for linearly magnetizable media (μ = const) is

$$(\omega - k_y U_a)^2 \rho_a \coth ka + (\omega - k_y U_b)^2 \rho_b \coth kb$$
$$= gk(\rho_b - \rho_a) + k^3\sigma - \left[\frac{k^2\mu_a\mu_b(H_z^a - H_z^b)^2}{\mu_b \tanh ka + \mu_a \tanh kb} - \frac{k_y^2 H_y^2(\mu_a - \mu_b)^2}{\mu_b \coth kb + \mu_a \coth ka}\right] \quad (7.93)$$

Equation (7.93) simplifies for thick layers ($a \to \infty$, $b \to \infty$) because the coth and tanh functions tend to 1 as their arguments grow without bound. The dispersion relationship becomes

$$(\omega - k_y U_a)^2 \rho_a + (\omega - k_y U_b)^2 \rho_b$$
$$= gk(\rho_b - \rho_a) + k^3\sigma - \left[\frac{k^2\mu_a(H_z^a - H_z^b)^2}{1 + \mu_a/\mu_b} - \frac{k_y^2 H_y^2(\mu_a - \mu_b)^2}{\mu_a + \mu_b}\right] \quad (7.94)$$

Setting U_a, U_b, ρ_a, $H_y = 0$, $\mu_a = \mu_0$, $\rho_b = \rho$, $\mu_b = \mu$ and recognizing that $H_z^a - H_z^b = M_z^b - M_z^a = M_z^b \equiv M_0$ recovers equation (7.76), describing the normal-field instability.

7.3 Rayleigh–Taylor problem

When a light fluid in contact with a dense fluid is accelerated in the direction of the latter, the plane interface between them may become unstable. An illustration of the instability is given by liquid spilling from an inverted bottle even though atmospheric pressure is equivalent to a water column 10 m high. Photographs reveal bubbles of air invading the liquid and spikes of the liquid extending into the air. The instability is prevented in a familiar demonstration by the placement of a layer of stiff paper over and in contact with the mouth of the bottle. When the system is inverted, the paper mechanically prevents the growth of any disturbance on the interface. The magnetic field can exert a comparable stabilizing influence in a nonmechanical fashion (but see Section 7.5 for an important caveat and its resolution).

The appropriate form of the dispersion relation is obtained by restricting (7.94) to the absence of mean flow ($U_a = U_b = 0$) and the presence of a tangential applied field ($H_z^a = H_z^b = 0$):

$$\omega^2(\rho_a + \rho_b) = gk(\rho_b - \rho_a) + \sigma k^3 + k_y^2 \frac{\mu_0 M_0^2}{\chi + 2} \quad (7.95)$$

where $\mu_0 = \mu_b$ and $\chi = \mu_a/\mu_0 - 1$. In the nonmagnetic case

$$\omega^2 = \frac{gk(\rho_b - \rho_a) + \sigma k^3}{\rho_a + \rho_b} \quad (7.96)$$

When $\rho_b > \rho_a$ (i.e., when the denser layer is on the bottom), ω is real and so $z_0 = \hat{z}_0 \cos(\omega t - k_x x - k_y y)$. The solution then describes traveling waves that are neutrally stable.

With dense fluid overlaying less dense fluid, $\rho_a < \rho_a$, and the right side of (7.96) can now assume negative as well as positive values. When the

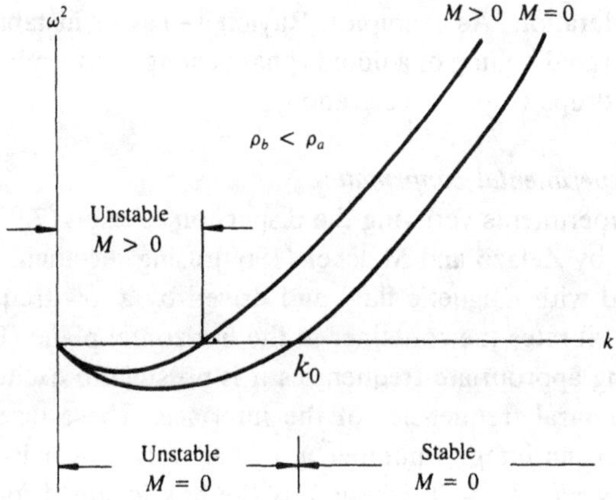

7.12 Dispersion in the ω–k plane for the Rayleigh–Taylor instability in magnetized and unmagnetized fluids.

right side of (7.96) is negative, ω is imaginary and $z = \hat{z}_0 e^{\nu t} \cos(k_x x + k_y y)$, corresponding to static waves that are unstable. This represents the Rayleigh–Taylor instability.

Incipient instability corresponds to the value $k = k_0$, obtained from (7.96) when $\omega = 0$:

$$k_0 = \sqrt{g(\rho_a - \rho_b)/\sigma} \tag{7.97}$$

$\lambda_0 = 2\pi/k_0$ is the *Taylor wavelength*, and k_0 is known as the *capillary constant*. With the aid of (7.97), (7.96) can be written

$$\omega^2 = \frac{\sigma k}{\rho_a + \rho_b}(k^2 - k_0^2) \tag{7.98}$$

As illustrated in Figure 7.12, wave numbers greater than k_0 or, equivalently, wavelengths shorter than λ_0 are stabilized by interfacial tension.

As seen from the last term of (7.95), magnetization provides a stabilizing (stiffening) influence for disturbances propagating along the field lines, and self-field effects are absent in perturbations propagating across the lines of field intensity. The stiffening influence shifts the range of instability to lower wave numbers, as also illustrated in Figure 7.12.

The instability occurs in more situations than might be thought, for kinematic acceleration or deceleration can take the place of gravitational acceleration. As examples, Rayleigh–Taylor instability tends to develop in rapid boiling of a liquid (phase change), in explosions, and in splattered drops (rapid deceleration).

Experimental confirmation

Experiments verifying the dispersion relation (7.95) have been performed by Zelazo and Melcher (1969) using rectangular containers partly filled with magnetic fluid and driven by a low-frequency transducer that vibrates the container in the horizontal plane (Figure 7.13). By choosing appropriate frequencies it is possible to excite resonances near the natural frequencies of the interface. These occur when the box contains an integral number n_s of half waves over its length such that $k_y = n_s\pi/l_0$, $k_z = 0$, where l_0 is the box length. From (7.95) the relative frequency shift needed to produce standing waves of a given wavelength is

$$\frac{\omega_m^2 - \omega_0^2}{\omega_0^2} = \frac{(n_s\pi/l_0)\mu_0 M^2/(\chi + 2)}{g(\rho_b - \rho_a) + (n_0\pi/l_0)^2\sigma} \tag{7.99}$$

The experimental results, which are displayed in Figure 7.13, are in general agreement with this theoretical expression.

7.4 Kelvin–Helmholtz instability

Classical Kelvin–Helmholtz instability relates to the behavior of a plane interface between moving fluid layers. A basic situation in FHD is the inviscid wave behavior at the interface between layers of magnetized fluid having permeabilities μ_a and μ_b. Consider the case of an applied magnetic field of intensity H_y oriented parallel to the unperturbed surface. Again gravity is oriented normal to the interface, and the fluid layers move at speeds U_a and U_b relative to fixed boundaries, as shown in Figure 7.11. The dispersion relationship is obtainable as a special case of equation (7.93):

$$(\omega - k_y U_a)^2\rho_a \coth ka + (\omega - k_y U_b)^2\rho_b \coth kb$$

$$= gk(\rho_b - \rho_a) + k^3\sigma + \frac{k_y^2 H_y^2(\mu_a - \mu_b)^2}{\mu_b \coth ka + \mu_a \coth kb} \tag{7.100}$$

Two familiar phenomena closely related to this instability are wind-generated ocean waves and the flapping of flags. A variation of the Kelvin–Helmholtz instability is also displayed when a gas emerges as a jet through a nozzle into ambient air at rest, as shown in Figure 7.14a

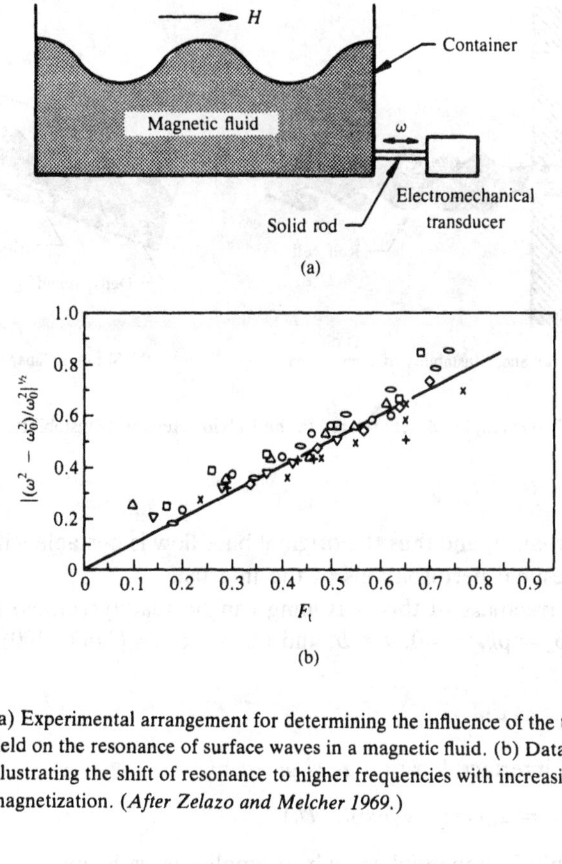

7.13 (a) Experimental arrangement for determining the influence of the tangential field on the resonance of surface waves in a magnetic fluid. (b) Data illustrating the shift of resonance to higher frequencies with increasing fluid magnetization. (*After Zelazo and Melcher 1969.*)

and in the wave machine shown in Figure 7.14b, where the fluids are immiscible.

To develop an appreciation of this instability, consider the situation depicted in Figure 7.15a. Initially the two fluid layers are moving with velocities $U_a = -U_b = U$; i.e., the two velocities are equal and opposite. Now suppose that the interface separating the two fluids becomes slightly wavy, as shown in Figure 7.15b. Because of this deformation of the interface, the fluid at A, A', and A'' moves slightly faster than before, and the fluid at B, B', and B'' moves slightly slower. Now, according to the Bernoulli equation, the pressure rises where the fluid velocity decreases and falls where the fluid velocity increases. However, the pressure gradients are in directions producing amplification of

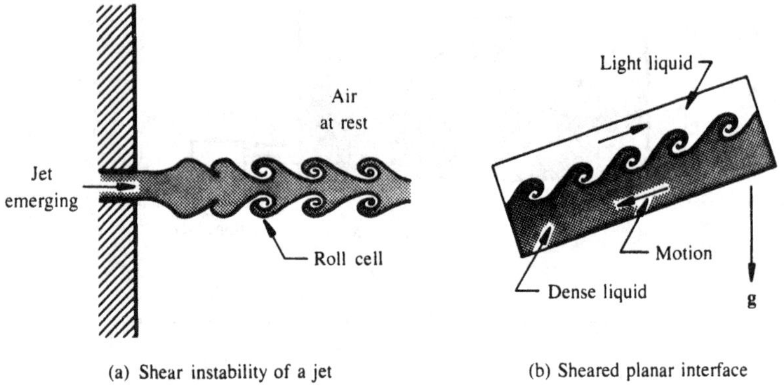

7.14 Two examples closely related to the Kelvin–Helmholtz instability.

the disturbance, and thus the original base flow is unstable with respect to infinitesimal perturbations of the interface.

The correctness of this reasoning can be readily verified by taking $H_y = 0$, $\rho_a = \rho_b$, $\sigma = 0$, $a = b$, and $U_a = -U_b = U$ in (7.100), with the result

$$\omega = \pm i k_y U \tag{7.101}$$

Thus the interface location z_0 is given by

$$z_0 = z_0 \cos(k_y y) \exp(k_y U t) \tag{7.102}$$

and exhibits exponential growth in amplitude with time.

Criterion for Kelvin–Helmholtz instability in a ferrofluid

For simplicity, consider the limiting form of (7.100) at $k = k_y$ for thick layers; that is, let $a, b \to \infty$. It is found that ω is given by

$$\omega = \frac{k(U_a\rho_a + U_b\rho_b)}{\rho_a + \rho_b} \pm \frac{k}{(\rho_a + \rho_b)^{1/2}} \beta_k^{1/2} \tag{7.103}$$

where

$$\beta_k \equiv \frac{g(\rho_b - \rho_a)}{k} + k\sigma + \frac{(\mu_a - \mu_b)^2 H_y^2}{\mu_a + \mu_b} - \frac{\rho_b\rho_a(U_b - U_a)^2}{\rho_b + \rho_a} \tag{7.104}$$

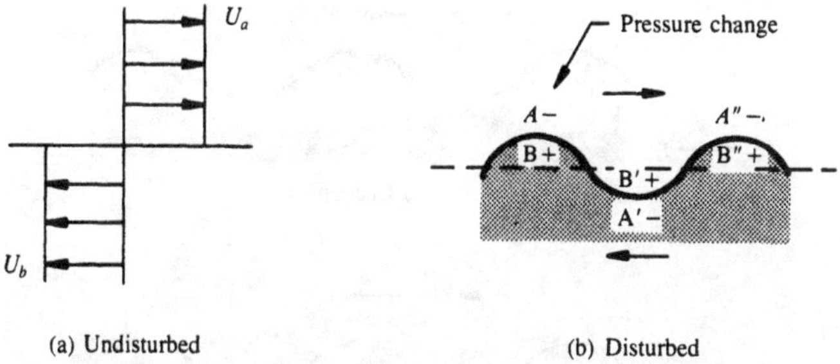

7.15 Velocity profile and pressure variation in Kelvin–Helmholtz instability.

The flow is stable as long as $\beta_k > 0$. Thus the flow is incipiently unstable when the following two conditions are satisfied:

$$\beta_k = 0 \tag{7.105}$$

$$\frac{d\beta_k}{dk} = -\frac{g(\rho_b - \rho_a)}{k^2} + \sigma = 0 \tag{7.106}$$

The critical wave number is found from (7.106):

$$k = k_c = [g(\rho_b - \rho_a)/\sigma]^{1/2} \tag{7.107}$$

which is equivalent to the critical wave number of (7.97) found in the Rayleigh–Taylor problem. Eliminating k from (7.105) with the aid of (7.107) and (7.104) then produces a criterion for instability in the magnetic Kelvin–Helmholtz problem:

$$(U_b - U_a)^2 > \frac{\rho_b + \rho_a}{\rho_b \rho_a}\left\{2[g(\rho_b - \rho_a)\sigma]^{1/2} + \frac{(\mu_a - \mu_b)^2 H_y^2}{\mu_a + \mu_b}\right\} \tag{7.108}$$

The larger the applied field and the larger the difference in permeability across the interface, the greater is the velocity difference that can be accommodated before instability occurs. Low density of a layer also promotes stability.

When the tangential applied magnetic field is in a direction normal to the direction of wave propagation, as shown in Figure 7.16a, no poles are created and H_x never pierces the surface. Consequently there is no change in field energy as the interface deforms, and hence there is no coupling. Thus a uniform magnetic field oriented normal to the wave offers no stabilization. However, when the field is collinear with the

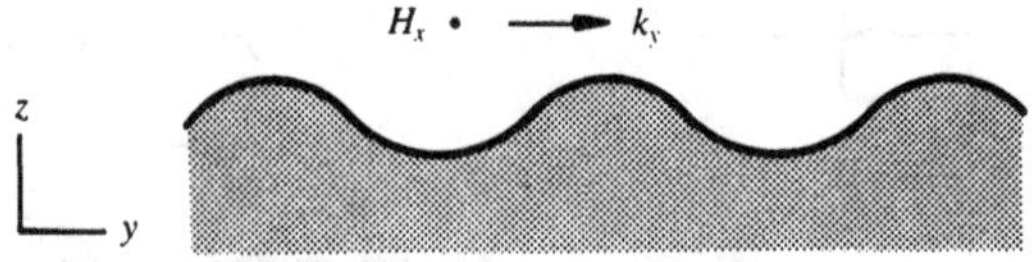

(a) Uncoupled

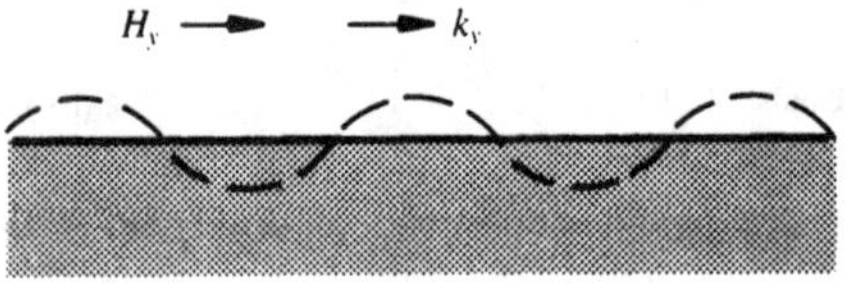

(b) Coupled

7.16 Magnetic stabilization of the Kelvin–Helmholtz instability. (a) The field normal to the direction of the wave yields no coupling. (b) Coupling occurs when the field is collinear with the wave.

direction of wave propagation, as shown in Figure 7.16b, the field lines tend to concentrate at the peaks of the disturbance as they follow the easiest flux path. The perturbed magnetic field then causes Maxwell stresses to act in the direction shown, causing the interface to flatten.

7.5 Gradient-field stabilization

Whereas a uniform tangential magnetic field stiffens the fluid interface for wave propagation along the field direction, waves propagating normal to the field remain uninfluenced and are able to grow in amplitude. However, an imposed magnetic field having a gradient of intensity is capable of stabilizing the fluid interface against growth in the amplitude of waves having any orientation. The theory for gradient-field stabilization is more complex than that for uniform-field self-interaction and includes the self-field influence as a special case (Zelazo and Melcher 1969). To be stabilizing, the field intensity must increase in the direction of the magnetizable fluid, whether the magnetizable layer is denser than the underlaying fluid or less dense and underlays the nonmagnetic fluid, in which case buoyant mixing should be prevented.

A normal field possessing a gradient of intensity is less satisfactory for this purpose than a tangential field having the requisite gradient owing to the destabilizing tendency of uniform field oriented normal to the interface.

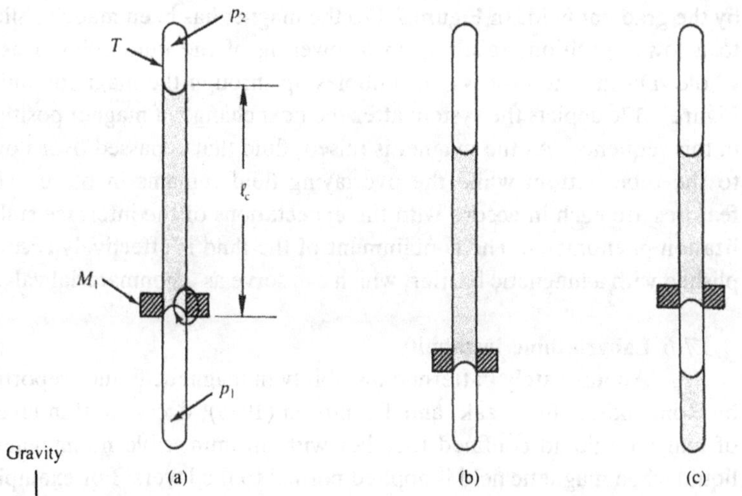

7.17 Field-gradient stabilization of the Rayleigh–Taylor instability, illustrating the mechanism of a magnetic-fluid contactless valve. (*After Rosensweig 1979a.*)

The following expression gives the interface criterion for a tangential gradient field to prevent the Rayleigh–Taylor instability:

$$\mu_0 M(dH_y/dn) > g(\rho_a - \rho_b) \tag{7.109}$$

Surface tension, which was neglected, only further increases the stability. Zelazo and Melcher (1969) tested (7.109) in adverse gravitational acceleration using wedge-shaped steel pole pieces to provide the gradient in imposed field intensity; they report quantitative agreement between theory and experiment. However, it appears that the gradient field extended over the whole volume of the magnetic fluid, so the experiment was unable to distinguish between field-gradient support of the liquid and gradient-field stabilization of the interface.

A demonstration to illustrate the effectiveness of employing a gradient field localized at the magnetic fluid interface is shown in Figure 7.17, which illustrates the stably supported liquid column. In one apparatus a sealed glass tube T of 8-mm i.d. and 330 mm in length contained magnetic fluid with $\rho = 1200\ \text{kg} \cdot \text{m}^{-3}$ and $\mu_0 M = 0.012\ \text{Wb} \cdot \text{m}^{-2}$. The local gradient-containing field is furnished by a ring magnet M_1 slid over the tube, having face-to-face magnetization, and made of oriented barium ferrite, of 25-mm o.d. by 7.5 mm thickness. As shown in Figure 7.17a, the fluid column of length l_c is initially supported against gravity by a pressure difference p_1 less p_2, the lower interface being stabilized by the gradient field. In Figure 7.17b the magnet has been made to slide to a lower position, resulting in a lowering of the fluid column as a whole. During the process, air bubbles up through the magnetic fluid. Figure 7.17c depicts the system after the next change of magnet position in this sequence. As the magnet is raised, fluid that is passed over flows to the tube bottom while the overlaying fluid remains in place. The features are each in accord with the expectations of the interface stabilization phenomena. The containment of the fluid is effectively accomplished with a magnetic barrier, which can serve as a nonmaterial valve.

7.6 Labyrinthine instability

An intricately patterned instability in magnetic liquid, reported by Romankiw, Slusarczak, and Thompson (1975), occurs in thin layers of magnetic liquid confined together with an immiscible nonmagnetic liquid when magnetic field is applied normal to the layers. For example, the fluids may be contained between closely spaced horizontal glass plates (Hele–Shaw cell). The marvelous equilibrium static pattern that appears exhibits a maze, or labyrinthine structure, having walls or lanes of opaque magnetic liquid separated by analogous paths of clear immiscible fluid. The phenomenon is best described with reference to a vertical cell such that gravity establishes a flat interface between the liquid layers in the absence of applied field. Figure 7.18 illustrates a sequence of patterns that evolves in response to a ramp increase in the applied field (Rosensweig 1982a,b).

The ferrofluid in these photographs consists of the magnetite form of iron oxide suspended in kerosene. The specific gravity of the ferrofluid is 1.22, the clear fluid is aqueous with specific gravity of 1.0, the glass cell is 75 mm on a side with a spacing between plates of 1 mm, and the applied field has a final intensity of 0.0535 T. Initially, in the absence of a magnetic field (as shown in the first photograph) or at low magnetic-field intensity, the denser magnetic liquid remains indefinitely at the bottom of the cell with the interface flat. With the application of a sufficiently intense magnetic field, small perturbations suddenly develop and grow into invasive fingers (seen in the second and third photographs). The fourth photograph was made at the end of 90 s, when equilibrium had been reached. It may be seen that the pattern grows by a process of multiple bifurcations, with two new lanes emanating from a given node. The nodal angles approximate 120°, as would be expected

7.18 Photographs of the labyrinthine instability developed in a thin vertical layer in response to an increasing applied magnetic-field intensity. Gravity acts downward, and the applied field, which is uniform, is horizontal and perpendicular to the page. (*Rosensweig 1982a,b.*)

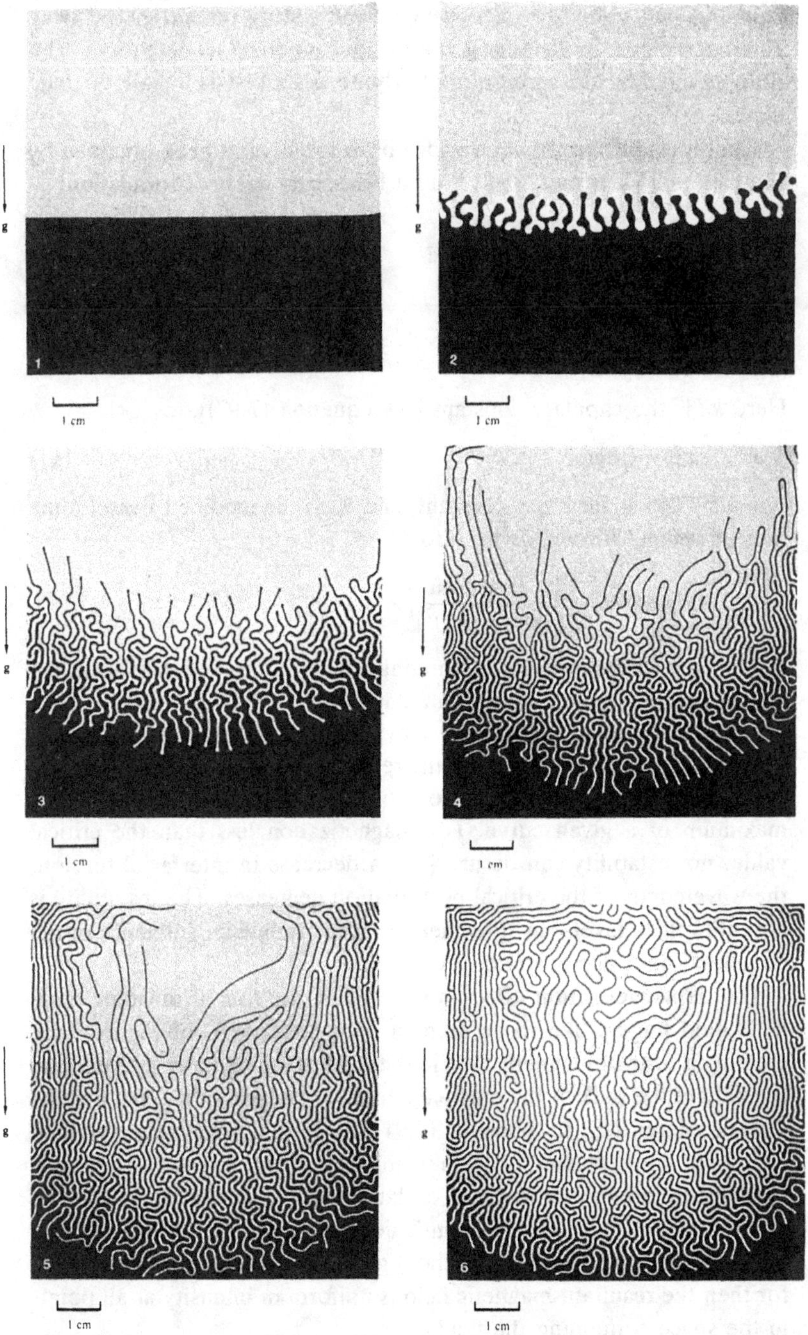
1
2
3
4
5
6
g
1 cm

from Plateau's rule. Throughout the maze the strips remain spaced away from each other, as each strip is a magnet repelling its neighbors. The ultimate equilibrium pattern can fairly be described as a state of *static chaos*.

The threshold for the appearance of instability has been obtained by Tsebers and Maiorov (1980) from a linearized energy formulation:

$$\frac{\sigma}{\mu_0 M_0^2 h} = \frac{4J(kh)}{(kh)^2 + (k_0 h)^2} \tag{7.110a}$$

where

$$J(kh) \equiv \gamma + \ln \tfrac{1}{2}kh + K_0(kh) \tag{7.110b}$$

Here k_0 is the capillary constant [see equation (7.97)],

$$k_0^2 \equiv g\,\Delta\rho/\sigma \tag{7.111}$$

$\gamma = 0.577215$ is the Euler constant, and K_0 is the modified Bessel function of second kind of order zero:

$$K_0(kh) = \int_0^\infty \frac{\cos\alpha\, d\alpha}{[\alpha^2 + (kh)^2]^{1/2}}$$

For a convenient reference containing definitions and tables of the Bessel functions, see Abramowitz and Stegun (1964).

Neutral curves of the vertical labyrinthine instability are plotted in Figure 7.19 for several values of the reduced capillary constant $k_0 h$. The most dangerous perturbations correspond to the wave number at the maximum of a given curve. For magnetization less than the critical value, no instability can occur. With a decrease in interfacial tension, the wavelength of the critical perturbation decreases. The instability is absent in thick layers, in agreement with the stabilizing influence of the tangential field demonstrated by (7.95).

The demagnetization effect is reduced at the tips of invading magnetic-fluid fingers like those seen in the photographs of Figure 7.18, creating a zone of magnetic field in the fluid at the tip more intense than the field in the bulk of the magnetic fluid. Correspondingly, the magnetostatic energy U_m, as shown by (7.91), is reduced when magnetic fluid can flow into regions of higher magnetic-field intensity, and this mechanism drives the process of labyrinth formation. It should be remarked, however, that labyrinth development will not occur if the magnet pole faces are placed without gap directly against the test fluids, for then the resultant magnetic field is uniform in intensity at all points in the space containing the fluids.

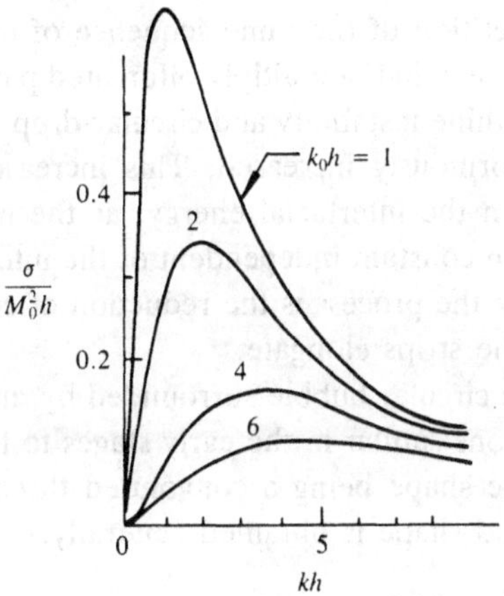

7.19 Neutral curves of the vertical labyrinth instability. (*After Tsebers and Mairov 1980.*)

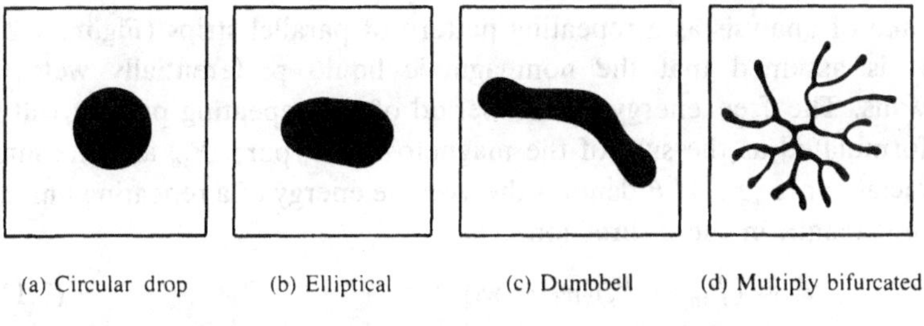

(a) Circular drop (b) Elliptical (c) Dumbbell (d) Multiply bifurcated

7.20 Experimental transition with increasing field of a circular cylindrical drop. (*After Tsebers and Maiorov 1980.*)

A phenomenon related to labyrinth formation is the response of a circular drop of magnetic fluid confined between closely spaced walls (see Figure 7.20a). At least one magnetic pole piece must be distant. The circular drop shape is stable up to a critical value of magnetization, beyond which the shape becomes elliptical, in accord with theory (Figure 7.20b). The elliptical shape elongates with further increase of the applied field; analysis of a long strip indicates instability with respect to a serpentine shape (but not with respect to a sausage shape), a prediction supported by a transition to the dumbbell shape illustrated in Figure 7.20c: Note the thickening of the pattern at the ends and on the convex side of the center, which is due to the relatively higher field in these regions of lower demagnetization. The thickened zones are susceptible to a repetition of the same sequence of instabilities and ultimately, as in Figure 7.20d, a multiply bifurcated pattern is established.

In both labyrinthine instability and circular-drop instability the interfacial area is enormously increased. This increase entails a proportionate increase in the interfacial energy, as the interfacial tension is believed to remain constant independent of the intensity of the applied field. What drives the process is the reduction of configurational magnetic energy as the strips elongate.

A nonmagnetic circular bubble surrounded by magnetic fluid undergoes transformations similar in the early stages to those of the circular drop, the ultimate shape being a convoluted thread. It is not known whether the thread shape is obtained generally.

Labyrinth spacing

When a labyrinth is established in a horizontal plane, the pattern is uniform in average opacity unlike the vertical labyrinths of Figure 7.18, which have a graded appearance because of gravity force.

The well-developed horizontal labyrinth may be idealized for the sake of analysis as a repeating pattern of parallel strips (Figure 7.21). It is assumed that the nonmagnetic liquid preferentially wets the walls. The free energy of one period of the repeating pattern will be formulated as the sum of the magnetostatic part F_m and the interfacial part F_s. If F denotes the average energy of a repeating unit per unit length in the z direction,

$$F = (F_m + F_s)(w_f + w_l)^{-1} \tag{7.112}$$

From the geometry of Figure 7.21, the interfacial energy per unit depth is

$$F_s = 2\sigma t + 2\sigma'(w_f + w_l) \tag{7.113}$$

where σ denotes the interfacial tension between the liquids and σ' between liquid and wall. The magnetic field energy associated with the configuration can be written within a constant [see (7.91)] as

$$F_m = -\tfrac{1}{2}\mu_0 \int_V MH_0\, dV = -\tfrac{1}{2}\mu_0 MH_0 t w_f \tag{7.114}$$

where it is assumed that the magnetization is uniform at all points within the ferrofluid. The magnetization $M = \chi H$, where H, the field within a strip, is reduced from the intensity H_0 of the applied field by the demagnetization field H_d due to all the strips in the system. Thus $H = H_0 - H_d = H_0 - DM$, where D is the demagnetization coefficient, and $M =$

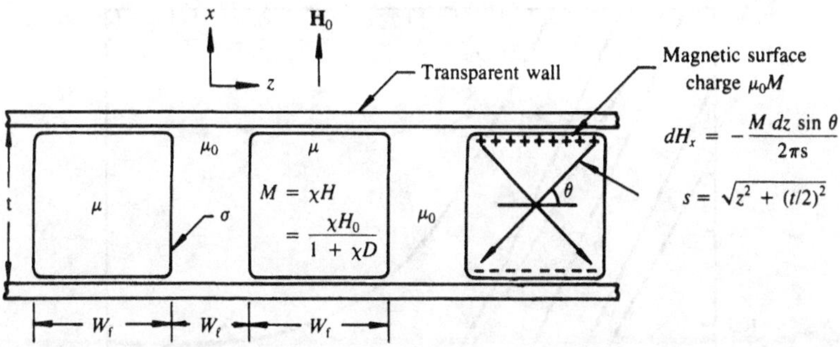

7.21 The labyrinth, idealized as a periodic system of parallel strips. σ denotes the interfacial tension of the fluids. The nonmagnetic fluid preferentially wets the walls. (*After Rosensweig, Zahn, and Shumovich 1983.*)

$\chi H_0/(1 + \chi D)$, so F_m from (7.114) may be rewritten

$$F_m = -\frac{\mu_0}{2}\frac{\chi H_0^2 t w_f}{1 + \chi D} \tag{7.115}$$

Substituting (7.115) and (7.113) into (7.112) gives, on rearrangement

$$N_F = -\frac{\chi N_{Bo}}{1 + \chi D} + \frac{2}{z} + 2\frac{\sigma'}{\sigma}(1 + r) \tag{7.116}$$

where the magnetic Bond number $N_{Bo} \equiv \mu_0 H_0^2 t/2\sigma$ and where

$$N_F \equiv F(1 + r)/\sigma, \qquad r \equiv w_l/w_f, \qquad z \equiv w_f/t \tag{7.117}$$

Minimization of the energy in (7.116) at constant phase ratio r yields the expression

$$\left(\frac{\partial N_F}{\partial z}\right)_r = \frac{\chi^2 N_{Bo}}{(1 + \chi D)^2}\left(\frac{\partial D}{\partial z}\right)_r - \frac{2}{z^2} = 0 \tag{7.118}$$

The demagnetization coefficient at the midline of a strip is found by integrating the demagnetizing field of the assumed-uniform magnetic surface charge $\pm\mu_0 M$ of a given strip and N neighbor strips on either side of the given strip, as illustrated in Figure 7.21. The result is

$$D = \frac{2}{\pi}\left\{\tan^{-1} z + \sum_{n=0}^{N}\left[\tan^{-1}\frac{(n+1)\,r + (n+\frac{3}{2})}{1/2z} - \tan^{-1}\frac{(n+1)\,r + (n+\frac{1}{2})}{1/2z}\right]\right\} \tag{7.119}$$

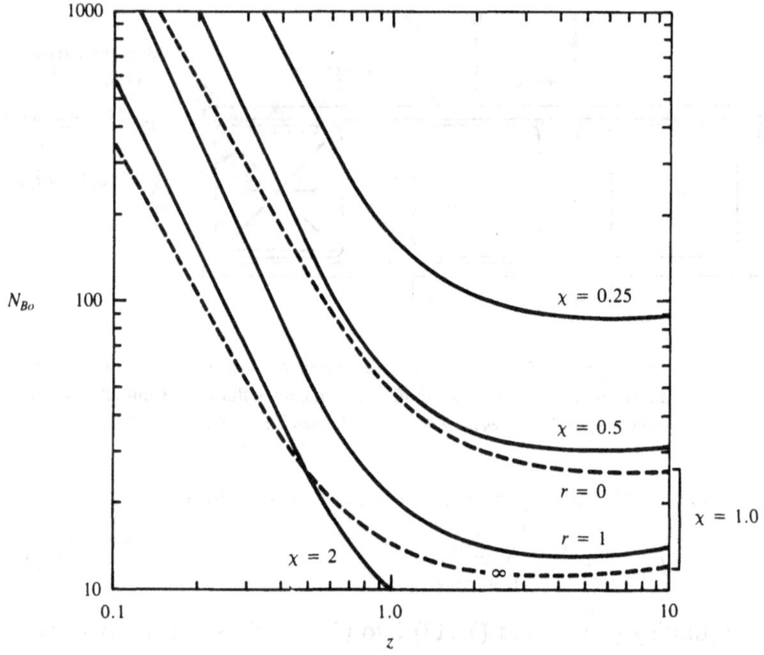

7.22 The dependence of the equilibrium magnetic Bond number N_{Bo} on the normalized magnetic strip thickness z for various values of the magnetic susceptibility χ. The phase ratio $r = 1$ unless otherwise indicated by a dashed line. (*After Rosensweig, Zahn, and Shumovich 1983.*)

From (7.118) and (7.119) the following relationship is attained at equilibrium

$$N_{\text{Bo}} = \frac{\pi}{\chi^2 z^2}\left[1 + \frac{2\chi}{w}\left(\tan^{-1} z + \sum_{n=0}^{N} \{\tan^{-1} 2z[(n+1)r + n + \tfrac{3}{2}] - \tan^{-1} 2z[(n+1)r + n + \tfrac{1}{2}]\}\right)\right]^2$$
$$\times \left(\frac{1}{1+z^2} + \sum_{n=0}^{N}\left\{\frac{(n+1)r + n + \frac{3}{2}}{1 + 4z^2[(n+1)r + n + \frac{3}{2}]^2} - \frac{(n+1)r + n + \frac{1}{2}}{1 + 4z^2[(n+1)r + n + \frac{1}{2}]^2}\right\}\right)^{-1} \tag{7.120}$$

The general dependence of N_{Bo} on z, x, and r given by (7.120), displayed in Figure 7.22, illustrates that a minimum value of the Bond number is required to establish the labyrinthine pattern. A curious

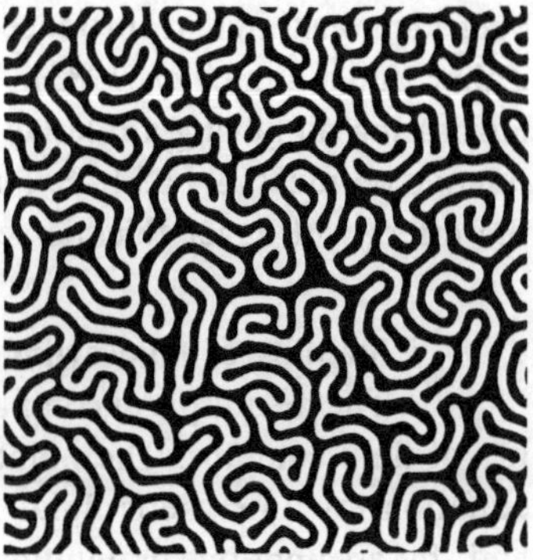

.010 T .035 T

10 mm

7.23 Labyrinthine instability in a horizontal cell, illustrating the closer-spaced pattern resulting from an increase in the magnetic-field intensity. The layer thickness is 0.9 mm. Thinner layers produce yet more finely convoluted patterns. (*From Rosensweig, Zahn, and Shumovich 1983.*)

feature is that z is double valued in the range of N_{Bo} just above the threshold; a more exact analysis is needed to determine if this feature is an artifact. Neglecting neighbors by setting $N = 0$ in (7.120) simplifies the relation greatly in that the summations disappear, and the calculated Bond number is reduced typically by only 15%–20%. However, in computing the curves of Figure 7.22 the value $N = 100$ was used.

Experiments with horizontal cells in which the wall spacing and applied field were systematically varied resulted in spacings that showed a reasonably close comparison with the theoretically computed values over a three-decade range of the magnetic Bond number (Rosensweig, Zahn, and Shumovich 1983). Representative photographs are shown in Figure 7.23.

Note that the magnetic-fluid labyrinth is closely analogous to the labyrinthine structure of band domains in thin films of ferromagnets (Thiele 1969). There also appears to be an analogy to structures in type-I superconductors in reverting from the superconducting to the normal conducting state (Faber 1958). Along the same lines, the normal-field instability of a magnetic fluid bears analogy to the hexagonal patterns observed in the transition of type-II superconductors (Essmann and Trauble 1967).

Dielectric fluids subjected to an intense applied electric field have also been demonstrated to produce the labyrinthine pattern (Rosensweig, Zahn, and Shumovich 1983). The dielectric fluids employed were a lubricating oil having a relative permeability of 2.33 and castor oil of a relative permeability of 4.48. An applied field of 16 $kV \cdot cm^{-1}$ produced the effect with layer thicknesses of 0.8 and 1.6 mm.

7.7 Stability of fluid cylinders

The next topic encompasses the flow of round jets and the stability of fluid columns. The point of departure is (7.22), the momentum equation for a magnetic fluid. Because the viscosity term is neglected and, by assumption, gravitational force is absent, the relationship has the form of an Euler equation:

$$\rho(\partial \mathbf{v}/\partial t + \mathbf{v}\cdot\nabla\mathbf{v}) = -\nabla\Pi \tag{7.121}$$

where $\Pi = p + p_s$ for a magnetic fluid and $\Pi = p$ for a nonmagnetic fluid. Applying the curl operation to each term of the Euler equation gives a relationship for the vorticity $\mathbf{\Omega}$ in incompressible flow:

$$\frac{D\mathbf{\Omega}}{Dt} = \frac{\partial\mathbf{\Omega}}{\partial t} + \mathbf{v}\cdot\nabla\mathbf{\Omega} = \mathbf{\Omega}\cdot\nabla\mathbf{v} \tag{7.122}$$

This has the property that

$$\text{if} \quad \mathbf{\Omega} = \mathbf{0}, \qquad \text{then} \quad D\mathbf{\Omega}/Dt = \mathbf{0} \tag{7.123}$$

Thus, if an element of inviscid magnetic fluid has no vorticity at some initial instant of time, then it can never acquire any. Many realizable flows are initially of this type – for example, the potential flow field surrounding a viscous boundary layer. All the instability problems treated up to this point are also essentially inviscid. Such motion is termed irrotational and

$$\mathbf{\Omega} = \nabla \times \mathbf{v} = \mathbf{0} \tag{7.124}$$

throughout the flow. Accordingly, a velocity potential Φ may be defined such that

$$\mathbf{v} \equiv -\nabla\Phi \tag{7.125}$$

because (7.124) is satisfied identically.

Combining the continuity equation $\nabla\cdot\mathbf{v} = 0$ with (7.125) gives

$$\nabla^2\Phi = 0 \tag{7.126}$$

showing that the velocity potential satisfies Laplace's equation. The analysis leading to (7.126) applies to both steady and unsteady flow, yet no time variation appears explicitly in the equation. Techniques for solving Laplace's equation are highly developed as a result of its importance in a variety of physical contexts. Laplace's equation is applied below to the analysis of stability of a cylindrical jet of magnetic fluid.

Magnetic-fluid jet in a uniform field

The interaction of a field with an infinitely deep fluid, ensuring the stability of surface perturbations, should be manifested in jet flows as well. The critical value of the field and the critical wavelength will also be affected by the curvature of the surface and the finite thickness of the round jet. The classical stability problem of an unmagnetized round jet was treated by Rayleigh (1878), who showed that such a flow is always unstable; any perturbation with a wavelength greater than the perimeter of the jet increases with time. The instability is driven by surface-tension forces.

In this analysis the velocity of the jet motion is assumed to be small so that the effect of the surrounding air on the disintegration of the stream may be neglected, and gravity is assumed to be absent. It is clear that the uniform translational motion of the jet has no influence on the dynamics, and hence the analysis may conveniently be done in a reference frame that moves with the undisturbed velocity of the jet. A uniform field H_0 is applied to the jet by a coaxially positioned solenoid. The problem is formulated in a cylindrical coordinate system with the z axis directed along the stream axis.

The velocity potential Φ of the fluid satisfies the Laplace equation, as seen in (7.126). In cylindrical coordinates this is

$$\frac{1}{r}\frac{\partial}{\partial r}\left(r\frac{\partial \Phi}{\partial r}\right) + \frac{1}{r^2}\frac{\partial^2 \Phi}{\partial \theta^2} + \frac{\partial^2 \Phi}{\partial z^2} = 0 \qquad (7.127)$$

where r is the radius, z the axial distance, and θ the angle. It will be assumed that the jet is axisymmetric and hence that Φ does not depend on the angle θ. A solution is sought in the form

$$\Phi = \hat{\Phi} R(r) e^{i(\omega t - kz)} \qquad (7.128)$$

where $\hat{\Phi}$ is a small quantity, k is the wave number, and ω is the frequency and can be complex. It should be understood that the real part of the exponential function is intended.

When the derivatives of Φ that appear in (7.127) are found from

(7.128) and substituted into (7.127), the following total differential equation is obtained:

$$\frac{d^2R}{dr^2} + \frac{1}{r}\frac{dR}{dr} - k^2R = 0 \tag{7.129}$$

This is a Bessel equation and has the general solution

$$R(r) = c_1 I_0(kr) + c_2 K_0(kr) \tag{7.130}$$

where $I_0(kr)$ is the modified Bessel function of the first kind of order zero, and $K_0(kr)$ is the modified Bessel function of the second kind of order zero. $K_0(kr)$ becomes infinitely large for $kr = 0$, but the velocity and hence Φ must remain finite, so the constant c_2 must be taken as zero for this problem; that is, $R(r) = c_1 I_0(kr)$.

If the radius of the unperturbed stream is r_0, the equation of the deformed surface of the stream may be written in the form

$$r = r_0 + \zeta(z, t) \tag{7.131}$$

where

$$\zeta = a_0 e^{i(\omega t - kz)} \tag{7.132}$$

This describes a sausage shape for the perturbed fluid jet. At the surface of the stream the normal component of the fluid velocity must equal the rate of displacement of the surface, according to the linearized kinematic condition [compare (7.21)]. Thus

$$v_n = v_r = \partial\zeta/\partial t \qquad \text{at} \quad r = r_0 \tag{7.133}$$

Computing $v_r = -\partial\Phi/\partial r$ from (7.128), evaluating $\partial\zeta/\partial t$ from (7.132), and noting that $dI_0(kr)/dr = kI_1(kr)$, where $I_1(kr)$ is the modified Bessel function of the first kind of order 1, one finds from (7.133) that

$$\Phi = -\frac{i\omega a_0 I_0(kr)}{kI_1(kr_0)} e^{i(\omega t - kz)} \tag{7.134}$$

Up to this point what has been determined is the product $\hat{\Phi}R(r)$ appearing in (7.128). A relationship between k and ω has yet to be established. This will be accomplished using force balance at the interface. As a preliminary, a solution is needed for the magnetic field. The field is described by Maxwell's equations in the magnetostatic limit:

$$\nabla \times \mathbf{H} = 0, \qquad \nabla \cdot \mu\mathbf{H} = 0 \tag{7.135}$$

For a magnetic fluid having constant permeability μ, it follows from (7.135) that $\mathbf{H}$ has a potential ψ such that $\mathbf{H} = -\nabla\psi$ and such that ψ satisfies Laplace's equation $\nabla^2\psi = 0$. (Compare this to the development of the scalar potential in free space, seen in Section 3.1.) The potential ψ may be represented

$$\psi \equiv H_0 z + \psi' \tag{7.136}$$

where the first term on the right side describes the uniform unperturbed field inside the solenoid, and ψ' is the superposed field that arises owing to perturbations in the shape of the fluid column constituting the jet. Just as in the case of the velocity potential, Laplace's equation for the magnetic potential may be expressed in cylindrical co-ordinates and solutions found for the region inside the jet and outside. This gives

$$\psi_1' = a_1 I_0(kr)e^{i(\omega t - kz)} \tag{7.137}$$

$$\psi_2' = [a_2 I_0(kr) + b_2 K_0(kr)]e^{i(\omega t - kz)} \tag{7.138}$$

where the subscript 1 refers to the inner region and 2 to the outer region. These solutions must satisfy the magnetic-field boundary conditions at the interface between the regions. These conditions express the continuity of the normal induction field and of the tangential magnetic field at $r = r_0$; they are

$$(\mu H_n)_1 = (\mu H_n)_2 \tag{7.139a}$$

$$(H_t)_1 = (H_t)_2 \tag{7.139b}$$

or, in terms of the potentials at $r = r_0$, with $\mu_1 = \mu$ and $\mu_2 = \mu_0$,

$$\mu \mathbf{n}\cdot\nabla\psi_1 = \mu_0 \mathbf{n}\cdot\nabla\psi_2 \tag{7.140a}$$

$$\psi_1 = \psi_2 \tag{7.140b}$$

On the surface of the solenoid, where $r = R_0$, the potential perturbation ψ_2' is zero:

$$\psi_2' = 0 \quad \text{at} \quad r = R_0 \tag{7.140c}$$

The solutions for the perturbation potentials of (7.137) and (7.138) contain the three unknowns a_1, a_2, and b_2, and the field boundary conditions of equations (7.140) provide an equal number of constraints. Thus it is possible to evaluate the constants using these conditions.

Evaluation of the terms in (7.140a) requires an expression for the normal $\mathbf{n}$ in the cylindrical coordinates. The technique introduced in Section 7.1 is used. The interface of the jet may thus be represented as

$r = f(z)$, so $r - f(z) =$ const represents contours near the interface and $\nabla[r - f(z)]$ yields a vector oriented in the direction of $\mathbf{n}$. When this operation is carried out using the cylindrical-coordinates expression for the gradient operator ($\nabla = \mathbf{i}_r \partial/\partial r + \mathbf{i}_\theta(\partial/\partial\theta)/r + \mathbf{i}_z \partial/\partial z$) recognizing that $\partial/\partial\theta = 0$ in this problem, and the resulting vector is divided by its length $|\nabla[r - f(z)]|$, one obtains the following result, correct to terms of the first order:

$$\mathbf{n} = \mathbf{i}_r - \mathbf{i}_z \frac{\partial\zeta}{\partial z} \tag{7.141}$$

where the right side of (7.131) has been substituted for $f(z)$.

The values of the constants are found after algebraic manipulation to be

$$a_1 = ia_0 H_0(\mu - \mu_0)[\mu I_1(kr_0) - \mu_0 A I_0(kr_0)]^{-1} \tag{7.142a}$$

$$a_2 = \frac{ia_0 H_0(\mu - \mu_0)\, I_0(kr_0) K_0(kR_0)}{[\mu I_1(kr_0) - \mu_0 A I_0(kr_0)][I_0(kr_0)K_0(kR_0) - I_0(kR_0)K_0(kr_0)]} \tag{7.142b}$$

$$b_2 = -\,a_2 I_0(kR_0)/K_0(kR_0) \tag{7.142c}$$

where

$$A \equiv \frac{I_1(kr_0)K_0(kR_0) - I_0(kR_0)K_1(kr_0)}{I_0(kr_0)K_0(kR_0) - I_0(kR_0)K_0(kr_0)}$$

With the assumption that the solenoid is distant, so that $r_0/R_0 \ll 1$, A reduces to

$$A = -K_1(kr_0)/K_0(kr_0)$$

Next the force-equilibrium condition at the stream interface is introduced, as in the development of (5.24). The condition can be written

$$\Pi + \mu_0 \int_0^H M\, dH + \frac{\mu_0}{2} M_n^2 = p_0 + 2\mathscr{H}\sigma \tag{7.143}$$

where $\Pi = p + p_s$ for the magnetic fluid, and p_0 is the environmental pressure. From (7.11) for $2\mathscr{H}$ and with $\mathbf{n}$ from (7.141), it is found that

$$\begin{aligned} 2\mathscr{H} = \nabla\cdot\mathbf{n} &= \frac{1}{r}\frac{\partial}{\partial r}(rn_r) + \frac{1}{r}\frac{\partial n_\theta}{\partial\theta} + \frac{\partial n_z}{\partial z} \\ &= \frac{1}{r} - \frac{\partial^2\zeta}{\partial z^2} \end{aligned} \tag{7.144}$$

From (7.132) for ζ and because to linear terms $r^{-1} \simeq r_0^{-1}(1 - \zeta/r_0)$ the expression for mean curvature becomes

$$2\mathscr{H} = \left\{\frac{1}{r_0} - \frac{\zeta}{r_0^2}[1 - (kr_0)^2]\right\} \tag{7.145}$$

The magnetic terms in (7.143), correct to first order in small quantities, are

$$\mu_0 \int_0^H M\,dH = (\mu - \mu_0)(\tfrac{1}{2}H_0^2 + H_0 H_z'), \qquad \tfrac{1}{2}\mu_0 M_n^2 = 0 \tag{7.146}$$

where $H_z' = -d\psi'/dz = ik\psi'$. (Remember that ψ' is complex and only the real part of this expression is retained.)

Thus, the surface-force-balance expression (7.143) for undisturbed flow becomes

$$\Pi_0 + (\mu - \mu_0)(\tfrac{1}{2}H_0^2) = p_0 + \sigma/r_0 \tag{7.147}$$

where Π_0 is the value of Π in equilibrium flow.

When this relationship is subtracted from the complete surface-force-balance expression for the disturbed flow, the result is

$$(\Pi - \Pi_0) + (\mu - \mu_0)ikH_0\psi' = -\frac{\sigma\zeta}{r_0^2}(1 - k^2 r_0^2) \tag{7.148}$$

The term $\Pi - \Pi_0$ can be found in a similar manner from the unsteady form of the Bernoulli equation, (5.5), wherein the function $f(t) = \Pi_0 =$ const, permitting p and Φ by assumption to be spatially periodic. In the absence of a gravitational field and to first order, this yields

$$\Pi - \Pi_0 = \rho\,\partial\Phi/\partial t \tag{7.149}$$

Substituting (7.134) into (7.149) and the result, together with (7.138) and (7.132) into (7.148) produces the dispersion equation for waves propagating on the surface of the jet, a result obtained by Taktarov (1975):

$$\omega^2 = \frac{k^2 H_0^2(\mu - \mu_0)^2 I_0(kr_0)K_0(kr_0)}{\rho[\mu I_1(kr_0)K_0(kr_0) + \mu_0 I_0(kr_0)K_1(kr_0)]} - \frac{\sigma k I_1(kr_0)}{\rho r_0^2 I_0(kr_0)}(1 - k^2 r_0^2) \tag{7.150}$$

The quantity ω^2 may have both positive and negative values. In the first case ω is real and the corresponding motion is stable. In the second case,

ω has two imaginary values, one of which has a positive part; this leads to an increased wave amplitude and therefore to instability. In this case the stream breaks up into droplets.

When the field is absent, the first term on the right side of (7.150) disappears, and what remains is the classical Rayleigh result. The onset of instability occurs for $k_c r_0 < 1$ ($\lambda > 2\pi r_0$), i.e., when the wavelength of the disturbance exceeds the perimeter of the jet.

A graph of the dimensionless quantity $\omega^2(\sigma/\rho r_0^3)^{-1}$ as a function of $(kr_0)^{-1}$ in terms of μ/μ_0 is shown in Figure 7.24. It is seen that the zero crossing, which represents the minimum wavelength exhibiting instability, is shifted to longer wavelengths as the fluid permeability increases. The minimum in any one of the curves, representing the fastest-growing unstable wave, is shifted to longer wavelengths, indicating that larger droplets are formed. The length of the intact portion of the jet is of the order of $v/|\omega_m|$, where v is the jet velocity. The depth of the minimum $|\omega_m|$ decreases with increasing permeability, illustrating that the disturbance amplitude grows more slowly. Thus, a stream of magnetic fluid can in theory be stabilized by a longitudinal magnetic field so that the intact length of the jet increases, and the droplet size when breakup finally occurs is greater.

The normalized group $\mu_0 H_0^2(\sigma/r_0)^{-1}$ has the common value $5/4\pi$ for all the curves in Figure 7.24. This corresponds, for example, to $\mu_0 H_0 = 10^{-3}$ T, $\sigma = 0.02$ N·m^{-1}, $\rho = 10^3$ kg·m^{-3}, and $r_0 = 0.01$ m. These values are readily accessible using existing magnetic fluids and magnetic-field sources of modest intensity. Bashtovoi and Krakov (1978) report laboratory observation of the magnetic stabilization of a magnetic-fluid jet.

Cylindrical column in a radial gradient field

An interesting related problem that has been studied is the stability of a cylindrical column of magnetic fluid surrounding a straight, long current-carrying conductor. This system may be prepared in the laboratory as a stationary column and studied at leisure. A review of this work is given by Berkovsky and Bashtovoi (1980). The magnetic field induced by a cylindrical current-carrying conductor is the following exact solution of the Maxwell equations:

$$H_r = 0, \qquad H_\theta = H = I/2\pi r, \qquad H_z = 0 \tag{7.151}$$

where the z axis of the cylindrical coordinate system (r, θ, z) is directed along the conductor axis. Because the magnetic field has only an azimuthal component, it is oriented tangential to the free surface of the

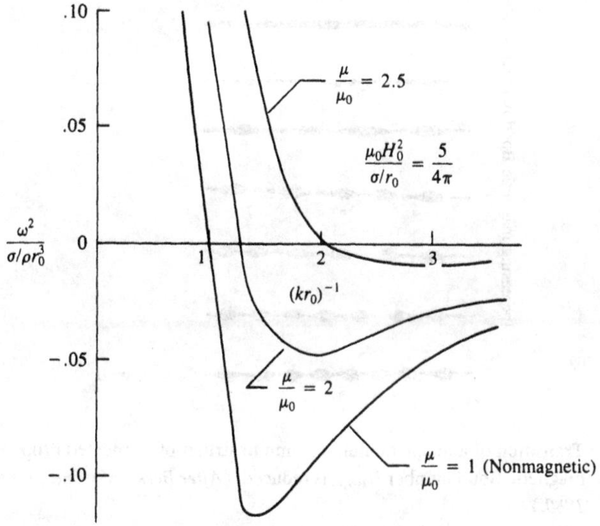

7.24 **Dispersion relationship for an axisymmetric jet in a uniform longitudinal magnetic field.**

magnetic fluid, and hence there is no tendency to establish normal-field instability. The magnetic-field influence is manifested as the existence of body force due to the field nonuniformity. That is, the field possesses the gradient

$$G \equiv |dH/dr| = I/2\pi r_0^2 \tag{7.152}$$

evaluated at the position of the free surface, r_0.

Stability analysis of the disturbances that may propagate along the free surface, in the linear approximation for a fluid of constant susceptibility $\chi = M/H$, yields the following dispersion equation, which is written in dimensionless form:

$$\frac{\rho r_0^3 \omega^2}{\sigma} = \left[\frac{I_1(kr_0)K_1(kR) - I_1(kR)K_1(kr_0)}{I_0(kr_0)K_1(kR) + I_1(kR)K_0(kr_0)} + \frac{\rho_e K_0(kr_0)}{\rho K_1(kr_0)}\right]^{-1} (kr_0)(N_{\mathrm{Bo,m}} - 1 + k^2 r_0^2) \tag{7.153}$$

where $N_{\mathrm{Bo,m}}$ is defined as a magnetic Bond number, R is the radius of the conductor, ρ_e is density of the external fluid, and ρ is density of the magnetic fluid.

$$N_{\mathrm{Bo,m}} \equiv \mu_0 M G d^2/\sigma = \mu_0 \chi I^2/4\pi^2 r_0 \sigma \tag{7.154}$$

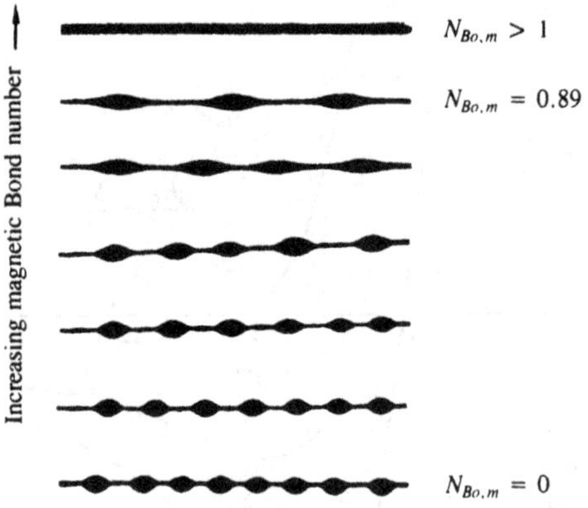

7.25 Transition of a magnetic-fluid column to strings of connected drops as the magnetic Bond number $N_{\mathrm{Bo,m}}$ is reduced. (*After Berkovsky and Bashtovoi 1980.*)

The onset of instability is at $\omega^2 = 0$. The sign of the right side of (7.153) is determined entirely by the expression $N_{\mathrm{Bo,m}} - 1 + k^2r_0^2$. A negative value of this expression is possible only for $N_{\mathrm{Bo,m}} < 1$. From this point of view the critical value of $N_{\mathrm{Bo,m}}$ is unity. Most dangerous are disturbances that have the shortest development time and hence correspond to the value of $N_{\mathrm{Bo,m}}$ giving from calculus, a minimum for the dimensionless frequency. This result will be quoted for the limiting case of a thin cylindrical layer, $r_0/R - 1 \ll 1$, surrounded by gas of negligible density $\rho_e = 0$. Under these conditions (7.153) reduces to

$$\frac{\rho r_0^3 \omega^2}{\sigma} = \left(\frac{r_0}{R} - 1\right)(kr_0)^2(N_{\mathrm{Bo,m}} - 1 + k^2r_0^2) \tag{7.155}$$

The critical wavelength λ_c is determined from (7.155) with the condition

$$\partial\omega^2/\partial k = 0 \tag{7.156}$$

from which it is found that

$$\lambda_c = 2\pi/k_c r_0 = 2\pi[\tfrac{1}{2}(1 - N_{\mathrm{Bo,m}})]^{1/2} \tag{7.157}$$

Experiments using a paraffin-based magnetic fluid immersed in an aqueous glycerine solution of equal density are illustrated in Figure 7.25. The current in the 1-mm stainless-steel conductor is preset at a supercritical value of the magnetic Bond number, $N_{\mathrm{Bo,m}} > 1$. Thus, the

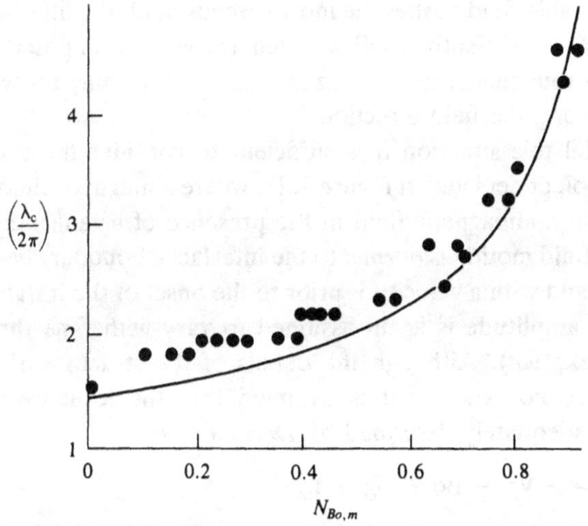

7.26 The theoretical wavelength of the most dangerous disturbances of the fluid column compared with the experimental points. (*After Berkovsky and Bashtovoi 1980.*)

liquid initially forms a stable cylindrical column around the conductor. As the current is decreased, the column is observed to undergo a beautiful series of transitions to strings of connected drops. Because of the finite length of the conductor, the transitions occur in discrete steps of current. This phenomenon is clearly evident in the data points for a fluid cylinder in air shown in Figure 7.26. The curve in Figure 7.26 is a plot of the theoretical prediction of (7.157).

The reader should note the similarity of this problem to the gradient-field stabilization mechanism discussed in Section 7.5.

7.8 Porous-medium flow: fingering instability

Consider flow through a porous medium in which a given fluid is driven through the voids by another fluid. It is found that the interface between the two fluids can be unstable if the more-viscous fluid is being driven by the less-viscous fluid but not vice versa. This *fingering instability* was analyzed by Saffman and Taylor (1958) using a linearized stability analysis. The phenomenon occurs frequently in oil production from reservoirs and in the flow of ground waters. Rosensweig, Zahn, and Vogler (1978) using a Hele–Shaw cell demonstrated that if a layer of magnetizable fluid pushes the more-viscous fluid, the interface can be stabilized for sufficiently small wavelengths with an imposed magnetic field. In porous media the stabilization is effective only for waves propagating along the field direction.

To model this situation it is sufficient to consider the simple two-region problem depicted in Figure 7.27, where a magnetic fluid is shown pushing the nonmagnetic fluid in the presence of a tangential applied field. The fluid motion is *normal* to the interfacial boundary between the two fluids and with a velocity V prior to the onset of the instability. The interfacial amplitude is again assumed to vary with time through the factor of $\exp(i\omega t)$. Although the details of the structure of a porous medium are not known, it is assumed that the local average fluid motion is adequately described by *Darcy's law*:

$$\mathbf{0} = -\nabla p - \beta \mathbf{q} + \rho \mathbf{g} + \mathbf{f}_{\mathrm{m}} \tag{7.158}$$

where p is the hydrodynamic pressure; $\beta = \eta/K$ is the ratio of the fluid viscosity η to the permeability K, which in turn depends on the geometry of the interstices; $\mathbf{q}$ is the local average fluid velocity; and $\mathbf{f}_{\mathrm{m}}$ is the effective magnetic fluid body force density. Inherent in (7.158) is the assumption that inertia is unimportant in this problem and hence can be neglected. Additional governing equations are the incompressible continuity relationship, $\nabla \cdot \mathbf{q} = 0$, and the magnetostatic field equations, $\nabla \times \mathbf{H} = \mathbf{0}$ and $\nabla \cdot \mathbf{B} = 0$. All these variables are considered to be averaged over a region of space. In fact, equations of multiphase flow are derived in Chapter 9, and it can be verified that (7.158) results as a special case of the fluid-phase momentum balance, (9.42a), combined with the constitutive relations, (9.45) and (9.46a), when inertial terms are negligible, the phase velocity is zero, and the void fraction is spatially uniform.

The following dispersion relation applies to this problem:

$$\omega = i\frac{k}{\beta_b + \beta_a}\left[\sigma k^2 + g(\rho_b - \rho_a) + (\beta_b - \beta_a)V + \frac{\mu_0 M_0^2 k_y^2/k}{1 + s_0}\right] \tag{7.159}$$

where

$$s_0 \equiv \frac{\mu}{\mu_0}\left(1 + \frac{k_y^2}{k^2}\left.\frac{\partial \ln \mu}{\partial \ln H}\right|_{H_0}\right)^{1/2} \tag{7.160}$$

Note that ω is pure imaginary; if the quantity inside the brackets in (7.159) is positive, any perturbation decays with time, and if it is negative, the system is unstable and any perturbation grows exponen-

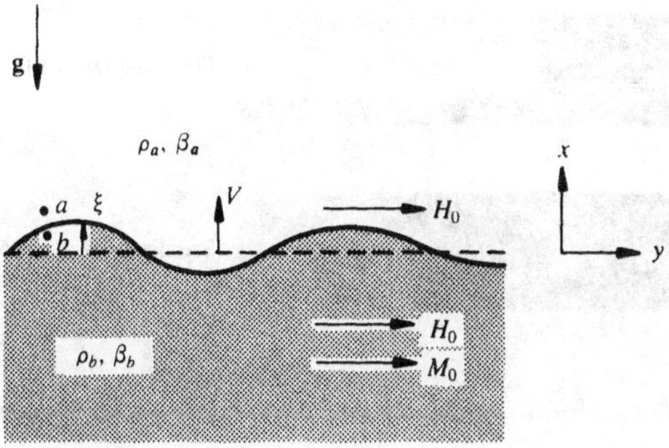

7.27 Perturbations on the interface separating two dissimilar fluids penetrating through a porous medium. V is the interfacial speed.

tially with time. Thus, surface tension and magnetization tend to stabilize the system; gravity also stabilizes the system if the more dense fluid is below ($\rho_b > \rho_a$). Unlike in the Rayleigh–Taylor instability, in fluid penetration it is viscous drag rather than inertia that controls the dynamics. Equation (7.159) confirms that the viscous fingering instability analyzed by Saffman and Taylor with no magnetic effects can occur when the more viscous fluid is being pushed ($\beta_a > \beta_b$).

The magnetic field stabilizes only those waves oriented along the direction of the field. Then, with $k_z = 0$, so that $k = k_y$, (7.159) can be written

$$\omega = i\frac{k_y}{\beta_b + \beta_a}\left(\sigma k_y^2 + \frac{\mu_0 M_0^2 k_y}{1 + r_0} - \Gamma\right) \tag{7.161}$$

where $r_0 = (\mu_c\mu_t/\mu_0^2)^{1/2}$, as given previously, and

$$\Gamma = (\beta_a - \beta_b)V + g(\rho_a - \rho_b) \tag{7.162}$$

From (7.161) and (7.162) it is seen that surface tension stabilizes the largest wave numbers (smallest wavelengths), and a magnetic field stabilizes intermediate wave numbers (intermediate wavelengths). However, if $\Gamma > 0$, the system is unstable over a range of small wave numbers. When $\Gamma \leqslant 0$, however, the system is stable for all wave numbers, whether magnetization is present or not.

In obtaining (7.161) from (7.159) s_0 is replaced by r_0. Because $\mu = B/H$, $(\partial \ln \mu/\partial \ln H)|_{H_0} = \mu_t/\mu_c - 1$, and substitution of this result into s_0 gives the desired reduction to r_0 when $k_z = 0$.

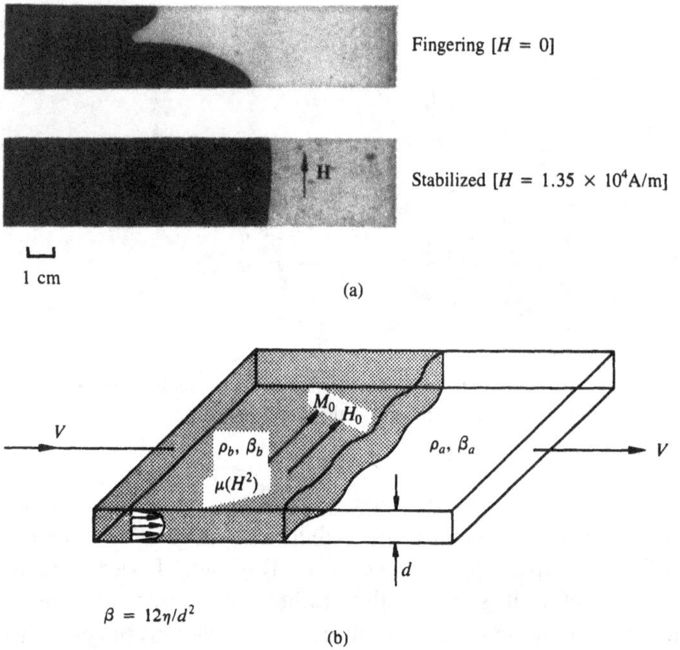

7.28 (a) Photographs of fluid penetration from left to right through a horizontal Hele–Shaw cell. The plate spacing $d = 0.52$ mm, the aqueous base magnetic fluid viscosity $\eta_b = 1.18$ mN·s·m^{-2}, the oil viscosity $\eta_a = 219$ mN·s·m^{-2}, the interfacial tension $\sigma = 30$ mN·m^{-1}, the velocity $V = 0.3$ mm·s^{-1}. (b) Schematic of the Hele–Shaw cell. (*After Rosensweig, Zahn, and Vogler 1978.*)

Experimental verification of the magnetic stabilization of fluid penetration has been carried out with a *Hele–Shaw cell*, which consists of two parallel plates separated by a small distance in the y direction, by Rosensweig, Zahn, and Vogler (1978) and Zahn and Rosensweig (1980); see Figure 7.28.

7.9 Thermoconvective stability

One of the most studied problems in transport phenomena is Rayleigh convection (often called the Bénard problem) sketched in Figure 7.29.

A liquid layer of depth d occupies the space between two parallel plates of infinite extent, as shown in Figure 7.29. The bottom plate is

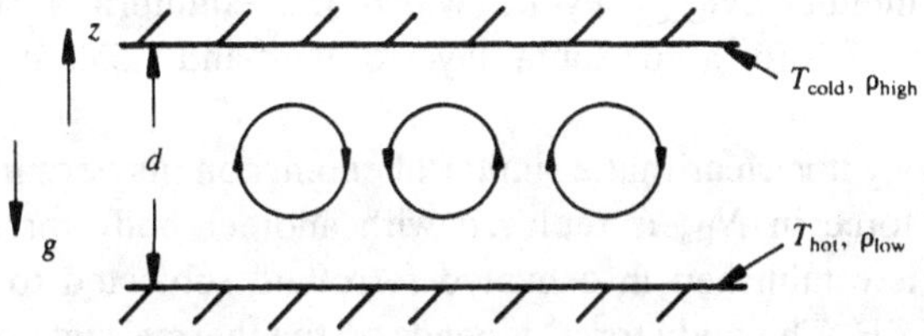

7.29 Description of and nomenclature for the thermal instability of a layer heated from below. Convection cells appear when the Rayleigh number slightly exceeds the critical value.

maintained at a temperature T_{hot} and the top plate at T_{cold}, with $T_{hot} - T_{cold} \equiv \Delta T$. If the temperature gradient across the gap is not too large, the fluid remains quiescent. In this steady state the temperature varies linearly across the gap, and consequently the temperature gradient in the fluid is everywhere uniform and given by

$$dT/dz = \Delta T/d \equiv A_0 \tag{7.163}$$

However, as the temperature difference between the two plates is increased, buoyancy effects start to become important. Owing to thermal expansion, the hotter portion of the fluid has a smaller body force acting on it per unit volume than does the colder fluid. The fluid may then be considered top-heavy, subject to a tendency to redistribute itself to offset this imbalance. The quiescent steady state is maintained until a dimensionless number, which is a function of the buoyancy, viscous forces, and heat transport by conduction and convection, exceeds a certain critical value. This dimensionless number is the *Rayleigh number*; in ordinary fluids acted upon only by the gravitational body force, it is given by

$$\begin{aligned} N_{\mathrm{Ra}} &= \frac{\text{buoyancy force}}{\text{viscous force}} \cdot \frac{\text{heat transport by convection}}{\text{heat transport by conduction}} \\ &= \frac{\beta_0 \rho g d\,\Delta T}{\eta u_0/d} \cdot \frac{u_0 c_0\,\Delta T}{k\,\Delta T/d} \\ &= \beta_0 \rho g c_0 A_0 d^4/\eta k \end{aligned} \tag{7.164}$$

where $\beta_0 \equiv -d \ln \rho/dT$ is the coefficient of thermal expansion, c_0 is the heat capacity at constant pressure per unit volume, u_0 is a characteristic velocity, and k is the thermal conductivity. The critical value of the

Rayleigh number $N_{Ra,0}$ beyond which the equilibrium solution is unstable is 1,708 for a horizontal layer of fluid and 1,558 for a vertical layer.

By analogy it is clear that a similar phenomenon may occur when the buoyancy force in N_{Ra} is replaced with another body force. Such a condition is established in a heated ferrofluid subjected to the body force $\mu_0 M \nabla H$. The body force depends on the thermal state of the fluid because $M = M(T, H)$ with $(\partial M/\partial T)_H < 0$. The physics can be readily understood with the aid of Figure 7.30, which illustrates the convective instability of a magnetic liquid. Let a magnetic field gradient be established across the gap, as shown in Figure 7.30, and an increase of temperature be established in the direction of ∇H. The colder fluid is more strongly magnetized, so it is drawn to the higher-field region, displacing the warmer fluid. The resultant flow is another example of magnetocaloric energy conversion, discussed in Chapter 6.

The thermoconvective instability of magnetic fluids can be investigated by a small-signal analysis through linearizing the equation of motion, the equation of continuity, the equation of heat conduction, and the appropriate Maxwell's equations. There have been numerous studies in this area since 1970; whenever analytical solutions are possible, a generalized Rayleigh number can be obtained that predicts the transition from a quiescent equilibrium state to a stationary convective motion. This generalized Rayleigh number is given by

$$N_{Ra} = \frac{c_0 A_0 d^4}{\eta k} \times \begin{cases} \beta_0 \rho g & [1] \\ \mu_0 M k_0 G_0 & [2] \\ \mu_0 M^2 k_0^2 A_0/(1 + \chi_t) & [3] \end{cases} \tag{7.165}$$

Here $G_0 = -dH/dz$ is the field gradient, which is constant throughout the fluid layer, $Mk_0 = -(\partial M/\partial T)_H$ is the pyromagnetic coefficient K encountered in Chapter 6, and $\chi_t \equiv (\partial M/\partial H)_T$. These results will now be summarized by considering each case in (7.165) individually.

1. *Only the buoyant force mechanism is operative.* This is the case discussed at the beginning of this section. The stimulus for 80 years of extensive research in convective instability can be traced to the classic experiments in 1900 of Bénard, who reported the cellular circulation patterns that occur in shallow (0.5–1 mm) pools of liquid heated from below. Inspired by Bénard's observations, Rayleigh (1916) analyzed the buoyancy-

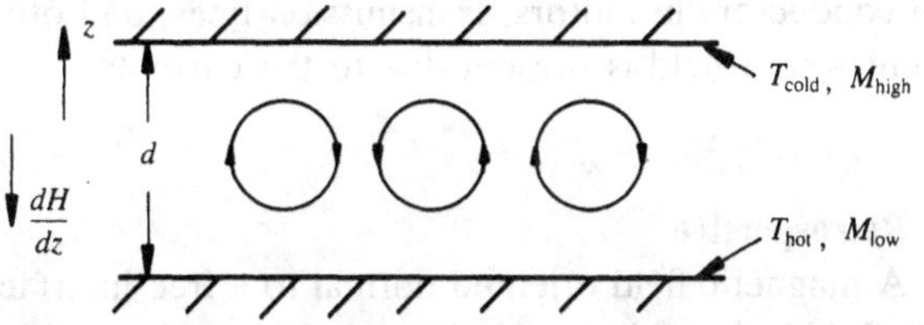

7.30 Convective instability of a magnetic liquid in the presence of an applied magnetic field gradient and the absence of a gravitational field.

driven instability by small-signal analysis and determined the critical Rayleigh number $N_{Ra,0}$ to be about 1700. However, it is now known that the cellular convective motions observed by Bénard were the result of surface-tension gradients rather than buoyancy forces (see, e.g., Pearson 1958; Sternling and Scriven 1959; Scriven and Sternling 1964) because the top boundary in his experiments was a free surface where surface tension forces could become operative owing to local variations in the surface temperature.

2. *The applied magnetic field has gradient G_0 that is constant throughout the fluid layer and the buoyancy has been turned off.* It is also assumed that adiabatic expansion and magnetocaloric cooling are negligible. This is the form of the modified Rayleigh number given by Lalas and Carmi (1971) and Curtis (1971). An analysis, including the effects of adiabatic expansion and magnetocaloric cooling, has been given by Shliomis (1974a). Instability is again indicated by the following criteria:

$$N_{Ra} > N_{Ra,0}$$

$$N_{Ra,0} = \begin{cases} 1{,}708 & \text{(horizontal layer)} \\ 1{,}558 & \text{(vertical layer)} \end{cases} \tag{7.166}$$

3. *The applied field is uniform but the magnetically induced temperature change is appreciable.* This problem has been analyzed by Finlayson (1970), who found that for values of $M > 3 \times 10^5\ \text{A} \cdot \text{m}^{-1}$ the critical value of the Rayleigh number approaches 1,708 for any value of χ_t.

It would appear that the enhancement of convective heat transfer in magnetic fluids could have applications in the cooling of current-carrying conductors in motors, transmission lines, and other electrical equipment where field is present due to the currents.

7.10 Retrospective

A magnetic field oriented normal to a free interface of a magnetizable fluid is destabilizing. In its pure form, this is evidenced by the sudden appearance of a uniform array of cells on a flat interface of large extent. In comparison, a magnetic field having a tangential orientation relative to an interface of great extent in the field direction is stabilizing. Examples of the stabilizing influence are the damping of standing waves in a vibrated vessel, the prevention of fingering in viscous-dominated flow, and the stability behavior of a fluid jet in a collinear magnetic field. Curiously, a tangential field is destabilizing to thin layers and leads to the formation of the labyrinthine and related instabilities. The behavior is a result of the field gradients accompanying the fringe field that exists at the edge of such layers.

Certain of the fluid configurations exhibit as an essential feature a critical point for the onset of instability. A critical point exists for the normal-field instability, the axisymmetric column stabilized by an azimuthal field, and the labyrinthine instability. In other cases, a dichotomy of wavelengths always exists, wavelengths in one range being stable and those in the other range being unstable, with the demarcation between the ranges mediated by the magnetic-field intensity. This occurs in Rayleigh–Taylor instability, porous-media flow, and the axisymmetric jet.

In every case, when the scale of the apparatus is smaller than that of the smallest unstable-disturbance mode, the system exhibits effective stability.

Another aspect of stabilization, the bulk instability of a variable-density medium, will be considered in Chapter 9, when the behavior of magnetized fluidized solids is discussed.

Addendum: representation of disturbance waves

The mathematical representation of a disturbance wave is frequently used in this work. Although the concept is itself simple, neglecting to define terms has perplexed generations of students. Interest in describing a wave stems from the assumption that initially a train of waves of infinitesimally small amplitude is suddenly produced in a fluid. A spectrum of such small disturbances will always be present as noise resulting from statistical fluctuations, flow irregularities, and other sources. The development of a single Fourier component or wave of the noise may be considered in isolation owing to the linearity of the equations. Then the motion of a wave through the fluid and its change in amplitude with respect to time may be computed analytically. If the ampitude increases exponentially, the flow is said to be unstable with respect to that disturbance mode.

The disturbance ξ may be represented by the following expression for a plane wave, sometimes referred to as a normal mode:

$$\xi(x, y, t) = \xi_0 \cos(\omega t - k_x x - k_y y) \tag{7.167}$$

For concreteness, $\xi(x, y, t)$ may be pictured as the surface elevation of a fluid interface relative to its undisturbed (flat) shape. The rectangular coordinate x, y in the reference plane identify a point on the surface, and t is time. ξ is the size of the disturbance at the point, ξ_0 is the amplitude or maximum value of disturbance, ω is the angular frequency, and k_x and k_y are wave-number components. In other contexts $\xi(x, y, t)$ may represent the variation of a velocity component, the pressure, a magnetic-field component, or another variable. Except when representing the surface position, the parameters mentioned may vary continuously with position in the direction z orthogonal to x and y. Thus ξ_0 may be taken as a function of z such that $\xi_0 = \xi_0(z)$.

The surface wave propagates in a particular direction. Representing a point in the plane of propagation of the wave by the position vector $\mathbf{r} = \mathbf{i}x + \mathbf{j}y$ and defining the wave vector $\boldsymbol{k} = k_x\mathbf{i} + k_y\mathbf{j}$, one sees that the scalar product of $\boldsymbol{k}$ with $\mathbf{r}$ gives $\boldsymbol{k}\cdot\mathbf{r} = k_x x + k_y y = ks$. Here s is the distance along the direction of propagation $\boldsymbol{k}$, and $k = |\boldsymbol{k}| = (k_x^2 + k_y^2)^{1/2}$ is the *wave number* and measures the number of cycles per unit length. When the wave number is measured in units of radians per meter, $\lambda \equiv 2\pi/k$ is the *wavelength* of the disturbance in meters. When the argument $\omega t - ks$ of the cosine function is constant, then $\xi(x, y, t)$ is constant for a particular value of z. Taking the constant to be zero corresponds to considering a position on the crest or antinode of the wave, for which $\omega t - ks = 0$ or $s/t = \omega k$. Thus, the expression for the wave represents a traveling disturbance having propagation velocity $v = \omega/k$.

Based on this discussion, the disturbance wave may alternatively be written

$$\xi(x, y, t) = \xi_0(z)\cos(\omega t - \boldsymbol{k}\cdot\mathbf{r}) = \xi_0(z)\cos(\omega t - ks) \tag{7.168}$$

Drastic simplification in the algebraic manipulation required results if exponentials are introduced in place of trigonometric functions in accord with *Euler's formula*, $\exp i\theta = e^{i\theta} = \cos\theta + i\sin\theta$, with the understanding that the real part of the exponential is to be retained. The symbol i denotes $\sqrt{-1}$, the *imaginary number*. With these conventions the disturbance may be represented as

$$\xi = \xi_0(z)\,\mathrm{Re}\exp[i(\omega t - \boldsymbol{k}\cdot\mathbf{r})] \qquad (7.169)$$

where Re denotes the real real part of a complex number.

If ω and k are real-valued constants then the traveling wave has temporally constant amplitude. However, if complex values $\omega = \gamma - i\nu$ with γ and ν both real are admitted, the expression for the wave acquires additional versatility. In this case

$$\xi = \xi_0\,\mathrm{Re}\exp[i(\omega t - \boldsymbol{k}\cdot\mathbf{r})] = \xi_0(z)e^{\nu t}\cos(\gamma t - ks) \qquad (7.170)$$

The factor $e^{\nu t}$ yields a time-varying amplitude of the wave form. Indeed, a main objective of instability theory is determining the time evolution of a test disturbance. It is of interest, therefore, to categorize the types of behavior that can arise. Because both ν and γ can be negative, zero, or positive, there are $3^2 = 9$ combinations arising. These cases are depicted in Table 7.1.

A wave for which $\nu < 0$ decays in amplitude with time and so connotes a stable system. The opposite behavior occurs for $\nu > 0$: The disturbance grows in amplitude with time, and the system is said to be unstable. With $\nu = 0$ the system is in a state of marginal stability.

The value of γ indicates the oscillatory nature of disturbances. Then $\gamma < 0$ yields waves propagating toward smaller values of the s-coordinate axis (left propagating waves), $\gamma > 0$ gives oppositely or right-propagating waves, and $\gamma = 0$ indicates a nonoscillatory disturbance.

With Fourier analysis any physical-disturbance form can be represented as a superposition of sine and cosine waves of varying wavelength or wave number. Normally it suffices to analyze the fate of one (general) harmonic, from which one can determine the wave number possessing the fastest rate of growth, as well as other information.

A special disturbance having physical interest is the superposition of oppositely traveling waves of the same wavelength. Such a superposition of progressive waves is described by the expression

$$\xi_0 e^{\nu t}[\cos(\gamma t - ks) + \cos(\gamma t + ks)] = 2\xi_0 e^{\nu t}\cos\gamma t\cos ks \qquad (7.171)$$

and represents a pattern of *standing waves*. The standing waves have no direction of propagation. Table 7.1 also illustrates the nature of standing

Table 7.1 *Classification of surface waves*[a]

	Oscillatory (left traveling) $\gamma < 0$	Nonoscillatory $\gamma = 0$	Oscillatory (right traveling) $\gamma > 0$	Standing (superposition of left- and right-traveling waves)
$\nu < 0$; decaying (stable)				
$\nu = 0$; persistent (neutrally stable)				
$\nu > 0$; growing (unstable)				

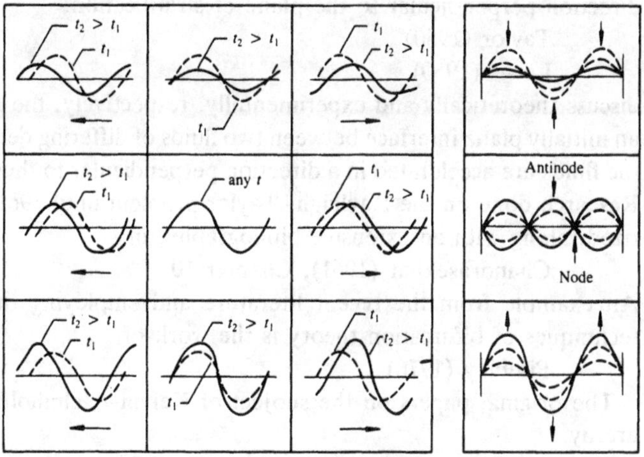

[a] $\xi = \xi_0 \operatorname{Re} \exp[i(\omega t - \boldsymbol{k}\cdot\mathbf{r})] = \xi_0 e^{\nu t} \cos(\gamma t - ks)$; $\omega = \gamma - i\nu$.

waves. There are three types, according to whether the amplitude increases, decreases, or remains constant with time.

Comments and supplemental references

An authoritative reference on the subject of hydrodynamic stability is the book by

Chandrasekhar (1961)

and an excellent treatment with recent developments is that of

Drazin and Reid (1981)

The analog of normal-field instability occurs at the free surface of a liquid dielectric in a constant vertical electric field and has been studied experimentally by

Taylor and McEwan (1965)

and theoretically by

Melcher (1963)

The patterns resulting from the normal-field instability of a ferrofluid offer a means for direct visual observations of phenomena related to two-dimensional melting, including the movement of lattice defects; see

Skjeltorp (1983)

The fundamental paper in the subject of Rayleigh–Taylor instability is that of

Rayleigh (1883)

which considers the stability of a heterogenous fluid accelerated in a direction perpendicular to the plane of stratification.

Taylor (1950)

Lewis (1950)

discuss theoretically and experimentally, respectively, the behavior of an initially plane interface between two fluids of differing densities when the fluids are accelerated in a direction perpendicular to their interface. Research done on the Rayleigh–Taylor problem until 1961 is summarized, along with an extensive bibliography, in

Chandrasekhar (1961), Chapter 10

An example from the recent literature and employing the powerful techniques of bifurcation theory is the work of

Pimbley (1976)

The original papers on the subject of Kelvin–Helmholtz instability are by

Helmholtz (1868)

Kelvin (1910)

Helmholtz's discussion is largely qualitative, whereas Kelvin's is analytical and exceptionally complete.

The magnetic Kelvin–Helmholtz problem was originally treated abstractly in the monograph of

Melcher (1963)

Magnetic influence of instabilities in boiling phase change are inherent in the study of

Papell and Faber (1966)

Many other stability problems of a magnetic fluid have been treated and can be culled from the extensive bibliographies of

Zahn and Shenton (1980)

Charles and Rosensweig (1983)

which serve also as sources for information on all aspects of ferrohydrodynamics.

8

MAGNETIC FLUIDS AND ASYMMETRIC STRESS

In treating dynamic flow it has been assumed up to this point that the magnetization **M** is collinear with the magnetic field **H**, as would be the case in static equilibrium. Collinearity is realized to a very good approximation for subdomain particles of sufficiently small size that superparamagnetic behavior is achieved. The direction of **M** then rotates freely within the solid particle. For particles in a larger size range, the magnetic moment is locked to the orientation of the particle. Then if **H** shifts its orientation, **M** responds by the slower process of particle rotation, which is resisted by fluid viscous-drag torque. The result is that the product $\mu_0 \mathbf{M} \times \mathbf{H}$ has some finite value and so constitutes a body couple.

It will be shown that when body couple is present the viscous stress tensor is no longer symmetric, and novel flow fields arise owing to a state of *asymmetric stress*.

8.1 **Phenomena**

Figure 8.1 illustrates a diversity of phenomena arising because of the rotation of magnetic particles relative to the liquid matrix of a colloidal ferrofluid. This may be observed, for example, if a beaker of magnetic fluid is subjected to a *rotating* magnetic field, with the result depicted in Figure 8.1c. The changing field orientation induces a swirling flow pattern, as was found by Moskowitz and Rosensweig (1967). Additional details and data resulting from the experiments are shown in Figures 8.2 and 8.3. In addition, in any flow possessing vorticity (i.e., fluid rotational motion) and subjected to a *steady* magnetic field, some degree of antisymmetric stress will be present; in viscometric uniform shear flow in a magnetic field (see Figure 8.1e), Rosensweig, Kaiser,

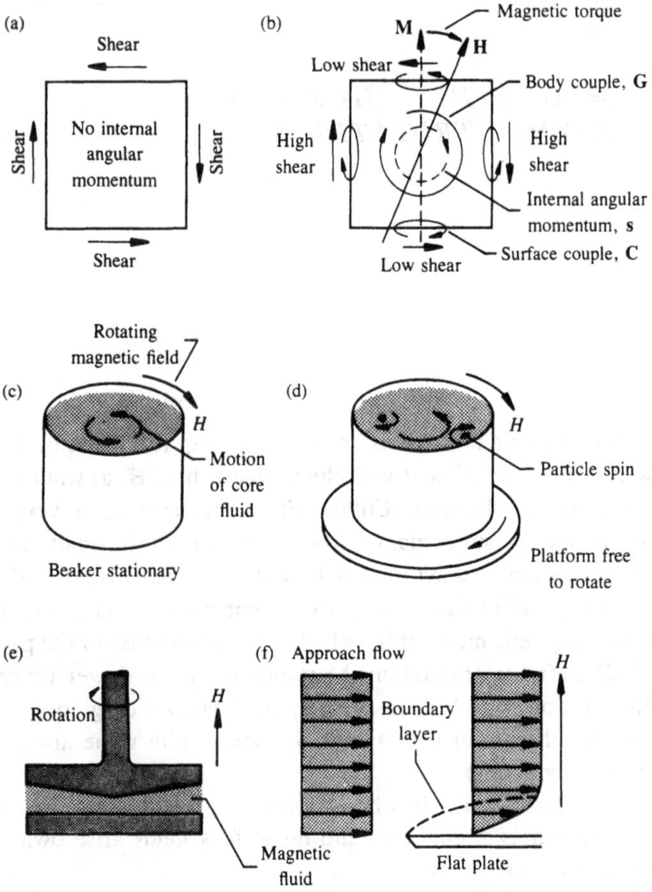

8.1 Asymmetric stresses due to the magnetic torque are set up in a ferrofluid when an imposed magnetic field and the fluid magnetization are oriented in different directions; see (a) and (b). The stresses are due to rotation of the magnetic particles relative to the fluid carrier. The torque may establish a viscous flow, as in (c) or (d). Alternatively, as in (e) and (f), a viscous flow may excite a torque of magnetic origin. The unequal surface shear forces in (b) are due to nonvanishing **A**, the rate of interchange of external and internal angular momentum.

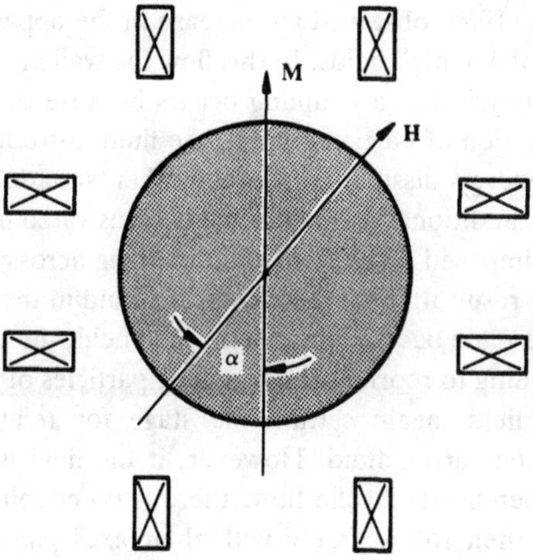

8.2 Top view of a ferrofluid in a beaker. Alternating current is supplied in phase quadrature to the two pairs of field coils to produce a uniform rotating field. The magnetization lags the applied field as shown.

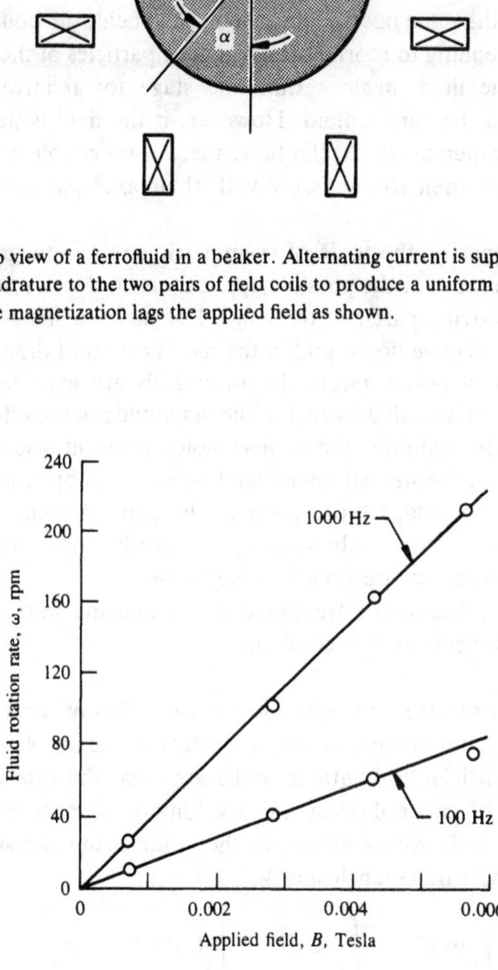

8.3 Data from the experiment of Moskowitz and Rosensweig (1967).

and Miskolczy (1969) observed an increase in the apparent viscosity of up to a factor of 4 in high fields. In this flow the wall moves and the field is stationary. Physically, a coupling occurs because the magnetic field pins the orientation of particles within the fluid, introducing additional friction and energy dissipation. Yet another situation in which the mechanism for additional dissipation acts is illustrated in Figure 8.1f. A *steady* field is imposed on the ferrofluid moving across a *stationary* flat plate, with the result that the velocity of the fluid in the boundary layer adjacent to the plate becomes coupled to the field; the boundary flow is rotational, tending to reorient the magnetic particles of the ferrofluid in the magnetic field, again setting the stage for relative rotation of particles and the carrier fluid. However, if the field is parallel to the plate and perpendicular to the flow, there is no coupling, because the particles may then rotate freely with their axes parallel to the field direction.

Technologically, the topic of asymmetric stress in magnetic fluids is intriguing because of the possible applications to devices such as pumps having no moving parts, in the control of heat- or mass-transfer processes in convective flows, and in the modification of drag in boundary layers. Certain polymers and liquid crystals are expected to display these effects to a small degree, but the magnitude of the effect is large in ferrofluids. In addition, the subject holds inherent scientific interest.

The present theoretical understanding in this subject is incomplete. This chapter develops those parts of the general theory that appear rigorous and correct. Subsequently, in treating particular problems, areas of uncertainty are noted as they arise.

Section 8.2 begins the treatment by introducing some fundamental ideas in the mechanics of continua.

8.2 Cauchy stress principle and conservation of momentum

The continuum-mechanical postulate extends Newton's second law from particles to continua and states that the rate of change of momentum of material occupying a volume V of space is given by the sum of the body forces acting on the volume and the surface forces acting on the surface enclosing V:

$$\frac{D}{Dt}\int_V \rho \mathbf{v}\, dV = \int_V \rho \mathbf{F}\, dV + \oint_S \mathbf{t}_n\, dS \tag{8.1}$$

where ρ is the density of the continuum, $\mathbf{v}$ is the velocity, $\mathbf{F}$ is the external body force per unit mass (p.u.m.), and $\mathbf{t}_n$ is the familiar traction

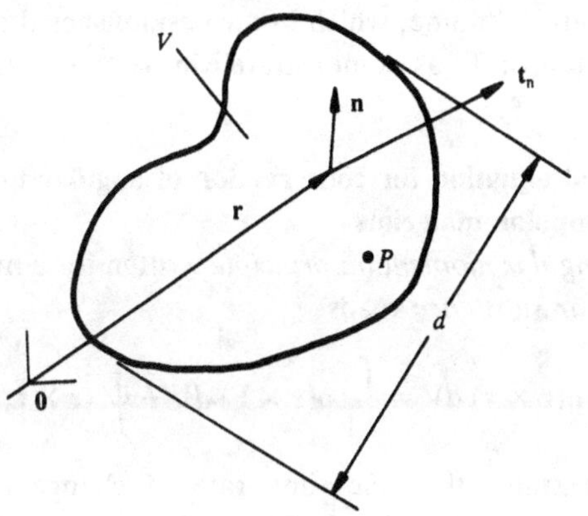

8.4 Volume V with characteristic linear dimension d.

vector. Equation (8.1) is called the *Cauchy stress principle* or *integral equation for conservation of linear momentum*.

Refer now to the volume shown in Figure 8.4. It is desired to obtain order-of-magnitude estimates for each of the terms in (8.1) as the volume is shrunk to a point P while its shape is preserved. If the characteristic linear dimension of V is d, then as $d \rightarrow 0$ the magnitude of each term in (8.1) apparently behaves as

$$\frac{D}{Dt}\int_V \rho\mathbf{v}\,dV \sim d^3 \qquad (8.2)$$

$$\int_V \rho\mathbf{F}\,dV \sim d^3 \qquad (8.3)$$

$$\oint_S \mathbf{t}_n\,dS \sim d^2 \qquad (8.4)$$

In the limit, the volume terms are negligible compared to the surface term. This permits the conclusion that

$$\lim_{d\rightarrow 0}\frac{1}{d^2}\oint \mathbf{t}_n\,dS = \mathbf{0} \qquad (8.5)$$

That is, the *stresses must be in local equilibrium. Local equilibrium of surface tractions holds for polar as well as nonpolar substances.* Equation (8.5) justifies writing a static force balance on an indefinitely small tetrahedral control volume, which in turn establishes the existence of a surface stress tensor **T**, as demonstrated in Section 1.7.

8.3 Integral equation for conservation of angular momentum for nonpolar materials

The *angular momentum principle* written for a material volume V of a *nonpolar substance* reads

$$\frac{D}{Dt}\int_V \rho(\mathbf{r}\times\mathbf{v})\,dV = \int_V \rho(\mathbf{r}\times\mathbf{F})\,dV + \oint_S (\mathbf{r}\times\mathbf{t}_n)\,dS \tag{8.6}$$

This equation states that the time rate of change of the angular momentum of the matter that is found at a given instant within the volume V is given as the sum of the moments exerted by the body force acting throughout the volume and the surface stress acting over the enclosing surface.

Carrying out an order-of-magnitude comparison between each of the terms in (8.6) and permitting the volume to shrink on a point leads to the apparent conclusion that

$$\lim_{d\to 0}\frac{1}{d^2}\oint_S (\mathbf{r}\times\mathbf{t}_n)\,dS = \mathbf{0} \tag{8.7}$$

Thus, *moments of surface stresses are locally in equilibrium in nonpolar fluids*. Equation (8.7) implies that the stress tensor **T**, regardless of its physical origin, is symmetric in a nonpolar fluid; i.e. $\mathbf{T} = \mathbf{T}^T$, or what is the same thing, $T_{ij} = T_{ji}$. A heuristic proof of the symmetry of the stress tensor is given next, and a more rigorous proof will follow.

The symmetry of the stress tensor can be demonstrated by the following line of reasoning. To simplify the proof (8.7) is written for a two-dimensional surface, but the proof can easily be extended to three dimensions.

Consider an infinitesimally small element of area of dimensions δx by δy in rectangular coordinates (Figure 8.5). Take the origin of the coordinate system to be at the center of the element. Taylor-series expansions of the stress components about the origin are given to first-order as

$$T_{xy}(\delta x/2, 0) = T_{xy}(0,0) + \frac{\delta x}{2}\left(\frac{\partial T_{xy}}{\partial x}\right)_{0,0} \tag{8.8}$$

$$T_{yx}(0, \delta y/2) = T_{yx}(0,0) + \frac{\delta y}{2}\left(\frac{\partial T_{yx}}{\partial y}\right)_{0,0} \tag{8.9}$$

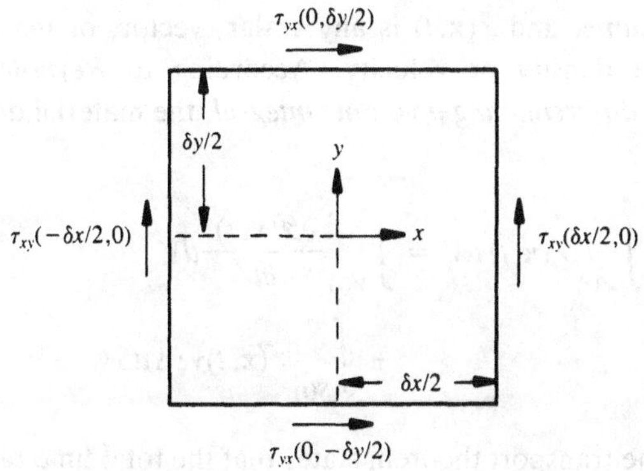

8.5 **Infinitesimal element over which the moment-of-momentum balance is made. Only those components of the stress tensor that contribute to the moment equation are shown.**

$$T_{xy}(-\delta x/2, 0) = T_{xy}(0,0) - \frac{\delta x}{2}\left(\frac{\partial T_{xy}}{\partial x}\right)_{0,0} \tag{8.10}$$

$$T_{yx}(0, -\delta y/2) = T_{yx}(0,0) - \frac{\delta y}{2}\left(\frac{\partial T_{yx}}{\partial y}\right)_{0,0} \tag{8.11}$$

Thus, for example, the force on the top face is $T_{yx}(0, \delta y/2)\,\delta x$, and the associated moment around the origin is $T_{yx}(0, \delta y/2)\,\delta x\,\delta y/2$; on the bottom face the force is $T_{yx}(0, -\delta y/2)\,\delta x$ and the associated moment is $T_{yx}(0, -\delta y/2)\,\delta x\,(-\delta y/2)$. When a balance of all the moments around the origin is formulated using the local equilibrium concept, the higher-order terms cancel, giving the result that

$$T_{xy} = T_{yx}$$

so the stress-tensor components are equal. The argument may be repeated for the other directions, and therefore the stress tensor is symmetric. Because the shear-stress components vary with position over a single face, a more elaborate derivation is needed to verify rigorously that the stress tensor is symmetric; the topic is treated in Section 8.6.

8.4 Reynolds' transport theorem

Before proceeding further it will be convenient to introduce a helpful theorem. Consider the integral $I \equiv \int_{V(t)} \mathscr{F}(\mathbf{x}, t)\, dV$, where $V(t)$ is a closed volume consisting of the same fluid particles at all times, i.e., a *material volume*, and $\mathscr{F}(\mathbf{x}, t)$ is any scalar, vector, or tensor function, such as the density or velocity. According to *Reynolds' transport theorem for differentiating a volume integral*, the material derivative of I is given by

$$\frac{D}{Dt}\int_{V(t)} \mathscr{F}(\mathbf{x}, t)\, dV = \int_{V(t)} \frac{\partial \mathscr{F}(\mathbf{x}, t)}{\partial t}\, dV + \oint_{S(t)} \mathscr{F}(\mathbf{x}, t)\mathbf{v}\cdot\mathbf{n}\, dS \tag{8.12}$$

In words, the transport theorem states that the total time rate of change of the integral over a volume of a quantity $\mathscr{F}$ is given by the sum of the local rate of change throughout the volume and the net rate of change due to the surface overtaking adjacent regions. The surface integral in (8.12) can be transformed into a volume integral with the aid of the divergence theorem:

$$\oint_{S(t)} \mathscr{F}(\mathbf{x}, t)\mathbf{v}\cdot\mathbf{n}\, dS = \int_{V(t)} \nabla\cdot(\mathscr{F}\mathbf{v})\, dV = \int_{V(t)} \mathscr{F}(\nabla\cdot\mathbf{v})\, dV + \int_{V(t)} \mathbf{v}\cdot\nabla\mathscr{F} dV \tag{8.13}$$

where the latter equality was obtained through the use of the vector identity $\nabla\cdot(\mathscr{F}\mathbf{v}) = \mathscr{F}(\nabla\cdot\mathbf{v}) + \mathbf{v}\cdot\nabla\mathscr{F}$ and hence is valid only when $\mathscr{F}$ is a scalar function. Substituting (8.13) into (8.12) and using (4.57) gives

$$\frac{D}{Dt}\int_{V(t)} \mathscr{F}(\mathbf{x}, t)\, dV = \int_{V(t)} \left[\frac{D\mathscr{F}}{Dt} + \mathscr{F}(\nabla\cdot\mathbf{v})\right] dV \tag{8.14}$$

If $\mathscr{F} = \rho\psi$, where ψ is an arbitrary function, (8.14) can be manipulated into the form

$$\frac{D}{Dt}\int_{V(t)} \rho\psi\, dV = \int_{V(t)} \left[\rho\frac{D\psi}{Dt} + \psi\left(\frac{D\rho}{Dt} + \rho\nabla\cdot\mathbf{v}\right)\right] dV \tag{8.15}$$

Putting $\psi = 1$ in (8.15) gives an expression for the rate of change of mass for a material control volume:

$$\frac{D}{Dt}\int_{V(t)} \rho\, dV = \int_{V(t)} \left(\frac{D\rho}{Dt} + \rho\nabla\cdot\mathbf{v}\right) dV \tag{8.16}$$

By its very definition, however, the mass of a material control volume is constant, and hence

$$\frac{D}{Dt}\int_{V(t)} \rho\, dV = 0 = \int_{V(t)} \left(\frac{D\rho}{Dt} + \rho \nabla\cdot\mathbf{v}\right) dV \qquad (8.17)$$

By the usual line of reasoning involving the arbitrariness of the volume, it is concluded that the integrand in the second equality must be identically zero:

$$D\rho/Dt + \rho\nabla\cdot\mathbf{v} = 0 \qquad (8.18a)$$

or

$$\partial\rho/\partial t + \nabla\cdot(\rho\mathbf{v}) = 0 \qquad (8.18b)$$

Equations (8.18) have recovered the *equation of continuity of mass.*

Substituting (8.18a) in (8.15) yields an equation of wide applicability:

$$\frac{D}{Dt}\int_{V(t)} \rho\psi\, dV = \int_{V(t)} \rho\frac{D\psi}{Dt}\, dV \qquad (8.19)$$

Because ψ can be any scalar, vector, or tensor component, (8.19) applies when ψ is considered to represent any scalar, vector, or tensor function.

8.5 Cauchy equation of motion

A very general equation valid for any kind of continuum is now derived, the *Cauchy equation of motion.* The surface integral in (8.1) can be rewritten as follows [see Section 1.7 and the remark following (8.5)]:

$$\oint_S \mathbf{t}_n\, dS = \oint_S \mathbf{n}\cdot\mathbf{T}\, dS = \int_V \nabla\cdot\mathbf{T}\, dV \qquad (8.20)$$

and from the Reynolds' transport theorem in the version given by (8.19), it is seen that the first term of (8.1) is equivalent to $\int_V \rho(D\mathbf{v}/Dt)\, dV$. When the latter result along with (8.20) is substituted back in (8.1), one finds, invoking once more the arbitrariness of the volume of integration, that

$$\rho(D\mathbf{v}/Dt) = \rho\mathbf{a} = \rho\mathbf{F} + \nabla\cdot\mathbf{T} \qquad (8.21)$$

where $\mathbf{a} \equiv D\mathbf{v}/Dt$ is the acceleration of a fluid element. The utility of the Cauchy equation of motion, (8.21), arises from the fact that it is free of constitutive assumptions.

8.6 Symmetry of the stress tensor for nonpolar fluids

With the aid of (8.19), letting ψ represent $\mathbf{r} \times \mathbf{v}$, the left side of equation (8.6) can be rewritten

$$\frac{D}{Dt}\int_V \rho(\mathbf{r} \times \mathbf{v})\, dV = \int_V \rho \frac{D}{Dt}(\mathbf{r} \times \mathbf{v})\, dV \tag{8.22}$$

Because

$$\frac{D}{Dt}(\mathbf{r} \times \mathbf{v}) = \frac{D\mathbf{r}}{Dt} \times \mathbf{v} + \mathbf{r} \times \frac{D\mathbf{v}}{Dt} = \mathbf{v} \times \mathbf{v} + \mathbf{r} \times \mathbf{a} = \mathbf{r} \times \mathbf{a}$$

(8.6) becomes

$$\int_V \rho(\mathbf{r} \times \mathbf{a})\, dV = \int_V \rho(\mathbf{r} \times \mathbf{F})\, dV + \oint_S (\mathbf{r} \times \mathbf{t}_n)\, dS \tag{8.23}$$

The surface integral can be cast into the following more useful form:

$$\begin{aligned}\oint_S (\mathbf{r} \times \mathbf{t}_n)\, dS &= \oint_S \mathbf{r} \times (\mathbf{n}\cdot\mathbf{T})\, dS \\ &= -\oint_S (\mathbf{n}\cdot\mathbf{T}) \times \mathbf{r}\, dS \\ &= -\oint_S \mathbf{n}\cdot(\mathbf{T} \times \mathbf{r})\, dS \\ &= -\int_V \nabla\cdot(\mathbf{T} \times \mathbf{r})\, dV \end{aligned} \tag{8.24}$$

In the following, use is made of the so-called *alternator* or *alternating unit tensor* $\boldsymbol{\varepsilon}$

$$\boldsymbol{\varepsilon} \equiv \mathbf{e}_i\mathbf{e}_j\mathbf{e}_k\, \varepsilon_{ijk} \tag{8.25}$$

where

$$\varepsilon_{ijk} = \begin{cases} +1 & \text{if } ijk = 123,\ 231, \text{ or } 312 \\ -1 & \text{if } ijk = 321,\ 132, \text{ or } 213 \\ 0 & \text{if } i = j,\ i = k, \text{ or } j = k \end{cases} \tag{8.26}$$

and $\mathbf{e}_i, \mathbf{e}_j, \mathbf{e}_k$ are unit vectors in the i, j, k directions, respectively. Thus

$$\boldsymbol{\varepsilon} = (\mathbf{e}_1\mathbf{e}_2\mathbf{e}_3 + \mathbf{e}_2\mathbf{e}_3\mathbf{e}_1 + \mathbf{e}_3\mathbf{e}_1\mathbf{e}_2 - \mathbf{e}_1\mathbf{e}_3\mathbf{e}_2 - \mathbf{e}_2\mathbf{e}_1\mathbf{e}_3 - \mathbf{e}_3\mathbf{e}_2\mathbf{e}_1)$$

The alternator is an example of a *polyadic*, and it is noted that the vector cross product is given by $\mathbf{e}_i \times \mathbf{e}_j = \varepsilon_{ijk}\mathbf{e}_k$.

Next it will be shown that

$$-\nabla\cdot(\mathbf{T}\times\mathbf{r}) = \mathbf{r}\times(\nabla\cdot\mathbf{T}) + \mathbf{A} \tag{8.27}$$

where **A** is a *pseudovector* given by

$$\mathbf{A} = \varepsilon_{ijk}\mathbf{e}_i T_{jk} \tag{8.28}$$

Proof: In indicial notation $\nabla\cdot(\mathbf{T}\times\mathbf{r})$ can be expanded as

$$\begin{aligned}
\nabla\cdot(\mathbf{T}\times\mathbf{r}) &= \mathbf{e}_i\frac{\partial}{\partial x_i}\cdot(\mathbf{e}_j\mathbf{e}_k T_{jk}\times\mathbf{e}_l x_l)\\
&= (\mathbf{e}_i\cdot\mathbf{e}_j)(\mathbf{e}_k\times\mathbf{e}_l)\left(x_l\frac{\partial T_{jk}}{\partial x_i} + T_{jk}\frac{\partial x_l}{\partial x_i}\right)\\
&= \delta_{ij}\varepsilon_{klm}\mathbf{e}_m\left(x_l\frac{\partial T_{jk}}{\partial x_i} + \delta_{li}T_{jk}\right)\\
&= \varepsilon_{klm}\mathbf{e}_m x_l\frac{\partial T_{jk}}{\partial x_j} + \varepsilon_{kim}\mathbf{e}_m T_{ik}\\
&= -\varepsilon_{lkm}\mathbf{e}_m x_l\frac{\partial T_{jk}}{\partial x_j} - \varepsilon_{mik}\mathbf{e}_m T_{ik}\\
&= -\mathbf{r}\times(\nabla\cdot\mathbf{T}) - \mathbf{A} \qquad \text{QED}
\end{aligned}$$

The nature of this curious vector **A** may be examined as follows:

$$\begin{aligned}
\mathbf{A} &= \varepsilon_{ijk}\mathbf{e}_i T_{jk}\\
&= \varepsilon_{1jk}\mathbf{e}_1 T_{jk} + \varepsilon_{2jk}\mathbf{e}_2 T_{jk} + \varepsilon_{3jk}\mathbf{e}_3 T_{jk}\\
&= \mathbf{e}_1(T_{23} - T_{32}) + \mathbf{e}_2(T_{31} - T_{13}) + \mathbf{e}_3(T_{12} - T_{21})
\end{aligned} \tag{8.29}$$

Thus, **A** is a vector with components $A_1 = T_{23} - T_{32}$, $A_2 = T_{31} - T_{13}$ and $A_3 = T_{12} - T_{21}$. As noted in Appendix 1, any second-order tensor **T** can be written as the sum of a symmetric $\mathbf{T}_s$ and antisymmetric $\mathbf{T}_a$ part:

$$\mathbf{T} = \mathbf{T}_s + \mathbf{T}_a, \qquad \mathbf{T}_s = \tfrac{1}{2}(\mathbf{T} + \mathbf{T}^T), \quad \mathbf{T}_a = \tfrac{1}{2}(\mathbf{T} - \mathbf{T}^T) \tag{8.30}$$

In matrix form $\mathbf{T}_a$ is given by

$$\mathbf{T}_a = \frac{1}{2}\begin{bmatrix} 0 & T_{12} - T_{21} & T_{13} - T_{31}\\ T_{21} - T_{12} & 0 & T_{23} - T_{32}\\ T_{31} - T_{13} & T_{32} - T_{23} & 0\end{bmatrix} \tag{8.31}$$

$$= \frac{1}{2}\begin{bmatrix} 0 & A_3 & -A_2\\ -A_3 & 0 & A_1\\ A_2 & -A_1 & 0\end{bmatrix} \tag{8.32}$$

Because

$$\mathbf{T} = \mathbf{e}_1\mathbf{e}_1 T_{11} + \mathbf{e}_1\mathbf{e}_2 T_{12} + \cdots + \mathbf{e}_3\mathbf{e}_3 T_{33} = \mathbf{e}_i\mathbf{e}_j T_{ij}$$

then

$$\begin{aligned} \text{vec } \mathbf{T} &= \mathbf{e}_1 \times \mathbf{e}_1 T_{11} + \mathbf{e}_1 \times \mathbf{e}_2 T_{12} + \cdots + \mathbf{e}_3 \times \mathbf{e}_3 T_{33} \\ &= \mathbf{e}_1 \times \mathbf{e}_j T_{ij} = \varepsilon_{ijk} T_{ij} \mathbf{e}_k = \varepsilon_{kij} \mathbf{e}_k T_{ij} \\ &= \mathbf{A} \end{aligned} \tag{8.33}$$

Thus, $\mathbf{A}$ is the *vector of the tensor* $\mathbf{T}$.

$\mathbf{A}$ may be found from $\mathbf{T}$, or $\mathbf{T}_a$ may be found from $\mathbf{A}$, using the following relationships employing the polyadic alternator $\boldsymbol{\varepsilon}$:

$$\mathbf{A} = -\boldsymbol{\varepsilon}:\mathbf{T} \tag{8.34}$$

$$\mathbf{T}_a = \tfrac{1}{2}\boldsymbol{\varepsilon}\cdot\mathbf{A} \tag{8.35}$$

Equations (8.34) and (8.35) may be verified by direct expansion following the rules of dyadic algebra and comparing with (8.28) or (8.33) and (8.32). The nesting convention (see Appendix 1) is followed in evaluating the multiple internal product. Note that $\mathbf{A} = \mathbf{0}$ if $\mathbf{T}_a = \mathbf{0}$ and vice versa. Also, $\mathbf{A} = \mathbf{0}$ if $\mathbf{T}$ is a symmetric tensor, and vice versa.

Example: Assuming $\mathbf{A}$ is oriented in the 3 direction so that $\mathbf{A} = A\mathbf{e}_3$, compute the dyadic expression for $\mathbf{T}_a$ from equation (8.35). Verify from (8.34) that $\mathbf{A}$ is recovered from this result.

Solution:

$$\begin{aligned} \mathbf{T}_a &= \tfrac{1}{2}\boldsymbol{\varepsilon}\cdot\mathbf{A} = \tfrac{1}{2}\mathbf{e}_i\mathbf{e}_j\mathbf{e}_k\varepsilon_{ijk}\cdot\mathbf{e}_3 A = \tfrac{1}{2}\mathbf{e}_i\mathbf{e}_j\varepsilon_{ij3}A \\ &= \tfrac{1}{2}(\mathbf{e}_1\mathbf{e}_2 - \mathbf{e}_2\mathbf{e}_1)A \end{aligned}$$

This expression is the dyadic representation of $\mathbf{T}_a$. Substituting the result into (8.34), one obtains

$$\begin{aligned} \mathbf{A} &= -\mathbf{e}_i\mathbf{e}_j\mathbf{e}_k\varepsilon_{ijk}:\tfrac{1}{2}(\mathbf{e}_1\mathbf{e}_2 - \mathbf{e}_2\mathbf{e}_1)A \\ &= -(\mathbf{e}_1\mathbf{e}_2\mathbf{e}_3 + \mathbf{e}_2\mathbf{e}_3\mathbf{e}_1 + \mathbf{e}_3\mathbf{e}_1\mathbf{e}_2 \\ &\quad - \mathbf{e}_1\mathbf{e}_3\mathbf{e}_2 - \mathbf{e}_2\mathbf{e}_1\mathbf{e}_3 - \mathbf{e}_3\mathbf{e}_2\mathbf{e}_1):\tfrac{1}{2}(\mathbf{e}_1\mathbf{e}_2 - \mathbf{e}_2\mathbf{e}_1)A \\ &= [(\mathbf{e}_3\mathbf{e}_2 - \mathbf{e}_2\mathbf{e}_3)\cdot\mathbf{e}_2 + (\mathbf{e}_3\mathbf{e}_2 - \mathbf{e}_1\mathbf{e}_3)\cdot\mathbf{e}_1](A/2) = A\mathbf{e}_3 \end{aligned}$$

Substituting (8.24) and (8.27) into (8.23) and isolating the integral involving the vector $\mathbf{A}$ gives the relationship

$$\int_V \mathbf{r} \times (\rho\mathbf{a} - \rho\mathbf{F} - \nabla\cdot\mathbf{T})\,dV = \int_V \mathbf{A}\,dV \tag{8.36}$$

The quantity inside the parentheses is zero by the Cauchy equation of motion, (8.21), so $\mathbf{A}$ vanishes for a nonpolar fluid. Because $\mathbf{A} = \mathbf{0}$, the antisymmetric stress tensor $\mathbf{T}_a$ is also zero, and from (8.30) it may be concluded that *the stress tensor is symmetric in a nonpolar fluid*, confirming the result obtained in Section 8.3.

What has been accomplished so far is preparation for treating polar fluids, which is the subject broached in the next section.

8.7 **Analysis for polar fluids**

Just as a system may exchange linear momentum with its surroundings, it may also exchange angular momentum. The rate of arrival of linear momentum from distant surroundings is accounted for by $\mathbf{F}$, which is the body force p.u.m. Similarly, the distant surroundings may transmit angular momentum to the system through a *body couple* $\mathbf{G}$ p.u.m. The contiguous surroundings exert a surface traction $\mathbf{t}_n = \mathbf{n} \cdot \mathbf{T}$ per unit area of the surface of the system, which may be accumulated within the system or removed to the surroundings as linear momentum. Analogously, the contiguous surroundings may also exert on the surface of the system a couple $\mathbf{c}_n$ per unit area.

If $\mathbf{L}$ denotes the total local density of angular momentum, then

$$\mathbf{L} = \mathbf{r} \times \mathbf{v} + \mathbf{s} \tag{8.37}$$

where $\mathbf{r} \times \mathbf{v}$ is the *external* or *orbital angular momentum* and $\mathbf{s}$ is the *internal angular momentum* p.u.m., or *spin*. Thus, spin field $\mathbf{s}$ is assigned to the rotation or spin of the magnetic-colloidal particles and viscous fluid that is locally entrained by the particles. There is coupling between the internal and external forms of angular momentum, and, as a first step in determining the nature of this exchange, let the *integral equation for the balance of total angular momentum for polar fluids* be written

$$\frac{D}{Dt}\int_V \rho(\mathbf{s} + \mathbf{r} \times \mathbf{v})\,dV = \int_V (\rho\mathbf{G} + \mathbf{r} \times \rho\mathbf{F})\,dV + \oint_S (\mathbf{c}_n + \mathbf{r} \times \mathbf{t}_n)\,dS \tag{8.38}$$

As mentioned, the term $\mathbf{c}_n$ accounts for the possible presence of a surface-couple density acting on an area dS. More can be learned about the quantity $\mathbf{c}_n$ as follows. Consider an infinitesimal tetrahedral control volume, shown in Figure 8.6, that is fixed in space and such that conditions are sensibly uniform within and on the surface. Application

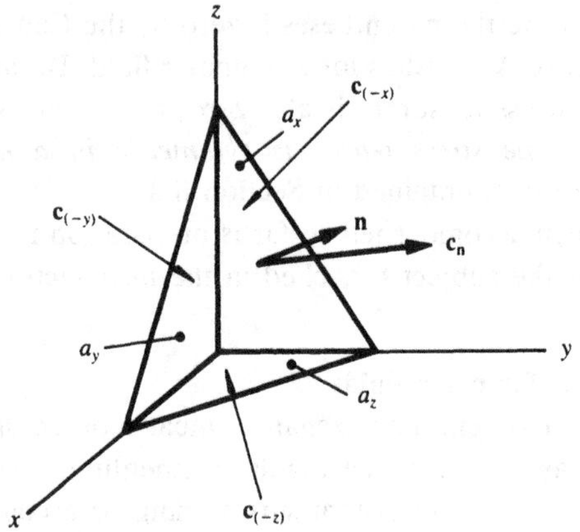

8.6 Tetrahedral control volume subjected to surface couple stress.

of (8.38) to the tetrahedral control volume gives, retaining only terms of highest order,

$$\oint_{S_{\text{tetrah}}} \mathbf{c}_n \, dS = \mathbf{0} \tag{8.39}$$

which can be written, with the aid of the notation introduced in Figure 8.6,

$$a_x \mathbf{c}_{(-x)} + a_y \mathbf{c}_{(-y)} + a_z \mathbf{c}_{(-z)} + a\mathbf{c}_n = \mathbf{0} \tag{8.40}$$

where a_x, a_y, a_z are the areas of the sides of the tetrahedral volume. By similarly applying (8.39) to a flat element of volume (a flake), one finds that

$$\mathbf{c}_{(-n)} = -\mathbf{c}_{(n)} \tag{8.41}$$

With the aid of (8.41) and $a_x = \mathbf{n}\cdot\mathbf{i}a$, $a_y = \mathbf{n}\cdot\mathbf{j}a$, $a_z = \mathbf{n}\cdot\mathbf{k}a$, (8.40) can be arranged into the following form:

$$\mathbf{c}_n = \mathbf{n}\cdot(\mathbf{i}\mathbf{c}_x + \mathbf{j}\mathbf{c}_y + \mathbf{k}\mathbf{c}_z) \tag{8.42}$$

The expression in parentheses is a dyadic called the *couple stress tensor* **C**. Physically, the couples represented by this tensor can arise from diffusive transport of internal angular momentum.

$$\mathbf{C} = \mathbf{i}\mathbf{c}_x + \mathbf{j}\mathbf{c}_y + \mathbf{k}\mathbf{c}_z \tag{8.43}$$

Thus, $\mathbf{c}_n = \mathbf{n}\cdot\mathbf{C}$, and now by the divergence theorem and with the introduction of (8.24), (8.38) may be written

$$\frac{D}{Dt}\int_V \rho(\mathbf{s} + \mathbf{r} \times \mathbf{v})\,dV = \int_V [\rho\mathbf{G} + \mathbf{r} \times \rho\mathbf{F} + \nabla\cdot\mathbf{C} - \nabla\cdot(\mathbf{T} \times \mathbf{r})]\,dV \tag{8.44}$$

or, with (8.27),

$$\frac{D}{Dt}\int_V \rho(\mathbf{s} + \mathbf{r} \times \mathbf{v})\,dV = \int_V [\rho\mathbf{G} + \mathbf{r} \times \rho\mathbf{F} + \nabla\cdot\mathbf{C} + \mathbf{r} \times (\nabla\cdot\mathbf{T}) + \mathbf{A}]\,dV \tag{8.45}$$

Application of (8.19) to the left side of (8.45) and invoking the arbitrariness of the volume of integration produces the *differential equation of change of the total angular momentum*:

$$\rho\frac{D}{Dt}(\mathbf{s}+\mathbf{r}\times\mathbf{v}) = \rho\mathbf{G}+\mathbf{r}\times\rho\mathbf{F}+\nabla\cdot\mathbf{C}+\mathbf{r}\times(\nabla\cdot\mathbf{T})+\mathbf{A} \tag{8.46}$$

Equation (8.46) applies to polar as well as nonpolar fluids.

Consider once more Figure 8.4, this time with the origin of the coordinate system located inside the volume V so that $|\mathbf{r}|$ is of order d. Now, if the volume V is shrunk on a point P as previously, then each term in (8.45) goes to zero as follows:

$$\frac{D}{Dt}\int_V \rho\mathbf{s}\,dV \sim d^3, \qquad \frac{D}{Dt}\int_V \rho(\mathbf{r} \times \mathbf{v})\,dV \sim d^4$$

$$\int_V \rho\mathbf{G}\,dV \sim d^3, \qquad \int_V (\mathbf{r} \times \rho\mathbf{F})\,dV \sim d^4, \qquad \int_V \nabla\cdot\mathbf{C}\,dV \sim d^3 \tag{8.47}$$

$$\int_V \mathbf{r} \times (\nabla\cdot\mathbf{T})\,dV \sim d^4, \qquad \int_V \mathbf{A}\,dV \sim d^3$$

As $d \to 0$, only terms of order d^3 survive, thus yielding the *differential equation of change of the internal angular momentum*:

$$\rho\frac{D\mathbf{s}}{Dt} = \rho\mathbf{G} + \nabla\cdot\mathbf{C} + \mathbf{A} \tag{8.48}$$

Subtraction of (8.48) from (8.46) yields a balance relationship of external angular momentum, which we shall obtain as (8.52b).

Equation (8.48) states that the rate of change of internal angular momentum **s** within a parcel of magnetic fluid has three causes: the exertion of a body couple **G** due to transmission from distant sources; the exertion of a surface couple $\nabla \cdot \mathbf{C}$, which may be regarded as the diffusion of internal angular momentum across the surface; and the exchange of angular momentum between the external and the internal types, a process represented by **A**. Referring to the comment following (8.36), it can be realized that with nonvanishing **A** the state of stress **T** is asymmetric.

Equation (8.48), derived in another manner by Dahler and Scriven (1961), is fundamental and deserving of a second proof, that due to those authors.

Proof: Form the cross product of **r** with the Cauchy equation of motion, (8.21):

$$\mathbf{r} \times \rho D\mathbf{v}/Dt = \mathbf{r} \times \rho\mathbf{F} + \mathbf{r} \times \nabla \cdot \mathbf{T} \tag{8.49}$$

However, as seen in Section 8.6,

$$\mathbf{r} \times \frac{D\mathbf{v}}{Dt} = \frac{D}{Dt}(\mathbf{r} \times \mathbf{v}) \tag{8.50}$$

Substituting (8.50) into (8.49) and subtracting the resulting equation from (8.46), which is the differential equation of change of total angular momentum, gives (8.48). QED

8.8 Summary of the basic laws of continuum mechanics

Cauchy linear-momentum equation:

$$\rho D\mathbf{v}/Dt = \rho\mathbf{F} + \nabla \cdot \mathbf{T} \tag{8.51}$$

External angular momentum equation:

$$\rho\frac{D}{Dt}(\mathbf{r} \times \mathbf{v}) = \mathbf{r} \times \rho\mathbf{F} + \mathbf{r} \times (\nabla \cdot \mathbf{T}) \tag{8.52a}$$

$$= \mathbf{r} \times \rho\mathbf{F} - \nabla \cdot (\mathbf{T} \times \mathbf{r}) - \mathbf{A} \tag{8.52b}$$

Total angular momentum equation:

$$\rho\frac{D\mathbf{L}}{Dt} = \rho\frac{D}{Dt}(\mathbf{s} + \mathbf{r} \times \mathbf{v})$$

$$= \rho \mathbf{G} + \mathbf{r} \times \rho \mathbf{F} + \nabla \cdot \mathbf{C} + \mathbf{r} \times (\nabla \cdot \mathbf{T}) + \mathbf{A} \qquad (8.53a)$$
$$= \rho \mathbf{G} + \mathbf{r} \times \rho \mathbf{F} + \nabla \cdot \mathbf{C} - \nabla \cdot (\mathbf{T} \times \mathbf{r}) \qquad (8.53b)$$

Internal angular momentum equation:

$$\rho \frac{D\mathbf{s}}{Dt} = \rho \mathbf{G} + \nabla \cdot \mathbf{C} + \mathbf{A} \qquad (8.54)$$

Equation (8.51) or (8.54) can be written in an alternative form that emphasizes the conservation–generation nature of the relationship. Multiplying the equation of continuity, (1.26), by **v** or **s** and adding the result to the left side of (8.51) or (8.54), respectively, expanded according to (4.57), and employing the tensor identity $\nabla \cdot (\mathbf{AB}) = \mathbf{A} \cdot \nabla \mathbf{B} + \mathbf{B} \nabla \cdot \mathbf{A}$ with $\mathbf{A} = \rho \mathbf{v}$ and $\mathbf{B} = \mathbf{v}$ or $\mathbf{s}$ yields

Cauchy balance of linear momentum:

$$\frac{\partial \rho \mathbf{v}}{\partial t} + \nabla \cdot (\rho \mathbf{v} \mathbf{v}) = \rho \mathbf{F} + \nabla \cdot \mathbf{T} \qquad (8.51')$$

Balance of internal angular momentum:

$$\frac{\partial \rho \mathbf{s}}{\partial t} + \nabla \cdot (\rho \mathbf{v} \mathbf{s}) = \rho \mathbf{G} + \nabla \cdot \mathbf{C} + \mathbf{A} \qquad (8.54')$$

where $\rho \mathbf{v} \mathbf{v}$ in (8.51′) is the flux of the linear-momentum volumetric density $\rho \mathbf{v}$ and $\rho \mathbf{v} \mathbf{s}$ in (8.54′) is the flux of the internal-angular-momentum volumetric density $\rho \mathbf{s}$, that is, $\rho \mathbf{v} \mathbf{s}$ is the spin flux.

The theoretical relationships describing the asymmetric-stress flow field developed up to this point are soundly based on broad principles and will be the bedrock of further advances.

8.9 Constitutive relations

In order to solve particular problems, it is necessary to complete the balance laws represented by equations (8.51)–(8.54) by the addition of constitutive relations. The variables that need to be constituted are $\mathbf{T} = \mathbf{T}_s + \mathbf{T}_a$, **C**, **A**, and **G**. Only problems in which linear acceleration of magnetic origin is absent will be considered; i.e., $\nabla \cdot \mathbf{T}_m = \mathbf{0}$. Thus **T** represents the pressure viscous-stress tensor. Following the approach of Condiff and Dahler (1964) developed for electrically polarized molecules, it will be assumed that the symmetric component $\mathbf{T}_s$ is independent of the field of internal angular-momentum density and is given by the usual expression for a Newtonian viscous fluid. The couple stress tensor **C** is assumed to be symmetric. It will be assumed to depend on the internal strain as represented by the angular spin rate $\boldsymbol{\omega}$, where $\mathbf{s} = I\boldsymbol{\omega}$, in which I is the average moment of inertia per unit mass. The arguments that produce the form of the symmetric viscous stress tensor $\mathbf{T}_v$ can then be taken over to yield an expression for the couple stress tensor $\mathbf{C}$. Thus, with $\mathbf{T}_s \equiv -p\mathbf{I} + \mathbf{T}_v$ and $\mathbf{T}_v$ as given previously by (4.62),

$$\mathbf{T} = \lambda(\nabla\cdot\mathbf{v})\mathbf{I} + \eta[\nabla\mathbf{v} + (\nabla\mathbf{v})^T] \tag{8.55}$$
$$\mathbf{C} = \lambda'(\nabla\cdot\boldsymbol{\omega})\mathbf{I} + \eta'[\nabla\boldsymbol{\omega} + (\nabla\boldsymbol{\omega})^T] \tag{8.56}$$

where by analogy η' and λ' are called the *shear* and *bulk coefficients of spin viscosity*, respectively.

As mentioned previously, **A** describes the rate of conversion of external angular momentum to internal angular momentum. Physically, the coupling arises when there is lack of synchronization between the rate of rotation of a fluid element and the rate of internal spin of the matter making up the fluid element. The effective rate of rotation of a fluid element is given by half the vorticity, $\frac{1}{2}\boldsymbol{\Omega} = \frac{1}{2}\nabla \times \mathbf{v}$, and the internal spin rate is $\boldsymbol{\omega}$. The conversion rate should be a function of the difference of these quantities, or what is equivalent, $\nabla \times \mathbf{v} - 2\boldsymbol{\omega}$. Assuming a linear relationship and introducing the phenomenological coefficient ζ allows the following expression to be written for **A**:

$$\mathbf{A} = 2\zeta(\nabla \times \mathbf{v} - 2\boldsymbol{\omega}) \tag{8.57a}$$

The coefficient ζ has been termed the *vortex viscosity*. The antisymmetric tensor $\mathbf{T}_a$ corresponding to **A** is, from (8.35) and (8.57a),

$$\mathbf{T}_a = \tfrac{1}{2}\boldsymbol{\varepsilon}\cdot\mathbf{A} = \zeta\boldsymbol{\varepsilon}\cdot(\boldsymbol{\Omega} - 2\boldsymbol{\omega}) \tag{8.57b}$$

Problem: In the equation of linear momentum, because $\mathbf{T} = \mathbf{T}_s + \mathbf{T}_a$ a body force $\nabla\cdot\mathbf{T}_a$ arises owing to the presence of antisymmetric stress. Using the relationship $\mathbf{T}_a = \frac{1}{2}\boldsymbol{\varepsilon}\cdot\mathbf{A}$ of (8.35), show that $\nabla\cdot\mathbf{T}_a = -\frac{1}{2}\nabla \times \mathbf{A}$.

Problem: Using the constitutive expression for **A** given by (8.57a), prove that the antisymmetric stress produces a total force on a volume of viscous fluid equal to

$$-\tfrac{1}{2}\oint_S \mathbf{n} \times \mathbf{A}\,dS = 2\zeta\oint_S \mathbf{n} \times \boldsymbol{\omega}\,dS$$

[*Hint:* Use the result of the preceeding problem and apply the divergence theorem for the curl of a vector, recognizing also that vorticity near a wall is oriented parallel to the wall.]

As another constitutive relation, it will be asserted that the body couple $\rho\mathbf{G}$ on a fluid element may be formulated analogously to the body couple on a small isolated magnetic body with the field taken as the local field $\mathbf{H}$ rather than the applied field:

$$\rho\mathbf{G} = \mu_0\mathbf{M} \times \mathbf{H} \tag{8.58}$$

Magnetization relaxation process

The magnetization $\mathbf{M}$ that appears in (8.58) will be perturbed from the equilibrium value $\mathbf{M}_0$ that is attained in a steady field $\mathbf{H}$ in motionless fluid. Before attempting to account for this effect, it is instructive to consider the mechanical response of a single magnetic particle that is sufficiently large not to be disturbed by thermal Brownian reorientation. The angular-momentum balance for a single particle with moment of inertia I_1 rotation rate ω_1, and magnetic moment μ_1 is

$$I_1\, d\omega_1/dt = \mu_1 \times \mathbf{H} - \Gamma_1\omega_1 \tag{8.59}$$

where Γ_1 is the viscous torque per unit angular rotation rate. If the applied field $\mathbf{H}$ rotates and a steady state is established, then the inertial term disappears in (8.59), and the particle tracks the field at the same frequency with a lag angle α_1, where

$$\sin\alpha_1 = \Gamma_1\omega_1/\mu_1 H \tag{8.60}$$

Thus, there can be no steady state when $\omega_1 > \mu_1 H/\Gamma_1$. Caroli and Pincus (1969) analyzed the response of such an isolated grain suspended in a liquid. Of further interest here is the proposal by Shliomis (1974b) of a phenomenological equation to describe the magnetic response of a *collection* of small magnetic particles.

As a model for this process, it is assumed that each subdomain particle is spontaneously magnetized to saturation with the magnetization locked rigidly to the mass. The orientation of the particles, and hence of the magnetic moments, will be distributed over all directions owing to thermal motion. Consider for a given volume element of the medium a local reference frame F' that moves and rotates with the average motion of the suspended particles. The magnetization in the F' system is described by the postulated *relaxation equation*,

$$D'\mathbf{M}/Dt = -(1/\tau)(\mathbf{M} - \mathbf{M}_0) \tag{8.61}$$

where τ is a relaxation time constant. Thus, the vector rate of change of magnetic moment in the sample is proportional to its vector displacement from equilibrium.

As a special case, the relaxation equation describes the transient magnetization of a motionless sample, for example, in response to a step increase in applied field. In this case **M** and **H**, and hence $\mathbf{M}_0$, are collinear, with the magnitude M_0 a constant. It is easily found from integration of (8.61) that the magnitude M of an initially unmagnetized sample evolves according to the expression $M/M_0 = 1 - \exp(-t/\tau)$.

More generally, the relaxation equation describes simultaneous magnetization and reorientation. Recall that the average spin rate of the particles in a unit volume of fluid was denoted $\boldsymbol{\omega}$. This rotation of particles will shift the magnetic vector of the fluid sample through an angle $\boldsymbol{\omega} \times \mathbf{M}$ in unit time. Then the time rate of change of magnetization in a moving but nonrotating frame is

$$D\mathbf{M}/Dt = D'\mathbf{M}/Dt + \boldsymbol{\omega} \times \mathbf{M} \tag{8.62}$$

Thus, the relaxation equation for the reference frame of a stationary observer becomes

$$D\mathbf{M}/Dt = \boldsymbol{\omega} \times \mathbf{M} - (1/\tau)(\mathbf{M} - \mathbf{M}_0) \tag{8.63}$$

The content of (8.63) has been illustrated by Shliomis (1975) for the response of a ferrofluid subjected to a uniform rotating magnetic field under the assumption that bulk fluid motion is absent. The average magnetization **M** is assumed to track the field with lag angle α:

$$\begin{aligned} H_1 &= H\cos\omega' t, \qquad & H_2 &= H\sin\omega' t \\ M_1 &= M\cos(\omega' t - \alpha), \qquad & M_2 &= M\sin(\omega' t - \alpha) \end{aligned} \tag{8.64}$$

The parameter ω' denotes the angular velocity of the applied field. Substituting these expressions into the relaxation equation (8.63) gives two relationships because there are two vectorial components. These relationships may be written

$$(\omega' - \omega)\tau = x \tag{8.65a}$$

$$M/M_0 = 1/\sqrt{1 + x^2} = \cos\alpha \tag{8.65b}$$

where $x \equiv \tan\alpha$. Note in (8.65a) that the field rotation rate ω' and the average particle rotation rate ω have different values. This is in contrast to the single-particle analysis and reflects the circumstance that not all the particles rotate in synchronization with the field. Nonetheless, the magnetization vector rotates in synchronization with the field, with the magnitude of the magnetization reduced as prescribed by (8.65b). Thus, when the lag angle $\alpha = 0$, $M = M_0$; whereas for $\alpha \to \pi/2$, $M \to 0$. At this stage α (or x) and ω are still unknown. These quantities can be determined, however, with the help of the continuum angular momentum equation, (8.54).

From equation (8.56), with $\boldsymbol{\omega}$ taken to be spatially uniform, it follows that $\nabla \cdot \mathbf{C}$ disappears. Then using $\mathbf{A}$ from (8.57a) with $\mathbf{v} = \mathbf{0}$ and replacing $\rho\mathbf{G}$ in (8.54) with $\mu_0 \mathbf{M} \times \mathbf{H}$ from (8.58), one obtains for the steady-state momentum balance

$$\mu_0 \mathbf{M} \times \mathbf{H} = 4\zeta\boldsymbol{\omega} \tag{8.66}$$

Using the definitions of the vector cross product $\mathbf{M} \times \mathbf{H} = MH \sin\alpha$, and $\sin\alpha = x/\sqrt{1 + x^2}$, together with the expression for M of (8.65b) then gives, from (8.66)

$$\mu_0 M_0(H) H / 4\zeta\omega = (1 + x^2)/x \tag{8.67}$$

Eliminating ω from (8.67) using (8.65a) gives the following cubic algebraic equation, which determines x:

$$x^3 - (\omega'\tau)x^2 + (P + 1)x - \omega'\tau = 0 \tag{8.68}$$

where $P \equiv \mu_0 M_0 H \tau / 4\zeta$. When x is determined from (8.68), the particle rotation rate ω can be found from (8.65a). Figure 8.7 presents calculated curves for the dependence of the lag angle on the field rotation rate; Figure 8.8 shows the mean particle rotation rate as a function of the field rotation rate. In both figures P is taken as a parameter of the curves. As $P \to 0$, $\tan\alpha \approx \omega'\tau$. More interestingly, it can be seen from Figure 8.7 that above a critical value of P that angle α experiences a discontinuity at some point ω'_m when the frequency of rotation of the field is increased. With a decrease of the field rotation frequency, a discontinuity occurs at $\omega' = \omega'_n < \omega'_m$, so the system exhibits *hysteresis*. In Figure 8.8 it is seen for ω/ω' that the rotation rate of particles also can exhibit hysteresis.

The responses described in this section are obviously rich in details. Nonetheless, the actual behavior of systems can be more complex than has been indicated; for example, the magnetic vector of a particle is not rigidly coupled to the particle mass under all circumstances (Néel relaxation). The interested reader is referred to the paper of Suyazov (1982) for further details of this more complex case.

8.10 Analogs of the Navier–Stokes equations for fluids with internal angular momentum

Combining the constitutive relationships with the balance equations will give the analogs of the Navier–Stokes equations. First it is

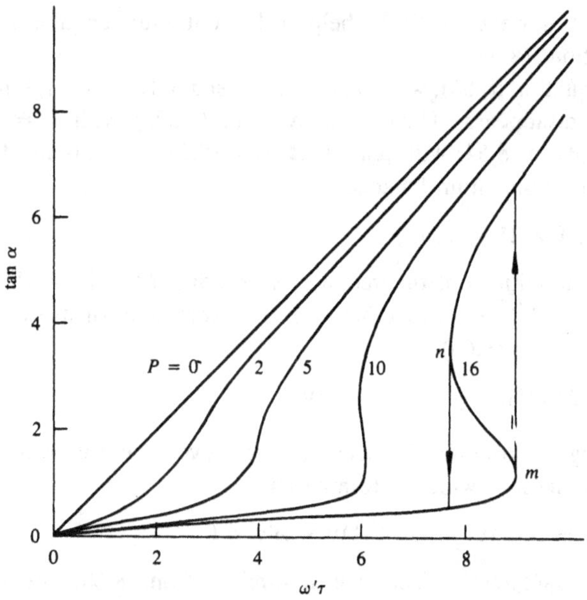

8.7 Graph of equation (8.68), illustrating the lag angle of mean magnetization in a motionless bulk fluid subjected to a uniform rotating magnetic field.

necessary to obtain divergence expressions for the dyadics $\mathbf{T}_v$, $\mathbf{C}$, and $\mathbf{T}_a$. From (8.55), (8.56), and (8.57b), these are, respectively,

$$\nabla \cdot \mathbf{T}_v = (\lambda + \eta)\nabla\nabla\cdot\mathbf{v} + \eta\nabla^2\mathbf{v} \tag{8.69a}$$

$$\nabla \cdot \mathbf{C} = (\lambda' + \eta')\nabla\nabla\cdot\boldsymbol{\omega} + \eta'\nabla^2\boldsymbol{\omega} \tag{8.69b}$$

$$\begin{aligned}\nabla \cdot \mathbf{T}_a &= -\tfrac{1}{2}\nabla \times \mathbf{A} \\ &= -\tfrac{1}{2}\nabla \times (\text{vec}\,\mathbf{T}) = \zeta\nabla\nabla\cdot\mathbf{v} + \zeta\nabla^2\mathbf{v} + 2\zeta\nabla\times\boldsymbol{\omega}\end{aligned} \tag{8.69c}$$

Substituting (8.57a), (8.58), and (8.69) into (8.51) and (8.54) gives the analogous relationships

$$\rho\, D\mathbf{v}/Dt = -\nabla p + \rho\mathbf{F} + 2\zeta\nabla\times\boldsymbol{\omega} + \beta\nabla\nabla\cdot\mathbf{v} + \eta_e\nabla^2\mathbf{v} \tag{8.70}$$

$$\rho I\, D\boldsymbol{\omega}/Dt = \rho\mathbf{G} + 2\zeta(\nabla\times\mathbf{v} - 2\boldsymbol{\omega}) + \beta'\nabla\nabla\cdot\boldsymbol{\omega} + \eta'\nabla^2\boldsymbol{\omega} \tag{8.71}$$

where

$$\eta_e \equiv \zeta + \eta, \qquad \beta \equiv \lambda + \eta - \zeta, \qquad \beta' \equiv \lambda' + \eta'$$

8.11 Ferrohydrodynamic-torque-driven flow

Based on the foregoing preparation a theoretical description of the experimentally discovered entrainment of fluid by a rotating mag-

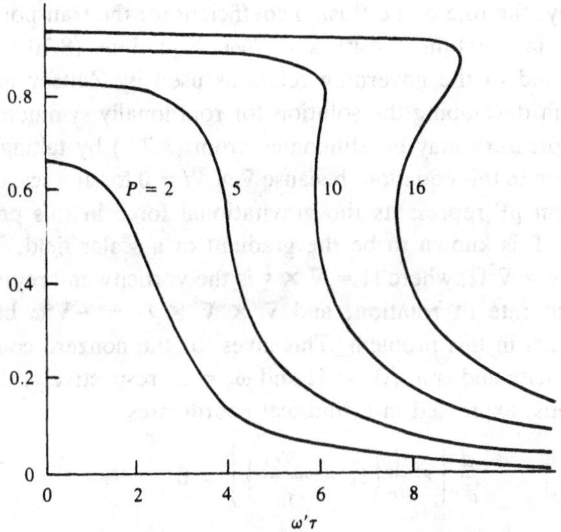

8.8 Dependence of the mean rotation rate of particles on the field rotation rate and applied field intensity.

netic field may now be given. It is desired to find a steady-state solution of the equations of motion (8.70) and (8.71) in a cylindrical coordinate system in which the fluid velocity has only an azimuthal (θ) component: $v_\theta = v(r)$, $v_r = v_z = 0$. It may be seen from (8.70) that such motion is possible if $\boldsymbol{\omega}$ is independent of time and $\omega_r = \omega_\theta = 0$, $\omega_z = \omega(r)$. It follows from (8.71) and (8.58) that a steady-state solution of this form exists only if $G_x = G_y = 0$ and $\rho G_z = \mu_0(M_x H_y - M_y H_x)$ is independent of time. This condition is satisfied when the magnetization vector rotates with the same angular velocity as the field while lagging the field at a constant phase angle α. Thus $\rho G_z = \mu_0 MH \sin\alpha$, where z denotes the vertical coordinate.

For an incompressible ferrofluid $\nabla \cdot \mathbf{v} = 0$ and, by cylindrical symmetry, $\nabla \cdot \boldsymbol{\omega} = 0$. Thus, in the steady state the equation set (8.70) and (8.71) reduces to the following linear momentum and internal angular-momentum balances, respectively:

$$\mathbf{0} = -\nabla p + \rho \mathbf{F} + 2\zeta \nabla \times \boldsymbol{\omega} + \eta_e \nabla^2 \mathbf{v} \tag{8.70$'$}$$

$$\mathbf{0} = \rho \mathbf{G} + 2\zeta(\nabla \times \mathbf{v} - 2\boldsymbol{\omega}) + \eta' \nabla^2 \boldsymbol{\omega} \tag{8.71$'$}$$

reception transformation diffusion

η'/I plays the role of a diffusion coefficient for the transport of internal angular momentum density $\mathbf{s} = I\rho\boldsymbol{\omega}$. Equations (8.70′) and (8.71′) correspond to the governing relations used by Zaitsev and Shliomis (1969) in developing the solution for rotationally symmetric flow.

The pressure may be eliminated from (8.70′) by taking the curl of each term in this equation, because $\nabla \times \nabla f = \mathbf{0}$ for any scalar function f. The term $\rho\mathbf{F}$ represents the gravitational force in this problem, and because $\mathbf{F}$ is known to be the gradient of a scalar field, $\nabla \times \mathbf{F} = \mathbf{0}$. $\nabla \times \nabla^2\mathbf{v} = \nabla^2\boldsymbol{\Omega}$, where $\boldsymbol{\Omega} = \nabla \times \mathbf{v}$ is the vorticity and represents twice the fluid rate of rotation, and $\nabla \times \nabla \times \boldsymbol{\omega} = -\nabla^2\boldsymbol{\omega}$ because $\nabla \cdot \boldsymbol{\omega}$ disappears in this problem. This gives for the nonzero components of the vorticity and spin ($\Omega_z = \Omega$ and $\omega_z = \omega$, respectively) the following equations, expressed in cylindrical coordinates:

$$\frac{d}{dr}\left[r\frac{d}{dr}\left(\Omega - \frac{2\zeta\omega}{\eta_e}\right)\right] = 0 \tag{8.72}$$

$$\frac{\eta'}{r}\frac{d}{dr}\left(r\frac{d\omega}{dr}\right) - 2\zeta(2\omega - \Omega) = -\mu_0 MH \sin\alpha \tag{8.73}$$

The boundary conditions at the surface of the cylinder where $r = R$ are

$$v(R) = 0 \tag{8.74a}$$

$$\omega(R) = 0 \tag{8.74b}$$

where the no-slip condition on the velocity v at the wall is conventially employed in ordinary fluids but has an uncertain status in the present context, and the no-spin condition on ω at the wall deserves further comment. For example, Brenner (1983) has developed an analysis of the motion of an isolated sphere rotating in viscous fluid near a wall. In the present context, the induced flow of the fluid surrounding the sphere results in translation of the sphere along the wall in a direction of rotation about the cylindrical-system axis opposite to that of the spin of the sphere. Assuming this mechanism is operative in a collection of suspended particles offers a physical explanation for the reverse flow of fluid observed in some experiments. For present purposes the simple boundary conditions of (8.74) are retained to illustrate the method of solution but not as an assertion that these simple conditions are physically sound in all instances.

Integration of (8.72) gives

$$\Omega = \frac{1}{r}\frac{d(rv)}{dr} = \frac{2\zeta\omega}{\eta_e} + C_1 \tag{8.75}$$

Eliminating Ω between (8.75) and (8.73) then gives

$$\frac{d^2\omega}{dr^2} + \frac{1}{r}\frac{d\omega}{dr} - \kappa^2\omega = -A \tag{8.76}$$

where

$$\kappa^2 \equiv 4\eta\zeta/\eta_e\eta' \tag{8.77a}$$
$$A \equiv (1/\eta')(\mu_0 MH \sin\alpha + 2\zeta C_1) \tag{8.77b}$$

Equation (8.76) may be transformed to a standard Bessel equation by the change of variables $y = \omega - A\kappa^{-2}$ and $x = \kappa r$. The solution for ω, obtained by imposing the condition that ω is finite at $r = 0$, is

$$\omega(r) = A\kappa^{-2} + C_2 I_0(\kappa r) \tag{8.78}$$

where I_0 is the modified Bessel function of the first kind of order zero. The fluid velocity is now found from (8.75) using ω from (8.78). The solutions for ω and v satisfying the boundary conditions (8.74) are

$$\omega(r) = \frac{\eta_e}{\eta(R)}\left(\frac{\mu_0 MH}{4\zeta}\sin\alpha\right)\left[1 - \frac{I_0(\kappa r)}{I_0(\kappa R)}\right] \tag{8.79}$$

$$v(r) = v_0\left[\frac{r}{R} - \frac{I_1(\kappa r)}{I_1(\kappa R)}\right] \tag{8.80}$$

where

$$\eta(R) \equiv \eta + \zeta\left[1 - \frac{2I_1(\kappa R)}{\kappa R I_0(\kappa R)}\right] \tag{8.81}$$

$$v_0 \equiv \frac{1}{2\kappa\eta(R)}(\mu_0 MH \sin\alpha)\frac{I_1(\kappa R)}{I_0(\kappa R)} \tag{8.82}$$

and M, if desired, is obtained as a solution from the relaxation equation, (8.63).

The solutions for ω and v depend markedly on κR. It may be argued that $l_D \equiv 1/\kappa$ represents a diffusion length with order of magnitude $l_D = l$, where l is the average distance between the solid particles, i.e., the distance between microeddy centers. Thus, asserting that η' depends only on the viscosity η and distance l gives dimensionally $\eta' \sim \eta l^2$. Likewise $\zeta \sim \eta$, and so it follows from (8.77a) that $l_D \sim l$. In a colloid with volume fraction of solids ϕ and particle diameter d, the ratio l/d is

$$l/d = (\pi/6\phi)^{1/3} \tag{8.83}$$

For example, a ferrofluid with volume fraction $\phi = 0.012$, corresponding to the fluid in the experiment of Moskowitz and Rosensweig (1967), yields $l/d = 3.5$. With $d = 100$ Å $= 10^{-6}$ cm and $R = 2$ cm, the distance $l = 3.5 \times 10^{-6}$ cm, and $\kappa R = R/l_D = 0.6 \times 10^6$. Hence, in the following it will be assumed that $\kappa R \gg 1$.

The Bessel functions appearing in (8.79) to (8.82) have the following limiting behavior:

$$I_0(x) = \begin{cases} 1, & x \to 0 \\ e^x/\sqrt{2\pi x}, & x \to \infty \end{cases}$$
$$I_1(x) = \begin{cases} x/2, & x \to 0 \\ e^x/\sqrt{2\pi x}, & x \to \infty \end{cases} \tag{8.84}$$

Consequently, for $\kappa R \gg 1$

$$\eta(R) = \eta + \zeta, \qquad v_0 = \frac{1/\kappa}{2(\eta + \zeta)} \mu_0 MH \sin\alpha \tag{8.85}$$

The profiles of ω and v from the solutions (8.79) and (8.80) display a marked boundary-layer nature (see Figure 8.9). Throughout the entire flow field except for a thin boundary layer adjacent to the wall, ω is constant, diminishing to zero over distances on the order of a particle diameter. The velocity profile is linear with radial distance up to the vicinity of the boundary layer; i.e., the bulk of the fluid rotates at constant angular velocity. This rotational rate is given by

$$\text{angular rate} = \frac{v_0}{R} = \frac{l_D/R}{2(\eta + \zeta)} \mu_0 MH \sin\alpha \tag{8.86}$$

When the coefficient $\eta' = 0$, and hence there is no diffusion of spin, it may be seen from (8.79) and (8.80) [allow $\kappa \to \infty$ and use the asymptotic expressions, (8.84)] that the spin field is spatially uniform and the flow field is motionless. If the spin field is spatially uniform even for $\eta' > 0$ and $\zeta > 0$, it may be found by integration of (8.72), using (8.74a) and the constraint that v is finite on the axis, that $v = 0$ or, again, the flow field is motionless. In this case the microeddies shear against each other and produce dissipation but no net flow, a circumstance that has led some authors to conclude that the generation of rotary motion in this system is not possible. However, we have seen that when a nonuniform field of spin is established owing to diffusion from the boundary, motion is indeed possible.

However, there is a serious discrepancy between theory and experi-

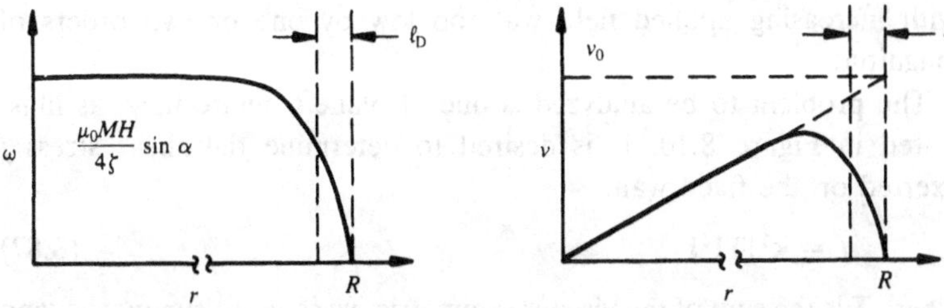

(a) Spin field (b) Velocity field

8.9 Theoretical solution for the flow of ferrofluid in a rotating field, displaying a very thin boundary layer having thickness on the order of the particle-to-particle separation distance l_D. (*After Zaitsev and Shliomis 1969.*)

ment, as can be appreciated by noting that the maximum value of $\sin\alpha$ [in (8.86)] is 1. In the experiment of Moskowitz and Rosensweig (1967). $\mu_0 H = 0.0060$ T, $\chi = 0.3$, $\eta = 0.0012$ kg·m^{-1}·s^{-1}, and $R \simeq 0.02$ m. With R/l_D, as computed following (8.83), and ζ taken to be equal to η, the maximum value of the angular-rotation rate is 0.0063 rad/s (0.06 rpm). This is smaller than the experimental value by a factor of 10^3–10^4. This variance is lethal not to the theory but to the details of how the theory is applied.* Glazov (1976) has suggested that azimuthal gradients in the applied field are responsible for the observed flow, basing an analysis on entrainment of the fluid in the asymmetric stress field of a traveling wave.

8.12 Effective viscosity of a magnetized fluid

A homogeneous magnetic field tends to orient the colloidal magnetic particles of a magnetic fluid in the direction of the field. If the fluid mixture is sheared, there is additional resistance because the oriented particles are impeded from rotating freely. At the same time, Brownian motion and hydrodynamic forces have a disorienting effect on the particles. The essential phenomenon brought into play is antisymmetric stress due to the relative motion between the grains and the surrounding liquid. The laws and relationships developed in this chapter can thus be used to determine the effective increase in viscosity. The present treatment is modeled on the work of Shliomis (1972). In earlier treatments, such as that of Hall and Busenberg (1969) neglecting Brownian rotation, the saturation exhibited by the increase in viscosity

*Free interface unsymmetric stress is responsible for the spin-up motion (see "Comments and supplemental references").

with increasing applied field was too low by one or two orders of magnitude.

The problem to be analyzed is one of plane Couette flow, as illustrated in Figure 8.10. It is desired to determine the shear stress f exerted on the fixed wall:

$$f = \mathbf{k}\cdot[\mathbf{T}]\cdot\mathbf{i} \tag{8.87}$$

where $\mathbf{T}$ is the sum of the viscous symmetric, viscous antisymmetric, and magnetic stress tensors, and [ ] denotes difference evaluated across the fluid–solid interface at the boundary wall:

$$\mathbf{T} = \mathbf{T}_s + \mathbf{T}_a + \mathbf{T}_m \tag{8.88}$$

The specified known conditions for this system may be stated as follows:

$$\mathbf{v} = \Omega z\mathbf{i} \tag{8.89a}$$
$$\mathbf{\Omega} = \Omega\mathbf{j} \tag{8.89b}$$
$$\mathbf{H} = H\mathbf{k} \tag{8.89c}$$

where the pressure p, field magnitude H, vorticity Ω, magnetization $\mathbf{M}$, and angular velocity $\boldsymbol{\omega}$ are constant. The magnetization may be decomposed as

$$\mathbf{M} = \mathbf{M}_0 + \mathbf{m} \tag{8.90a}$$
$$\mathbf{m} = m_1\mathbf{i} + m_2\mathbf{j} + m_3\mathbf{k} \tag{8.90b}$$

where $\mathbf{M}_0 = M_0\mathbf{k}$ is the unperturbed value and $\mathbf{m}$, which is not necessarily small, is the perturbed value due to the interactions of the flow, magnetic field, and thermal fluctuations.

The spin rate $\boldsymbol{\omega}$ is constant, so $\nabla\cdot\mathbf{C} = \mathbf{0}$, from (8.56), and the internal angular momentum equation (8.54) reduces to

$$\rho\mathbf{G} + \mathbf{A} = \mathbf{0} \tag{8.91}$$

Substituting for $\rho\mathbf{G}$ from (8.58) and using (8.57) gives

$$\mathbf{A} = 2\zeta(\mathbf{\Omega} - 2\boldsymbol{\omega}) = -\mu_0\mathbf{M}\times\mathbf{H} \tag{8.92}$$

It appears that no magnetic-stress difference can arise across the fluid-wall interface so evaluation of (8.88) depends on the nature of $\mathbf{T}_s$ and $\mathbf{T}_a$. These tensors are constant throughout the fluid layer, consistent with the linear-momentum equation (8.51). On the wall side of the interface both tensors disappear, so only the fluid side need be considered in evaluating f from (8.87). From the symmetric viscous-stress tensor of (8.55) and from (8.89a) it is readily found that the

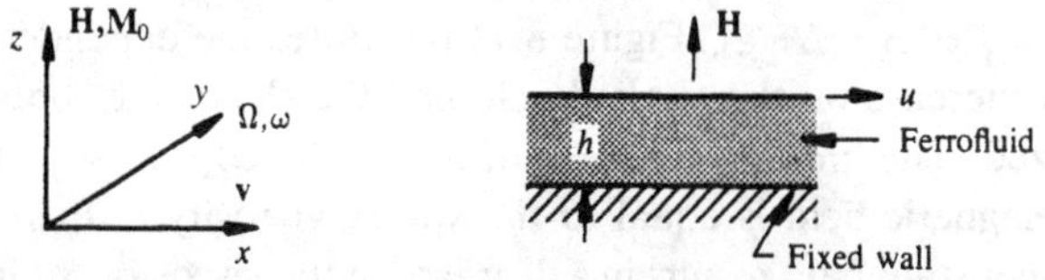

8.10 Nomenclature for the problem of sheared magnetized ferrofluid.

transpose dyadic $(\nabla\mathbf{v})^{\mathrm{T}}$ makes no contribution to the force in this problem and that $\mathbf{k}\cdot\mathbf{T}_s\cdot\mathbf{i} = \eta\Omega$. From (8.92), with the substitution of **H** from (8.89c) and **M** from (8.90), **A** may be expressed $\mathbf{A} = \mu_0 m_1 H\mathbf{j}$. Because $\mathbf{T}_a = \frac{1}{2}\boldsymbol{\varepsilon}\cdot\mathbf{A}$, this expression for **A** leads in a straightforward manner to $\mathbf{T}_a = \frac{1}{2}\mu_0 m_1 H(\mathbf{ki} - \mathbf{ik})$, which is the tensor of antisymmetric viscous stress for this problem. Carrying out the operation $\mathbf{k}\cdot\mathbf{T}_a\cdot\mathbf{i}$ yields the term $\mu_0 m_1 H/2$. Then taking the sum of terms yields the shear stress acting on the wall:

$$\begin{aligned} f &= \eta\Omega + \tfrac{1}{2}\mu_0 m_1 H \\ &= \Omega(\eta + \Delta\eta), \qquad \text{where} \quad \Delta\eta = \tfrac{1}{2}\mu_0 m_1 H/\Omega \end{aligned} \tag{8.93}$$

The product $\Omega\eta$ is the usual Newtonian viscous shear stress, and $\Delta\eta$ represents the rotational viscosity contribution, which is seen to depend solely on the m_1 component of the perturbed magnetization.

Determination of the component m_1 is necessary for use in equation (8.93). To begin, equation (8.92) may be used to eliminate $\boldsymbol{\omega}$ from the relaxation equation (8.63). For steady state this gives

$$\boldsymbol{\Omega}\times\mathbf{M} = (1/\tau)(\mathbf{M} - \mathbf{M}_0) + (\mu_0/2\zeta)\mathbf{M}\times(\mathbf{M}\times\mathbf{H}) \tag{8.94}$$

When the expressions for $\boldsymbol{\Omega}$ and **H** of (8.89), and **M** and **m** of (8.90) are introduced into (8.94), three algebraic equations result, one for each of the Cartesian component directions. From the equation for direction 2 it is found that $m_2 = 0$; that is, the magnetization **M** lies in the plane of the flow. The other two equations are

$$\Omega\tau(1 + y) = x + Px(1 + y) \tag{8.95a}$$
$$-\Omega\tau x = y - Px^2 \tag{8.95b}$$

where $x = m_1/M_0$, $y = m_3/M_0$, and $P = \mu_0 M_0 H\tau/4\zeta$. Eliminating y between these two equations gives a cubic algebraic equation that determines x or, equivalently, m_1 and hence $\Delta\eta$. This may be expressed

$$r^3 - 2r^2 + r[1 + (P + 1)/(\Omega\tau)^2] - P/(\Omega\tau)^2 = 0 \tag{8.96}$$

where $r \equiv Px/\Omega\tau = \Delta\eta/2\zeta$. Figure 8.11 illustrates the dependence of the viscosity increase on the applied field and the shear rate, both suitably normalized. The limiting value $\Delta\eta$ of the predicted viscosity increase in a high magnetic field is equal to the vortex viscosity ζ. Increasing the shear at constant field results in a decrease in the viscosity. At high shear in a high field, a hysteresis of viscosity is predicted, as seen in Figure 8.11.

The expressions given by Shliomis (1972) for the vortex viscosity and Brownian relaxation time τ [cf. (2.33)] for dilute ferrofluids correspond to

$$\zeta = \tfrac{3}{2}\eta\phi \tag{8.97}$$

$$\tau = 3V\eta/kT \tag{8.98}$$

where ϕ is the volume fraction of solids in the ferrofluid, V the volume of a single particle, and k the Boltzmann constant (1.38×10^{-23} $\mathrm{N \cdot m \cdot K^{-1}}$). Calculations for typical ferrofluids yield values of τ on the order of 10^{-6} s, and $\Omega\tau < 10^{-3}$ in experiments (Rosensweig, Kaiser, and Miskolczy 1969). For such small values of $\Omega\tau$, equation (8.96) reduces to $r = P/(1 + P)$, or

$$\frac{\Delta\eta}{\zeta} = \frac{\mu_0 M_0 H\tau/4\zeta}{1 + \mu_0 M_0 H\tau/4\zeta} \qquad (\Omega\tau \ll 1) \tag{8.99}$$

Because Ω^{-1} is a time for deformation of the fluid sample, $\Omega\tau$ represents the ratio of the magnetic relaxation time to the fluid deformation time. Thus, a small value of $\Omega\tau$ indicates that the sample magnetization can quickly reorient and closely track the field direction at all instants. As seen from (8.99), this results in the prediction that for small $\Omega\tau$ the viscosity is field dependent but not shear-rate dependent.

This section has considered the case of the vector $\mathbf{H}$ perpendicular to $\boldsymbol{\Omega}$. For an arbitrary orientation of these vectors it may be shown that (8.99) takes the form

$$\frac{\Delta\eta}{\zeta} = \frac{\mu_0 M_0 H\tau/4\zeta}{1 + \mu_0 M_0 H\tau/4\zeta}\sin^2\beta \tag{8.100}$$

where β is the angle between $\mathbf{H}$ and $\boldsymbol{\Omega}$. As seen from (8.100), when $\mathbf{H}$ is parallel to $\boldsymbol{\Omega}$ the viscosity is independent of the field. This result is understandable, because the orientation of the magnetic moment of the particle along the field does not prevent it from rotating with the angular velocity $\boldsymbol{\Omega}/2$ of the fluid.

These relationships take on a particularly simple form for a dilute

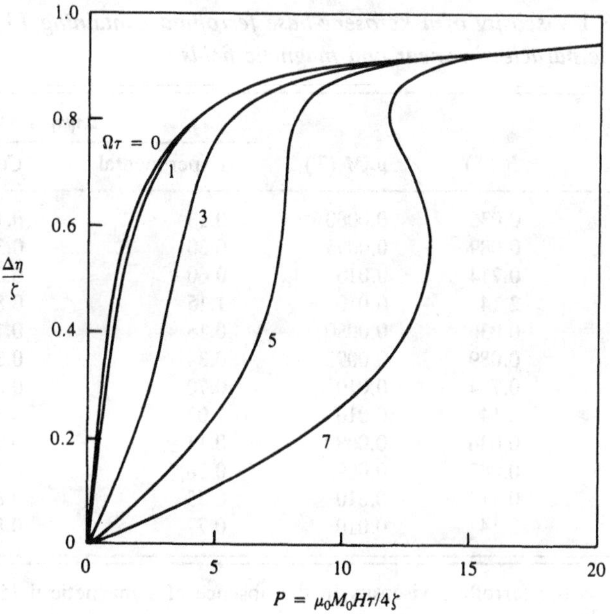

8.11 Prediction of equation (8.96) for the dependence of the viscosity on the applied field and shear rate (antisymmetric stress present).

ferrofluid. Substituting from (8.97) and (8.98) into (8.100) and recognizing that $M_0 = \phi M_d L(\alpha)$ – where ϕ is the volume fraction, M_d is domain magnetization and $L(\alpha)$ is the Langevin function, whose argument is $\alpha = \mu_0 M_d HV/kT$ [cf. (2.27)] – gives

$$\frac{\Delta\eta}{\eta} = \tfrac{3}{2}\phi \frac{\tfrac{1}{2}\alpha L(\alpha)}{1 + \tfrac{1}{2}\alpha L(\alpha)} \sin^2\beta \qquad (8.101)$$

Example – Correlation of ferrofluid viscosity data using the asymmetric stress concept: Data on the influence of the magnetic field and shear on concentrated magnetic fluids containing magnetite particles dispersed in mineral oil, kerosene, and fluorocarbon carriers are reported by Rosensweig, Kaiser, and Miskolczy (1968). A layer of ferrofluid at 30°C is subjected to the uniform shear of a cone-and-plate viscometer in the presence of a uniform magnetic field applied perpendicular to the layer. Inspection of the data indicates that in the case of the kerosene-base fluid the viscosity was nearly independent of the shear rate. The particle size for this fluid is reported as 13.6 nm, based

Table 8.1 *Viscosity of a kerosene-base ferrofluid containing 13.6-nm magnetic particles in shear and magnetic fields*

			$\Delta\eta/\eta$	
Ω (s^{-1})	B (T)	$\mu_0 M$ (T)	Experimental	Computed
46	0.036	0.0080	0.13	0.14
46	0.089	0.0095	0.30	0.30
46	0.714	0.010	0.60	0.82
46	2.14	0.010	1.15	0.83
115	0.036	0.0080	0.15	0.14
115	0.089	0.0095	0.34	0.30
115	0.714	0.010	0.70	0.82
115	2.14	0.010	1.02	0.83
230	0.036	0.0080	0.13	0.14
230	0.089	0.0095	0.26	0.30
230	0.714	0.010	0.55	0.82
230	2.14	0.010	0.77	0.83

Note: η is the ferrofluid viscosity in the absence of a magnetic field (0.0029 $N \cdot s \cdot m^{-2}$).

on electron microscopy. A magnetite particle of this size is expected to exhibit extrinsic superparamagnetism (see Table 2.2). Analyze the data of Table 8.1 to determine the vortex viscosity and the Brownian rotational time constant.

Analysis: The viscosity is little dependent on the shear rate, so it will be assumed that (8.99) applies. This equation may be rearranged into the following linear form, which is convenient for regression analysis:

$$y = a + bx$$

where y and x are variables known from the data and a and b are constants to be determined using all the data:

$$y = 1/\Delta\eta, \qquad x = 4/\mu_0 MH, \qquad a = 1/\zeta, \qquad b = 1/2\tau$$

Figure 8.12 is a plot of the data using y–x coordinates. The straight line is the least squares fit and represents the equation

$$y = 413 + 122{,}000x$$

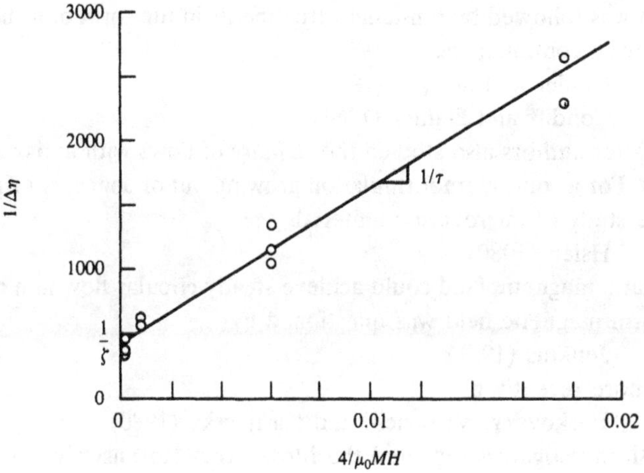

8.12 Correlation of the magnetoviscometric data for a kerosene-base magnetic fluid.

This yields the values

$$\zeta = 0.00242 \text{ N}\cdot\text{s}\cdot\text{m}^{-2} \qquad \text{(vortex viscosity)}$$
$$\tau = 4.10 \times 10^{-6} \text{ s} \qquad \text{(rotational time constant)}$$

An independent estimate of τ can readily be made using (8.98) with $\eta = 0.0029 \text{ N}\cdot\text{s}\cdot\text{m}^{-2}$, $D = 13.6$ nm, $V = (\pi/6)(1.36 \times 10^{-8})^3 = 1.32 \times 10^{-24}$ m^3, and $kT = (1.38 \times 10^{-23})(273 + 30) = 4.18 \times 10^{-21}$ N·m. Thus, $\tau = 3(1.32 \times 10^{-24})(0.0029)/(4.18 \times 10^{-21}) = 2.75 \times 10^{-6}$ s, in order-of-magnitude agreement with the value obtained from viscometry. A value of ϕ computed from (8.97) using ζ previously found gives $\phi = \frac{2}{3}(\zeta/\eta) = \frac{2}{3}(0.00242/0.00290) = 0.56$, which is reasonable for a *coated* particle.

Other data for concentrated mineral oil and fluorocarbon-base fluids exhibit greater dependence on the shear rate and correlate poorly with (8.99) or the more elaborate (8.96). The difference may be due to particle clustering, indicating that a different model is needed for reduction of those data.

Comments and supplemental references

The original continuum analysis of antisymmetric stress is in a short note by

Dahler and Scriven (1961)

which was followed by full-length treatments in the form of a theory of structured continua; see

Dahler and Scriven (1963)
Condiff and Dahler (1964)

The latter authors also studied the *stability* of flows with antisymmetric stress. For a continuum formulation growing out of concepts originated in the study of microelastic materials see

Hsieh (1980)

That a magnetic fluid could achieve steady circular flow in a rotating uniform magnetic field was questioned by

Jenkins (1971)

and more recently by

Berkovsky, Vislovich, and Kashevsky (1980)

British investigators repeated the Moskowitz–Rosensweig experiment with the beaker resting on a turntable that was free to rotate (see Figure 8.1d). They were astonished to see fluid rotating in one direction and the beaker and turntable in the other. Moreover, the fluid's direction of rotation was opposite that of the field, or "wrong way round"; see

Brown and Horsnell (1969)

Reversal of rotational direction with increasing magnetic-field intensity has been observed; the working fluid was a relatively coarse 0.1-μm suspension of ferrous particles and no doubt contained appreciable clustering of the particles, with effects difficult to predict. Kagan et al. report related measurements of the torque and rotation of dielectric cylindrical objects placed in a suspension of 0.2-μm ferrous particles; see

Calugaru, Badescu, and Luca (1976b)
Kagan, Rykov, and Yantovskii (1973)

Another experiment that strongly suggests the possibility of a non-symmetric stress tensor is reported in an early Russian reference:

Tsvetkov (1939)

The liquid crystal *p*-azoxyanisole was placed in a rotating magnetic field, which induced a torque on the boundary of the cylinder and a rotational velocity in the liquid crystal.

A suspension of gravitational dipoles, particles having off-center mass, furnishes an additional example of a polarizable fluid and is studied by

Brenner (1970)

An analysis of the effective viscosity in a magnetic field based on study of an orientational distribution function with the influence of rotary Brownian motion is developed by

Brenner and Weissman (1972)

Careful measurements show that while the pipe flow friction coefficient of a ferrofluid increases with magnetic field strength in the laminar flow regime, there is no magnetic-field effect in a turbulent flow regime.

Kamiyama, Koike, and Oyama (1983).

Enhancement of the heat-transfer rate in an antisymmetric stress field has been reported by

Shulman, Kordonskii, and Demchuk (1977).

Soviet investigators have become very active in investigating problems of ferrofluid antisymmetric stress. See papers by Tsebers, Berkovsky, Suyazov, and others in issues of *Magnetohydrodynamics* (translated from the Russian by Consultants Bureau, New York).

General expressions for entropy production have been developed in the study of irreversible thermodynamics and offer an elegant means of deducing the forms of the viscous stress tensor $\mathbf{T}_v$, the couple stress tensor $\mathbf{C}$, and the momentum-conversion vector $\mathbf{A}$. See

deGroot and Mazur (1962)

Berkovsky, Vislovich, and Kashevsky (1980).

That the driving force in spin-up flow is tangential stress arising at the free interfacial meniscus is shown experimentally and theoretically by Rosensweig, Popplewell and Johnston (1990). See *J. Magnetism Magnetic Mater.* 85 (1–3), 171–180.

9

MAGNETIC TWO-PHASE FLOW

9.1 Background

Two-phase flow is a special case of multiphase flow, a very broad topic that arises in connection with fluidized beds, the sedimentation of particles and droplets, the boiling of liquids, the motions of foams, and many other systems. The ferrohydrodynamics of these systems is in its infancy. The treatment in the following is specific to a two-phase mixture consisting of solids dispersed in a continuous phase that is fluid, and one particular application is studied in depth, that of magnetized fluidized solids. References to related studies in other systems are cited at the end.

Fluidization is an operation by which solid particles are transformed into a fluid-like state when placed in contact with a flowing gas or liquid. The size of the solid particles is large compared to that of particles in a colloidal ferrofluid and typically ranges from 50 to 1000 μm. The method of ensuring contact has a number of unusual features that have allowed it to be put to good use. A recent development is the use of FHD principles to make accessible a magnetically stabilized operating range in which turbulence is prevented. The stabilized beds yield advantages in moving-bed as well as stationary-bed contactors.

In the study of fluidization the dynamics of the two phases, solid and fluid, must be considered on an equal basis. Such was the case also for the magnetic fluid flows of Chapter 8 in which particle rotational motion relative to the fluid contributed a microstructure and led to consideration of antisymmetric stress. In the present context a dominant feature is that the *translational motion* of fluid relative to solid must be incorporated into the description. However, unlike in colloidal magnetic fluid, thermal Brownian motion of the particles is totally negligible in fluidized beds.

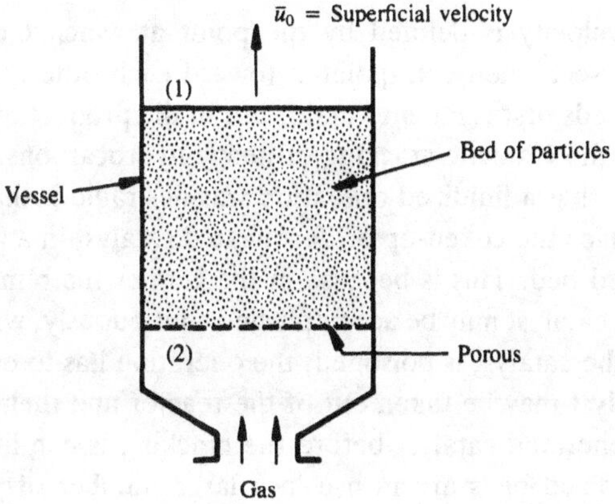

9.1 Fluidization vessel containing a bed of fine particles suitable for contacting fluids and solids. A batch operation is shown with the particles remaining in the vessel.

Following a brief review of ordinary fluidization, this chapter develops an approach for analyzing two-phase magnetized flows and then uses the results to determine the FHD stability behavior.

9.2 Ordinary fluidization

Depicted in Figure 9.1 is a bed of unmagnetized solid particles supported on a horizontal porous grid in a vertical tube. If a fluid is passed upward through the bed at a low flow rate, it merely percolates through the empty spaces between the solid particles; then, because the particles remain stationary, the bed is called a *fixed bed*. As the flow rate is increased and the pressure drop across the bed Δp_m is sufficient to support the weight of the particles, the bed is said to be fluidized. In an ordinary fluidized bed the particles are in chaotic motion, and the bed thus formed has many of the properties of a liquid: Objects float on the surface, and the addition or withdrawal of solid particles in processing equipment is also facilitated.

The superficial velocity, i.e., the fluid volumetric flow rate divided by the vessel cross-sectional area, at which *minimum fluidization* is achieved is denoted $\bar{u}_m$. The preferred method of determining $\bar{u}_m$ is to make a plot of pressure drop across the bed of solids versus superficial fluid velocity. As shown in Figure 9.2, the resulting plot contains two nearly linear portions. The translational region between them may be smooth (curve I) or may show a peak (curve II). The minimum fluidization velocity is defined by the point at which the two linear portions intersect when extrapolated toward each other.

Fluidized beds first came into major use in the production of aviation fuel through the catalytic cracking of heavy hydrocarbons. One of the main reasons that a fluidized operation was desirable is that it is much easier to replace the coked-up or deactivated catalyst in a fluidized bed than in a fixed bed. This is because in the former the removal and replacement of catalyst may be accomplished continuously, whereas in the latter, once the catalyst is poisoned, the operation has to be stopped so that the catalyst may be taken out of the reactor and then replaced by fresh or regenerated catalyst before the cracking is continued.

Today fluidized beds are in use in a large number of physical and chemical operations. The physical applications include transportation, drying and sizing solids, coating metal surfaces with plastic materials, and heating and drying operations; the chemical operations range from catalytic oxidation reactions to roasting sulfide ores, and more are expected in the areas of synthetic gas and fuel production.

An excellent introduction to the topic of ordinary fluidization is found in the book of Kunii and Levenspiel (1977).

9.3 Some fundamental problems in fluidization engineering

In a gas-fluidized bed, as the gas velocity is increased beyond the minimum fluidization velocity, most of the excess gas passes through the bed as bubbles (see Figure 9.3a). The bubbles agitate the bed in passing through and backmix the solid particles. This stirring is chaotic and creates an isothermal situation, which is desirable for control purposes if the reaction is temperature sensitive. On the other hand, the gas in the bubbles tends to bypass the bed, producing lower yields in heterogeneous catalytic reactors and generally reducing the contacting effectiveness. Also, motion produced by the bubbles creates impacts that cause *attrition*, leading to *entrainment* losses of smaller-size particles, which is undesirable. The entrainment of fine solids is, in addition, a pollution problem – these entrained particles must be removed by expensive filters, cyclones, or electrostatic precipitators.

It is known experimentally and theoretically (Rosensweig 1978b, 1979b,c) that with the use of bed solids that are magnetizable, and with application of a uniform field that is steady in time and oriented parallel to the direction of gas flow (see Figure 9.3b), bubble-free expanded beds in which particle motions, chaotic or otherwise, are totally absent are obtained over a substantial range of gas velocities. These systems,

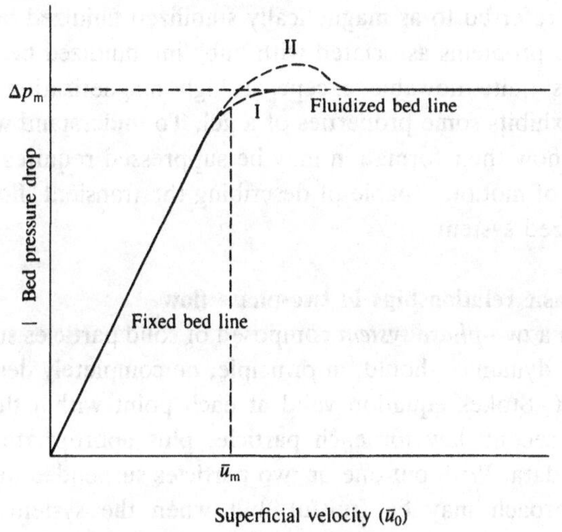

9.2 Graph of the bed pressure drop versus superficial velocity to illustrate features of fluidized solids behavior.

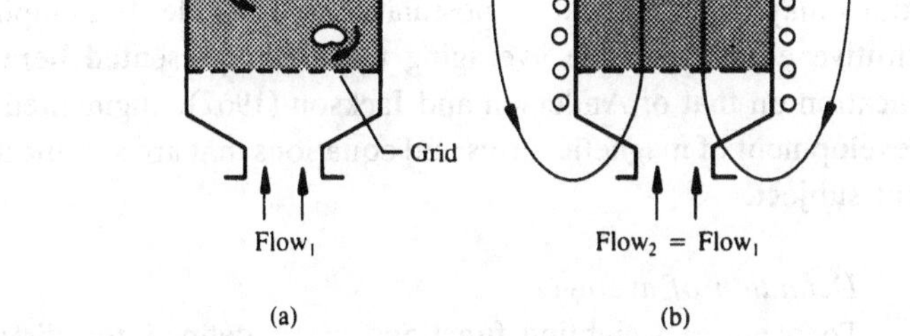

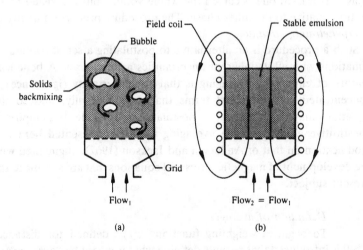

9.3 Comparison of (a) an ordinary fluidized solids with (b) a magnetically stabilized bed. The stable emulsion expands in length with increase of flow.

which are referred to as magnetically stabilized fluidized beds, offer a solution to problems associated with bubbling fluidized beds. The bed medium is quite flowable except at high magnetization, when the medium exhibits some properties of a gel. To understand why bubbles form and how their formation may be suppressed requires developing equations of motion capable of describing the transient, flow behavior of a fluidized system.

9.4 Basic relationships in two-phase flow

In a *two-phase system* composed of solid particles suspended in a fluid the dynamics should, in principle, be completely describable by the Navier–Stokes equation valid at each point within the fluid and Newton's second law for each particle, plus appropriate initial and boundary data. With but one or two particles suspended in a fluid this direct approach may be fruitful, but when the system of interest comprises a large number of closely spaced particles, as in a fluidized bed, it becomes intractable. Another approach is to replace the point-mechanical variables such as density and velocity, which vary rapidly on a scale comparable to the particle spacing, with smoothed variables obtained by averaging over regions large compared to the interparticle spacing but small compared to the characteristic dimension of the fluidized system. The resulting equations describe the motion of the two phases as if each one occupied the whole volume but is resisted by its motion relative to the other phase. The procedure produces a theory of *interpenetrating continua*.

Such a procedure is an alternative to postulating a set of continuum equations with all its inherent uncertainties at the outset. A beneficial feature of employing averaging is that the uncertainty (ignorance) is concentrated into identified terms that arise naturally, and so the equations may be completed by postulating reasonable, but empirical, constitutive equations. The averaging treatment presented here is a modification on that of Anderson and Jackson (1967), augmented with the development of magnetic terms and equations that are specific to the present subject.

Definition of averages

To begin, a weighting function $g(r)$ defined for distances $r > 0$ is introduced that permits defining what is meant by "average at a point." The exact form and scale of $g(r)$ are unimportant provided $g(r)$ possesses certain properties. Thus, $g(r)$ is defined to be positive for all r

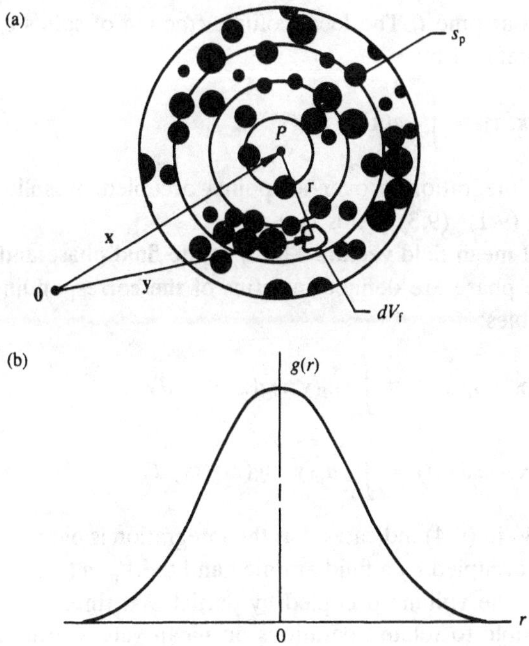

9.4 (a) The averaging region and (b) typical shape of a weighting function.

and to smoothly decrease with increasing r such that derivatives of all orders exist. Also, $g(r)$ is normalized so that

$$\int_{V_\infty} g(r)\, dV = 1 \tag{9.1}$$

where the V_∞ labeling the integral denotes integration over the entire system. In Figure 9.4a a field point P is identified by a position vector $\mathbf{x}$ drawn from an origin $\mathbf{0}$, and the location of a volume element dV_f is identified by a vector $\mathbf{y}$. Although $r = |\mathbf{x} - \mathbf{y}|$, the notation $g(r) = g(\mathbf{x} - \mathbf{y})$ is permissible and will be used for simplicity.

The local-mean voidage ε_f, or volume fraction of space occupied by fluid, is defined by

$$\varepsilon_f(\mathbf{x}, t) \equiv \int_{V_f} g(\mathbf{x} - \mathbf{y})\, dV \tag{9.2}$$

where the V_f denotes integration over all points $\mathbf{y}$ of the system occupied by the fluid at time t. The local volume fraction of solids ε_p is correspondingly defined by

$$\varepsilon_p(\mathbf{x}, t) \equiv \int_{V_p} g(\mathbf{x} - \mathbf{y})\, dV \tag{9.3}$$

where the integration is over all points occupied by solid at time t. Combining (9.1)–(9.3) yields $\varepsilon_f + \varepsilon_p = 1$.

The local-mean field vectors $\mathbf{a}_f(\mathbf{x}, t)$ of the fluid phase and $\mathbf{a}_p(\mathbf{x}, t)$ of the particle phase are defined in terms of the corresponding (primed) point variables:

$$\varepsilon_f(\mathbf{x}, t)\mathbf{a}_f(\mathbf{x}, t) \equiv \int_{V_f} \mathbf{a}_f'(\mathbf{y}, t) g(\mathbf{x} - \mathbf{y})\, dV \tag{9.4}$$

$$\varepsilon_p(\mathbf{x}, t)\mathbf{a}_p(\mathbf{x}, t) \equiv \int_{V_p} \mathbf{a}_p'(\mathbf{y}, t) g(\mathbf{x} - \mathbf{y})\, dV \tag{9.5}$$

where the V_f in (9.4) indicates that the integration is over all points $\mathbf{y}$ of the system occupied by a fluid at time t and the V_p in (9.5) that integration is over the volume occupied by particles at time t.

It is possible to relate operations on mean values to corresponding operations on point values using these definitions. Thus, from (9.4), with the notation ∇_x for the operator ∇ applied at position $\mathbf{x}$ and ∇_y for the operator applied at position $\mathbf{y}$,

$$\begin{aligned}
&\nabla_x \cdot [\varepsilon_f(\mathbf{x}, t)\mathbf{a}_f(\mathbf{x}, t)] \\
&= \int_{V_f} \mathbf{a}_f'(\mathbf{y}, t) \cdot \nabla_x g(\mathbf{x} - \mathbf{y})\, dV \\
&= -\int_{V_f} \mathbf{a}_f'(\mathbf{y}, t) \cdot \nabla_y g(\mathbf{x} - \mathbf{y})\, dV \\
&= \int_{V_f} g(\mathbf{x} - \mathbf{y}) \nabla_y \cdot \mathbf{a}_f'(\mathbf{y}, t)\, dV - \int_{V_f} \nabla_y \cdot [\mathbf{a}_f'(\mathbf{y}, t) g(\mathbf{x} - \mathbf{y})]\, dV \\
&= \int_{V_f} g(\mathbf{x} - \mathbf{y}) \nabla_y \cdot \mathbf{a}_f'(\mathbf{y}, t)\, dV - \int_{S_f} \mathbf{n}_f \cdot \mathbf{a}_f'(\mathbf{y}, t) g(\mathbf{x} - \mathbf{y})\, dS
\end{aligned} \tag{9.6}$$

where the divergence theorem was employed to obtain the last line, the S_f on the last integral denotes that it is taken over the whole boundary of the region occupied by a fluid, and $\mathbf{n}_f$ is the unit normal directed outward from the fluid boundary. The surface S_f bounding all the fluid consists of a number of disjoint parts, so the last term of (9.6) may be written

$$\int_{S_f} \mathbf{n}_f \cdot \mathbf{a}_f'(\mathbf{y}, t) g(\mathbf{x} - \mathbf{y})\, dS = \int_{S_{f\infty}} \mathbf{n}_f \cdot \mathbf{a}_f'(\mathbf{y}, t) g(\mathbf{x} - \mathbf{y})\, dS - \Sigma \int_{s_p} \mathbf{n}_p \cdot \mathbf{a}_f'(\mathbf{y}, t) g(\mathbf{x} - \mathbf{y})\, dS \qquad (9.7)$$

where $S_{f\infty}$ is the outer bound of the system, s_p denotes surface of a separate solid particle, and the summation is over all the particles in the system. In the last term of (9.7), $\mathbf{n}_p = -\mathbf{n}_f$.

For points not too close to the system boundary, the first integral on the right side of (9.7) is negligibly small compared to the contributions from the particle surfaces as $g(\mathbf{x} - \mathbf{y})$ is small at such points, and (9.6) yields the desired relation

$$\int_{V_f} g(\mathbf{x} - \mathbf{y}) \nabla_y \cdot \mathbf{a}_f'(\mathbf{y}, t)\, dV = \nabla_x \cdot [\varepsilon_f(\mathbf{x}, t) \mathbf{a}_f(\mathbf{x}, t)] - \Sigma \int_{s_p} \mathbf{n}_p \cdot \mathbf{a}_f'(\mathbf{y}, t) g(\mathbf{x} - \mathbf{y})\, dS \qquad (9.8)$$

A similar relation between the time derivatives of point and local-mean variables can be obtained directly from (9.4):

$$\int_{V_f} g(\mathbf{x} - \mathbf{y}) \frac{\partial}{\partial t} \mathbf{a}'_f(\mathbf{y}, t)\, dV = \frac{\partial}{\partial t} [\varepsilon_f(\mathbf{x}, t) \mathbf{a}_f(\mathbf{x}, t)] + \Sigma \int_{s_p} \mathbf{a}_f'(\mathbf{y}, t) \mathbf{n}_p \cdot \mathbf{v}_p'(\mathbf{y}, t) g(\mathbf{x} - \mathbf{y})\, dS \qquad (9.9)$$

where use has been made of the Reynolds' transport theorem for differentiating a volume integral, recognizing that the fluid volume has moving boundaries at the surfaces of particles, the surfaces comprising a number of disjoint parts. In the last term of (9.9) $\mathbf{v}_p'$ is the velocity of the solid at the surface.

Averaged continuity equations

The method of obtaining the desired continuum equations from point relations in the fluid or particle phases is illustrated, as an example, for the case of the continuity equation for the fluid phase. If the fluid is assumed to be incompressible, its point velocity satisfies the continuity equation

$$\nabla_y \cdot \mathbf{v}_f'(\mathbf{y}) = 0 \tag{9.10}$$

at all points instantaneously occupied by the fluid. Multiplying this by $g(\mathbf{x} - \mathbf{y})$ and integrating over all such points gives

$$\int_{V_f} g(\mathbf{x} - \mathbf{y}) \nabla_y \cdot \mathbf{v}_f'(\mathbf{y})\, dV = 0 \tag{9.11}$$

By (9.8) with $\mathbf{v}_f$ in place of $\mathbf{a}_f$ and $\mathbf{v}_f'$ in place of $\mathbf{a}_f'$, this may also be written

$$\nabla_x \cdot [\varepsilon_f(\mathbf{x})\mathbf{v}_f(\mathbf{x})] = \Sigma \int_{S_p} \mathbf{n}_p \cdot \mathbf{v}_f'(\mathbf{y}, t) g(\mathbf{x} - \mathbf{y})\, dS \tag{9.12}$$

Using (9.9), setting $\mathbf{a}_f'$ equal to a constant unit vector, gives the result

$$\frac{\partial}{\partial t} \varepsilon_f(\mathbf{x}, t) = -\Sigma \int_{S_p} \mathbf{n}_p \cdot \mathbf{v}_p(\mathbf{y}, t) g(\mathbf{x} - \mathbf{y})\, dS \tag{9.13}$$

Because $\mathbf{v}_f' = \mathbf{v}_p'$ everywhere on the fluid surfaces, the right sides of (9.12) and (9.13) cancel when these relations are added. Thus

$$\partial \varepsilon_f / \partial t + \nabla \cdot \varepsilon_f \mathbf{v}_f = 0 \tag{9.14}$$

which is the desired continuity equation in terms of local-mean fluid properties. This equation has the form suggested by intuition. A corresponding equation of continuity for the particle phase may be constructed from the point equations in a similar way:

$$\partial \varepsilon_p / \partial t + \nabla \cdot \varepsilon_p \mathbf{v}_p = 0 \tag{9.15}$$

Here and below the subscript on ∇ is dropped when the expression is valid for a field location at either $\mathbf{x}$ or $\mathbf{y}$.

Averaged magnetostatic relationships

Similar results for the magnetic-field vectors are required and will now be developed. It will be recalled that the induction field satisfies $\nabla \cdot \mathbf{B} = 0$ everywhere, so for any point in the fluid or solid, respectively,

$$\nabla_y \cdot \mathbf{B}_f' = 0 \quad \text{and} \quad \nabla_y \cdot \mathbf{B}_p' = 0 \tag{9.16}$$

The definitions of the spatially averaged induction fields $\mathbf{B}_f(\mathbf{x})$ in the fluid and $\mathbf{B}_p(\mathbf{x})$ in the particles, from (9.4), are, respectively,

$$\varepsilon_f \mathbf{B}_f \equiv \int_{V_f} \mathbf{B}'_f(\mathbf{y}) g(\mathbf{x} - \mathbf{y})\, dV \tag{9.17}$$

$$\varepsilon_p \mathbf{B}_p \equiv \int_{V_p} \mathbf{B}'_p(\mathbf{y}) g(\mathbf{x} - \mathbf{y})\, dV \tag{9.18}$$

The divergence of $\varepsilon_f \mathbf{B}_f$ and $\varepsilon_p \mathbf{B}_p$ may be found in a manner analogous to the procedure employed for obtaining (9.8), with the results

$$\nabla_x \cdot (\varepsilon_f \mathbf{B}_f) = \int_{V_f} g(\mathbf{x} - \mathbf{y}) \nabla_y \cdot \mathbf{B}'_f(\mathbf{y})\, dV + \sum \int_{S_p} \mathbf{n}_f \cdot \mathbf{B}'_f(\mathbf{y}) g(\mathbf{x} - \mathbf{y})\, dS \tag{9.19}$$

$$\nabla_x \cdot (\varepsilon_p \mathbf{B}_p) = \int_{V_p} g(\mathbf{x} - \mathbf{y}) \nabla_y \cdot \mathbf{B}'_p(\mathbf{y})\, dV - \sum \int_{S_p} \mathbf{n}_p \cdot \mathbf{B}'_p(\mathbf{y}) g(\mathbf{x} - \mathbf{y})\, dS \tag{9.20}$$

Because $\mathbf{n}_f \cdot \mathbf{B}'_f = \mathbf{n}_p \cdot \mathbf{B}'_p$ at any point on an interface between phases, the last terms in (9.19) and (9.20) cancel on addition. Also, from (9.16) the first term on the right side of both equations is zero. Thus a vector $\mathbf{B}$ may be defined as

$$\mathbf{B} \equiv \varepsilon_f \mathbf{B}_f + \varepsilon_p \mathbf{B}_p \tag{9.21}$$

representing the volume-weighted average field in the mixture of fluid and solids. Summing equations (9.19) and (9.20) proves that

$$\nabla \cdot \mathbf{B} = 0 \tag{9.22}$$

which is the same form as given by Maxwell's equation for a continuum.

The magnetic field may be treated in a corresponding manner:

$$\nabla_y \times \mathbf{H}'_f = \mathbf{0} \qquad \text{and} \qquad \nabla_y \times \mathbf{H}'_p = \mathbf{0} \tag{9.23}$$

Definitions for the averaged fields, from (9.4), are

$$\varepsilon_f \mathbf{H}_f = \int_{V_f} \mathbf{H}'_f(\mathbf{y}) g(\mathbf{x} - \mathbf{y})\, dV \tag{9.24}$$

$$\varepsilon_p \mathbf{H}_p = \int_{V_p} \mathbf{H}'_p(\mathbf{y}) g(\mathbf{x} - \mathbf{y})\, dV \tag{9.25}$$

Using the definitions of (9.24) and (9.25)

$$\nabla_x \times \varepsilon_f \mathbf{H}_f = \int_{V_f} g(\mathbf{x} - \mathbf{y}) \nabla_y \times \mathbf{H}_f' \, dV - \int_{V_f} \nabla_y \times [g(\mathbf{x} - \mathbf{y}) \mathbf{H}_f'] \, dV$$

$$= \int_{S_p} g(\mathbf{x} - \mathbf{y}) \mathbf{n}_p \times \mathbf{H}_f' \, dS \tag{9.26}$$

and, likewise,

$$\nabla_x \times \varepsilon_p \mathbf{H}_p = -\int_{S_p} g(\mathbf{x} - \mathbf{y}) \mathbf{n}_p \times \mathbf{H}'_p \, dS \tag{9.27}$$

where a corollary of the divergence theorem is employed to convert from the volume integral to the surface integral containing the cross product with the surface unit-normal vector, and the relationships (9.23) are invoked. The tangential component of the magnetic field is continuous across an interface, and therefore only the tangential field components survive the operations $\mathbf{n}_p \times \mathbf{H}_f'$ and $\mathbf{n}_p \times \mathbf{H}_p'$, and addition of (9.26) and (9.27) yields

$$\nabla \times \mathbf{H} = \mathbf{0} \tag{9.28}$$

where $\mathbf{H}$ is the average vector in the two-phase mixture, defined as

$$\mathbf{H} = \varepsilon_f \mathbf{H}_f + \varepsilon_p \mathbf{H}_p \tag{9.29}$$

Drew (1981) has derived equations (9.22) and (9.28) with averaging that uses a phase-indicator function.

Averaged momentum balances

Momentum balances for the fluid and particle phases can also be constructed using the averaging technique. The influence of magnetism will be considered later. The point of departure is the Cauchy momentum equation of a phase; this is given by (8.21) for the fluid phase:

$$\rho_f(\partial \mathbf{v}_f'/\partial t + \mathbf{v}_f' \cdot \nabla \mathbf{v}_f') = \rho_f \mathbf{g} + \nabla \cdot \mathbf{T}_f' \tag{9.30}$$

where a prime denotes the point value. The key term is that containing the surface stress tensor $\mathbf{T}_f'$. Multiplying the term by $g(\mathbf{x} - \mathbf{y})\, dV$, employing the tensor identity $\nabla\cdot(g\mathbf{T}_f') = g\nabla\cdot\mathbf{T}_f' + (\nabla g)\cdot\mathbf{T}_f'$ where $g = g(\mathbf{x}-\mathbf{y})$, and indicating the integrating gives

$$\int_{V_f} g(\mathbf{x}-\mathbf{y})\nabla_y\cdot\mathbf{T}_f'(\mathbf{y})\,dV = \int_{V_f}\nabla_y\cdot[g(\mathbf{x}-\mathbf{y})\mathbf{T}_f'(\mathbf{y})]\,dV - \int_{V_f}[\nabla_y g(\mathbf{x}-\mathbf{y})]\cdot\mathbf{T}_f'(\mathbf{y})\,dV \qquad (9.31)$$

The first term on the right side can be transformed using the divergence theorem to give

$$\int_{V_f}\nabla_y\cdot[g(\mathbf{x}-\mathbf{y})\mathbf{T}_f'(\mathbf{y})]\,dV = \int_{S_f}\mathbf{n}_f\cdot g(\mathbf{x}-\mathbf{y})\mathbf{T}_f'(\mathbf{y})\,dS = \int_{S_{f\infty}}\mathbf{n}_f\cdot g(\mathbf{x}-\mathbf{y})\mathbf{T}_f'(\mathbf{y})\,dS + \sum\int_{S_p} g(\mathbf{x}-\mathbf{y})\mathbf{n}_f\cdot\mathbf{T}_f'(\mathbf{y})\,dS \qquad (9.32)$$

where in the last term the summation is taken over all particles in the system.

The integral over $S_{f\infty}$, representing the outer bound of the fluid, is negligible for $\mathbf{x}$ not too close to a boundary, as $g(\mathbf{x}-\mathbf{y})$ is small at such points. In the summation term, $\mathbf{n}_f\cdot\mathbf{T}_f'(\mathbf{y})$ represents the stress $\mathbf{t}_f'$ experienced by the fluid at a particle boundary, and the entire summation represents the local mean value of the total force per unit volume exerted by the particles on the fluid, i.e., the local mean solid–fluid interaction force $\mathbf{f}_{pf}$:

$$\mathbf{f}_{pf} \equiv \sum\int_{S_p} g(\mathbf{x}-\mathbf{y})\mathbf{t}_f'\,dS \qquad (9.33)$$

From Newton's third law, $\mathbf{f}_{pf} + \mathbf{f}_{fp} = \mathbf{0}$, where $\mathbf{f}_{fp}$ is the corresponding interaction force that the fluid exerts on the particles.

From the relationship $\nabla_y g = -\nabla_x g$, the second term on the right side of (9.31) can be written

$$-\int_{V_f}[\nabla_y g(\mathbf{x}-\mathbf{y})]\cdot\mathbf{T}_f'(\mathbf{y})\,dV = \int_{V_f}[\nabla_x g(\mathbf{x}-\mathbf{y})]\cdot\mathbf{T}_f'(\mathbf{y})\,dV = \nabla_x\int_{V_f} g(\mathbf{x}-\mathbf{y})\mathbf{T}_f'(\mathbf{y})\,dV = \nabla_x\cdot(\varepsilon_f\mathbf{T}_f) \qquad (9.34)$$

where $\mathbf{T}_f$ is the spatially averaged fluid stress tensor. Substituting these results into (9.31) gives the final form of the averaged surface stress term:

$$\int_{V_f} g(\mathbf{x}-\mathbf{y})\,\nabla_y\cdot\mathbf{T}_f'(\mathbf{y})\,dV = \nabla_x\cdot(\varepsilon_f\mathbf{T}_f) - \mathbf{f}_{fp} \tag{9.35}$$

The gravity term of (9.30) transforms directly using the definition (9.4) to give the term $\varepsilon_f\rho_f\mathbf{g}$. The time-dependent term of (9.30) transforms directly using (9.9) to give the term $\partial(\varepsilon_f\mathbf{v}_f)/\partial t$ plus a term summing over the particle surfaces. With the incompressible continuity relationship $\nabla\cdot\mathbf{v}_f' = 0$, the convective term $\mathbf{v}_f'\cdot\nabla\mathbf{v}_f'$ of (9.30) becomes $\nabla\cdot(\mathbf{v}_f'\mathbf{v}_f')$. This term transforms upon averaging to give the term $\nabla\cdot(\varepsilon_f\mathbf{v}_f\mathbf{v}_f)$, plus a summation term that cancels the summation term arising from the time-dependent term, plus a fluctuation correlation term that arises from the quadratic term in a similar manner as in the Reynolds' stress theory of turbulence. In the present case, however, the correlation term can arise solely from spatial variations in the flow field due to the presence of obstacles presented by the other phase. Following common practice, the term is assumed negligible in the following, although in principle it could be represented by a constitutive relationship.

Collecting the various surviving terms gives the averaged fluid phase momentum balance:

$$\rho_f\varepsilon_f(\partial\mathbf{v}_f/\partial t + \mathbf{v}_f\cdot\nabla\mathbf{v}_f) = \nabla\cdot(\varepsilon_f\mathbf{T}_f) - \mathbf{f}_{fp} + \varepsilon_f\rho_f\mathbf{g} \tag{9.36}$$

where the averaged fluid-phase continuity equation (9.14) was introduced. Equation (9.36) differs from the form of the Cauchy point equation (9.30) from which it is derived in containing the solid–fluid interaction force as an additional term and in containing the parameter ε_f, which is a characteristic property of a two-phase mixture.

An averaged momentum balance for the disperse phase may be formulated in the same manner as for the continuous phase, basing the analysis on a Cauchy momentum balance written for an element of substance within a particle of the dispersed phase. The steps of the derivation need not be written, as they are essentially identical with those of the continuous-phase derivation except for the interchange of the subscripts f and p; a particle-phase momentum balance in the form

$$\rho_p\varepsilon_p(\partial\mathbf{v}_p/\partial t + \mathbf{v}_p\cdot\nabla\mathbf{v}_p) = \nabla\cdot(\varepsilon_p\mathbf{T}_p) - \mathbf{f}_{pf} + \varepsilon_p\rho_p\mathbf{g} \tag{9.37}$$

is obtained, where $\mathbf{T}_p$ is a particle-phase local-mean stress tensor analogous to $\mathbf{T}_f$. In deriving (9.36) and (9.37) the relationships governing rotational motion have been neglected, so that antisymmetric stresses such as those considered in Chapter 8 have been assumed absent.

Alternative forms of the momentum balances appear in the literature with compensating terms lumped into the interaction-force term. The form of the momentum balances presented here has the merit of a simple but logical derivation, symmetry of appearance, and expression in terms of physically meaningful averaged variables. However, there is arbitrariness in the choice of a constitutive form of $\mathbf{f}_{pf}$, and different forms appear in the literature.

Constitutive relations

To close the set of continuity equations and equations of motion, it is necessary to relate the interaction force $\mathbf{f}_{fp}$ and the stress tensors $\mathbf{T}_f$ and $\mathbf{T}_p$ to the local-mean voidage, velocity, and pressure fields. A possible assumption in constituting the stress tensors is that the form be the same as in a Newtonian fluid when expressed in terms of the local-mean pressure and velocity fields, i.e., the form of equation (4.62). However, for subsequent work in this chapter a drastic simplifying assumption is made to approximate conditions in a two-phase system, namely,

$$\mathbf{T}_f = -p_f\mathbf{I}, \qquad \mathbf{T}_p = -p_p\mathbf{I} \tag{9.38}$$

These relationships assert that the bulk fluid phase behaves in an inviscid manner and that stress in the dispersed phase is characterized as a hydrostatic pressure. Readers familiar with fluid dynamics are familiar with the concept of pressure in fluids. For a solid disperse phase the concept is less familiar. In this case the point stress, as in an elastic solid, is thought of as made up of a spherical part acting equally in all directions plus an extra stress. The spherical part when averaged yields the pressure p_p, and the extra stress is considered negligible.

It will be assumed that the fluidizing fluid is a low-density gas. The constitutive relation for the fluid–particle interaction force is then taken to be of the form proposed by Ishii (1975) and Drew (1983):

$$\mathbf{f}_{fp} = \varepsilon_f\beta(\varepsilon_f)(\mathbf{v}_f - \mathbf{v}_p) - p_f\nabla\varepsilon_f \tag{9.39}$$

where a virtual-mass term depending on the relative acceleration of the phases is negligible for low-density fluid and has been omitted. The first term on the right side represents a drag force that is proportional to the relative local mean velocity of the phases, with a local-mean drag coefficient $\beta(\varepsilon_f)$ that is voidage dependent. A linear dependence of the drag force on the relative velocity is appropriate for low-speed flows, in analogy to Darcy flow through a porous medium. The form of the last term in (9.39) reflects the fact that fluid exerts pressure on the particle phase, whereas particle pressure is not transmitted to the fluid phase. Thus the term $-p_f\nabla\varepsilon_f$ in (9.39) ensures that pressure in the continuous phase will be hydrostatic in a system in which both phases are motionless and unaccelerated.

The results of applying the two-phase flow equations to various steady flow and hydrostatic problems are summarized in Table 9.1.

In a fluidized system it will be assumed that $p_p = p_f$, i.e., that the pressure in particles of the disperse phase is in hydrostatic equilibrium with the surrounding bulk fluid. This is consistent with conditions on a system in which particles are free of forceful contact with each other.

The influence of particle magnetization may be incorporated into the particle-phase momentum balance by noting that each term on the right side of (9.37) is in the form of a force density. Thus, what is desired is an additional term representing the spatially averaged magnetic-force density. Now, from equation (1.10), the force per particle is given by the term $\nu\mu_0 M_p \nabla H_c$, where M_p denotes the particle magnetization, ν the particle volume, and subscript c indicates the mean cavity field present at the position of the particle. The process of averaging is indicated by

$$\varepsilon_p(\mathbf{x}, t)\langle \mu_0 M_p \nabla H_c \rangle = \sum \mu_0 M_p \nabla H_c \nu g(\mathbf{x} - \mathbf{x}_p)$$

where $\langle\ \rangle$ denotes the spatial average evaluated at the point $\mathbf{x}$, the summation is taken over all particles in the system, and $\mathbf{x}_p$ denotes the position of a particle center of mass. The nonlinearity due to the presence of a product of field variables creates a difficulty. Thus, to use this expression it is necessary to constitute a relationship for $\langle \mu_0 M_p \nabla H_c \rangle$ in terms of the bed magnetization M, which in this case equals $\varepsilon_p M_p$, and the mean magnetic field H. The following constitutive relationship, one of the simplest possible, is introduced for $\mathbf{f}_{mp}$, which represents the spatially averaged magnetic force per unit volume:

$$\mathbf{f}_{mp} = \varepsilon_p \langle \mu_0 M_p \nabla H_c \rangle = \mu_0 M \nabla H \tag{9.40}$$

If the fluid phase were magnetizable, it would be appropriate also to introduce a magnetic-force density $\mathbf{f}_{mf}$ acting on it.

Table 9.1 *Application of the steady-state two-phase momentum equations*

	Conditions	Relationships	Comments
Flow through a fixed bed	$\mathbf{v}_p = \mathbf{0}, \quad \mathbf{g} = \mathbf{0}$ $\nabla\varepsilon_f = \nabla\varepsilon_p = \mathbf{0}$	$\Delta p_f = \beta(\varepsilon_f) v_f L$	Permits the drag coefficient to be determined from experimental measurement of the pressure drop.
Batch sedimentation	$\varepsilon_p \mathbf{v}_p + \varepsilon_f \mathbf{v}_f = \mathbf{0}$ $\nabla\varepsilon_f = \nabla\varepsilon_p = \mathbf{0}$ $p_f = p_p$	$\Delta p_f = (\varepsilon_f \rho_f + \varepsilon_p \rho_p) g L$ $v_p = -\dfrac{\varepsilon_f \varepsilon_p (\rho_p - \rho_f) g}{\beta(\varepsilon_f)}$	Relates the pressure to the average density of the mixture. $\beta(\varepsilon_f)$ determined from the settling velocity.
Fluidized solids	$\mathbf{v}_p = \mathbf{0}$ $\nabla\varepsilon_f = \nabla\varepsilon_p = \mathbf{0}$ $p_f = p_p$	$\Delta p_f = (\varepsilon_f \rho_f + \varepsilon_p \rho_p) g L$ $v_f = \dfrac{\varepsilon_p (\rho_p - \rho_f) g}{\beta(\varepsilon_f)}$	Pressure drop is independent of the flow rate. Bed voidage is calculable from the flow rate.
Inhomogeneous settled bed	$\mathbf{v}_f = \mathbf{v}_p = \mathbf{0}$ $\nabla\varepsilon_p \neq \mathbf{0}$	$\Delta p_f = \rho_f g L$ $\Delta[\varepsilon_p (p_p - p_f)] = \varepsilon_p (\rho_p - \rho_f) g L$	Fluid pressure is the usual hydrostatic. $\varepsilon_p (p_p - p_f)$ is the force supported by the solids because of the solids buoyant weight.[a]

[a] The particle pressure can vary independently of the fluid pressure. In the simple case of an evacuated bed ($p_f = 0$, $\rho_f = 0$) supporting a constant compressive load W in the absence of gravity ($g = 0$), $\Delta\varepsilon_p p_p = 0$ and $\varepsilon_p p_p = W$, showing that the particle pressure p_p varies with position in the inhomogeneously packed bed. The particle phase equation found in the literature, and derived under approximations in which terms of the form $\nabla \cdot (\varepsilon \mathbf{T})$ are reduced to the form $\varepsilon \nabla \cdot \mathbf{T}$, is unable to recover this result.

9.5 Summary of the averaged equations

The spatially averaged equation set governing the dynamics of magnetized two-phase flow as developed in the preceding sections may be summarized as follows:

Continuity equations:

$$\partial \varepsilon_f / \partial t + \nabla \cdot \varepsilon_f \mathbf{v}_f = 0 \qquad (9.41a)$$

$$\partial \varepsilon_p / \partial t + \nabla \cdot \varepsilon_p \mathbf{v}_p = 0 \qquad (9.41b)$$

Momentum balances:

$$\rho_f \varepsilon_f (\partial \mathbf{v}_f / \partial t + \mathbf{v}_f \cdot \nabla \mathbf{v}_f) = \nabla \cdot (\varepsilon_f \mathbf{T}_f) - \mathbf{f}_{fp} + \varepsilon_f \rho_f \mathbf{g} + \mathbf{f}_{mf} \qquad (9.42a)$$

$$\rho_p \varepsilon_p (\partial \mathbf{v}_p / \partial t + \mathbf{v}_p \cdot \nabla \mathbf{v}_p) = \nabla \cdot (\varepsilon_p \mathbf{T}_p) - \mathbf{f}_{pf} + \varepsilon_p \rho_p \mathbf{g} + \mathbf{f}_{mp} \qquad (9.42b)$$

Definitions:

$$\mathbf{f}_{pf} + \mathbf{f}_{fp} = \mathbf{0} \qquad (9.43a)$$

$$\varepsilon_f + \varepsilon_p = 1 \qquad (9.43b)$$

Magnetostatic equations:

$$\nabla \cdot \mathbf{B} = 0, \quad \text{where} \quad \mathbf{B} = \varepsilon_f \mathbf{B}_f + \varepsilon_p \mathbf{B}_p \qquad (9.44a)$$

$$\nabla \times \mathbf{H} = \mathbf{0}, \quad \mathbf{H} = \varepsilon_f \mathbf{H}_f + \varepsilon_p \mathbf{H}_p \qquad (9.44b)$$

$$\mathbf{B} = \mu_0 (\mathbf{H} + \mathbf{M}), \quad \mathbf{M} = \varepsilon_f \mathbf{M}_f + \varepsilon_p \mathbf{M}_p \qquad (9.44c)$$

The idealized constitutive relations closing the set of governing equations are summarized as follows:

Constitutive equations:

fluid–particle interaction force

$$\mathbf{f}_{fp} = \varepsilon_f \beta(\varepsilon_f)(\mathbf{v}_f - \mathbf{v}_p) - p_f \nabla \varepsilon_f \qquad (9.45)$$

inviscid bulk fluid

$$\mathbf{T}_f = -p_f \mathbf{I} \qquad (9.46a)$$

stress-free solids

$$\mathbf{T}_p = -p_p \mathbf{I} \qquad (9.46b)$$

nonmagnetic fluid phase

$$\mathbf{f}_{mf} = \mathbf{0} \qquad (9.47a)$$

magnetic body force

$$\mathbf{f}_{mp} = \mu_0 M \nabla H = \mu_0 \varepsilon_p M_p \nabla(\varepsilon_f H_f + \varepsilon_p H_p) \qquad (9.47b)$$

negligible interparticle contact force

$$p_p = p_f \qquad (9.48)$$

$\beta(\varepsilon_f)$ is a drag coefficient expressing the viscous force per unit volume due to relative motion between the fluid and a solid phase of uniform porosity. In the absence of magnetization $\beta(\varepsilon_f)$ has been a well-studied

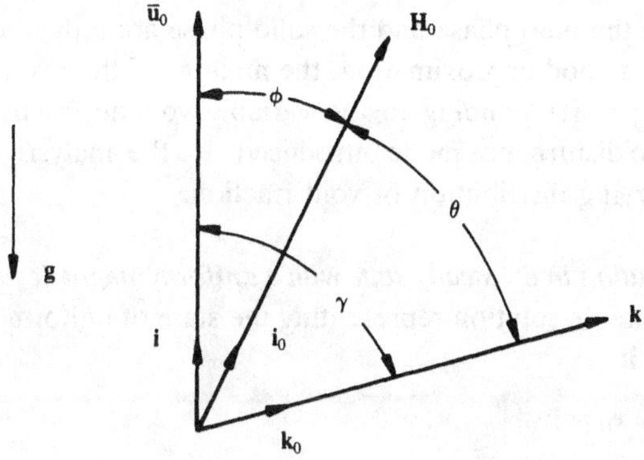

9.5 Nomenclature for the stability analysis of fluidized magnetized particles.

parameter in fixed-bed mechanics; a particular form for $\beta(\varepsilon_f)$ is introduced below. As noted in Section 7.8, the multiphase equation set reduces as a special case to the Darcy flow of a magnetic fluid in a porous medium.

Substituting the constitutive relationships (9.45)–(9.48) into the momentum balances with $\rho_f = 0$ yields the forms to be used in the analysis of magnetized solids that are fluidized with a gas of negligible density:

$$\nabla p_f + \beta(\varepsilon_f)(\mathbf{v}_f - \mathbf{v}_p) = \mathbf{0} \tag{9.49}$$

$$\varepsilon_p \rho_p(\partial \mathbf{v}_p/\partial t + \mathbf{v}_p \cdot \nabla \mathbf{v}_p) = -\varepsilon_p \nabla p_f + \varepsilon_f \beta(\varepsilon_f)(\mathbf{v}_f - \mathbf{v}_p) + \varepsilon_p \rho_p \mathbf{g} + \mu_0 \varepsilon_p M_p \nabla H \tag{9.50}$$

9.6 Magnetized fluidized solids

It is now possible to begin analyzing the dynamics and stability of magnetized fluidized beds. The analysis extends a treatment developed by Anderson and Jackson (1968) for unmagnetized beds. The nomenclature for the analysis is illustrated in Figure 9.5. For simplicity the analyzed system will be assumed to be of infinite extent in all directions. Flow of the fluidizing fluid is upward, opposite to the direction of gravity. A disturbance wave propagates through the system in the direction specified by its wave vector **k**, which is oriented at an angle γ relative to the flow direction. A uniform applied magnetic field $\mathbf{H}_0$ is impressed upon the system at an arbitrary orientation, specified by angle θ, taken relative to the direction of wave propagation.

Although the fluid phase and the solid phase are individually incompressible to a good approximation, the mixture of the two has variable mass density corresponding to the variable volume fraction of solids present. The disturbance mode introduced into the analysis represents a spatially varying distribution of void fraction.

Solution in the steady state with a uniform magnetic field

A simple solution representing the state of uniform fluidization is obtained if

$$\begin{aligned} &\mathbf{v}_f = \mathbf{v}_0 = \mathbf{i}v_0, \qquad \mathbf{v}_p = \mathbf{0} \\ &\varepsilon_f = \varepsilon_0, \qquad \varepsilon_p = 1 - \varepsilon_0 \\ &\mathbf{H} = \mathbf{H}_0 = \mathbf{i}_0 H_0, \qquad \mathbf{M} = \mathbf{M}_0 = \mathbf{i}_0 M_0, \qquad \mathbf{B} = \mathbf{B}_0 = \mathbf{i}_0 B_0 \end{aligned} \tag{9.51}$$

where $\mathbf{i}$ is the unit vector in the upward direction (opposite to gravity), $\mathbf{i}_0$ is a unit vector in the direction of the applied field, and u_0, ε_0, H_0, and M_0 are constants. In words, equations (9.51) represent the case of a local-mean fluid velocity that is constant in both space and time and directed vertically upward, a local-mean particle velocity that is everywhere zero, and voidage that is uniform in space and independent of time and the magnetic-field intensity. The magnetic induction and magnetization are everywhere uniform. The continuity equations (9.41) are satisfied identically, as are the magnetostatic field equations (9.44), and the fluid and solids momentum equations (9.49) and (9.50) reduce, respectively, to

$$\nabla p_{f,0} + \beta_0 \mathbf{v}_0 = \mathbf{0} \tag{9.52}$$

$$-(1 - \varepsilon_0)\nabla p_{f,0} + \varepsilon_0 \beta_0 \mathbf{v}_0 + (1 - \varepsilon_0)\rho_p \mathbf{g} = \mathbf{0} \tag{9.53}$$

Eliminating $p_{f,0}$ from equations these and noting that $\mathbf{v}_0$ is directed oppositely to $\mathbf{g}$ gives

$$\beta_0(\varepsilon_0)v_0 - (1 - \varepsilon_0)\rho_p g = 0 \tag{9.54}$$

When β is known as a function of ε, (9.54) determines the uniform voidage ε_0 corresponding to any fluid speed $\mathbf{v}_0$. Elimination of the velocity from (9.52) and (9.53) permits the fluid-pressure gradient to be calculated as

$$|\nabla p_{f,0}| = (1 - \varepsilon_0)\rho_p g \tag{9.55}$$

where it is seen that the gradient of pressure is uniform and equals the weight of bed solids per unit cross-sectional area of bed. This result recovers the experimental observation that at incipient fluidization the pressure drop across the bed is just sufficient to support the weight of the bed particles. It is also seen that the presence of the uniform applied magnetic field contributes no terms to the equilibrium solution, regardless of the field direction or magnitude.

Before proceeding further, it is necessary to specify the drag coefficient $\beta_0(\varepsilon_0)$ by assuming a model for the particle–fluid drag. For this purpose the well-known Ergun relationship will be adopted representing the frictional drag in the flow of fluid through a bed of packed particles:

$$\beta(\varepsilon_0) = K\left[\frac{(1-\varepsilon_0)^2}{\varepsilon_0^2} + \frac{1-\varepsilon_0}{k\varepsilon_0^2}N_{\mathrm{Re}}\right] \tag{9.56}$$

where

$$N_{\mathrm{Re}} \equiv D_{\mathrm{p}}v_0\rho_{\mathrm{f}}/\eta_{\mathrm{f}} \tag{9.57}$$

$$K \equiv 150\eta_{\mathrm{f}}/D_{\mathrm{p}}^2 \tag{9.58}$$

and k is an empirical constant whose value is 85.7. N_{Re} is the Reynolds number based on the particle diameter D_{p}, v_0 is the interstitial fluid velocity, the fluid density is ρ_{f}, and the fluid viscosity is η_{f}. Consistent with the approximation of a massless gas, N_{Re} is identically zero, so the second term in the brackets of (9.56) disappears.

9.7 Stability of the steady-state solution

The stability of the equilibrium solution with respect to small disturbances can now be examined by a linearized stability analysis (Rosensweig 1979c). Voidage perturbations that grow represent possible precursors of bubbles in fluidized beds. The qualification "possible" is made because a nonlinear analysis is needed to determine the actual fate of a disturbance that begins to grow. Nonetheless, linear stability analysis should give a necessary condition for stabilization to be achieved. The presence of voidage perturbations will perturb the uniform applied magnetic field initially present, so it may be anticipated that magnetic body forces will be generated and play a role in determining stability.

Derivation of the linearized equations

Each variable is written as the sum of its value in the uniform fluidization state and a small perturbation:

$$\begin{aligned}
&\varepsilon_{\mathrm{f}} = \varepsilon_0 + \varepsilon_1, \quad p_{\mathrm{f}} = p_{\mathrm{f},0} + p_1\\
&\mathbf{v}_{\mathrm{f}} \equiv \mathbf{v}_0 + \mathbf{v}_1 = \mathbf{i}v_0 + \mathbf{v}_1, \quad \mathbf{v}_{\mathrm{p}} \equiv \mathbf{0} + \mathbf{w}_1 = \mathbf{w}_1\\
&\mathbf{H} \equiv \mathbf{H}_0 + \mathbf{H}_1 = \mathbf{i}_0 H_0 + \mathbf{H}_1, \quad \mathbf{M} \equiv \mathbf{M}_0 + \mathbf{M}_1 = \mathbf{i}_0 M_0 + \mathbf{M}_1,\\
&\mathbf{B} \equiv \mathbf{B}_0 + \mathbf{B}_1 = \mathbf{i}_0 B_0 + \mathbf{B}_1
\end{aligned} \tag{9.59}$$

The disturbances are assumed to be small and also smoothly varying functions of both space and time, so any of their derivatives are also small. The linearized equations can now be obtained by substituting the relations (9.59) into (9.41a), (9.41b), (9.49), and (9.50) and using the steady solution given by (9.54).

Example – linearization: From (9.41a), the general spatially averaged continuity equation for the fluid phase is

$$\partial\varepsilon_f/\partial t + \nabla\cdot(\varepsilon_f \mathbf{v}_f) = 0$$

The individual terms are linearized to first-order terms as follows:

$$\frac{\partial\varepsilon_f}{\partial t} = \frac{\partial(\varepsilon_0 + \varepsilon_1)}{\partial t} = \frac{\partial\varepsilon_1}{\partial t}$$

$$\varepsilon_f \mathbf{v}_f = (\varepsilon_0 + \varepsilon_1)(\mathbf{v}_0 + \mathbf{v}_1) = \varepsilon_0\mathbf{v}_0 + \varepsilon_0\mathbf{v}_1 + \varepsilon_1\mathbf{v}_0$$

Recognizing that the term $\varepsilon_0\mathbf{v}_0$ is constant leads to

$$\nabla\cdot(\varepsilon_f\mathbf{v}_f) = \varepsilon_0\nabla\cdot\mathbf{v}_1 + \mathbf{v}_0\cdot\nabla\varepsilon_1 = \varepsilon_0\nabla\cdot\mathbf{v}_1 + v_0\,\partial\varepsilon_1/\partial x$$

Substitution of the linearized terms into the original equation gives the linearized fluid-continuity equation:

$$\partial\varepsilon_1/\partial t + v_0\,\partial\varepsilon_1/\partial x + \varepsilon_0\nabla\cdot\mathbf{v}_1 = 0 \tag{9.60}$$

Similarly, the linearized particle continuity equation is easily shown to be

$$-\partial\varepsilon_1/\partial t + (1 - \varepsilon_0)\nabla\cdot\mathbf{w}_1 = 0 \tag{9.61}$$

The small deviation velocity $\mathbf{w}_1$ such as occurs when a wave travels through the bed, enters into this equation.

The linearized fluid-momentum equation resulting from (9.49), although not as easily obtained as the linearized continuity relations, is

$$\nabla p_1 + \beta_0(\mathbf{v}_1 - \mathbf{w}_1) + \varepsilon_1\beta_0'\mathbf{v}_0 = \mathbf{0} \tag{9.62}$$

where $\beta_0 = \beta(\varepsilon_0)$ and the notation β_0' is used to denote the derivative $\beta_0' \equiv (d\beta/d\varepsilon)_0$.

The linearized solids-momentum equation corresponding to (9.50) is the most difficult one of the four to derive because of the final magnetic term. In order to transform this term into the desired form, H is expanded with the aid of the binomial theorem as follows:

$$\begin{aligned} H &= |\mathbf{H}_0 + \mathbf{H}_1| = [(H_0\mathbf{i}_0 + \mathbf{H}_1)\cdot(H_0\mathbf{i}_0 + \mathbf{H}_1)]^{1/2} \\ &= (H_0^2 + 2H_0\mathbf{H}_1\cdot\mathbf{i}_0)^{1/2} + \text{higher-order terms} \\ &= H_0 + \mathbf{H}_1\cdot\mathbf{i}_0 \end{aligned} \tag{9.63}$$

Similarly,

$$M = M_0 + \mathbf{M}_1 \cdot \mathbf{i}_0 \tag{9.64}$$

The following linearized solids-momentum equation is obtained after extensive algebraic manipulation:

$$\rho_p(1 - \varepsilon_0)\,\partial \mathbf{w}_1/\partial t = \mathbf{i}g\rho_p\varepsilon_1 + \beta_0(\mathbf{v}_1 - \mathbf{w}_1) + \mathbf{v}_0\beta_0'\varepsilon_1 + \mu_0 M_0 \nabla(\mathbf{H}_1 \cdot \mathbf{i}_0) \tag{9.65}$$

where use was made of (9.62) to eliminate the pressure gradient ∇p_1.

Form of the linearized equations for plane waves

In order to solve the simultaneous set of linearized equations of motion (9.60)–(9.62) and (9.65) it is necessary to relate the magnetic term of (9.65), which contains the unknown quantity $\mathbf{H}_1$, to independent parameters of the system. The relationship, to be derived, appears as (9.84). To begin, field equations are written for the small perturbations $\mathbf{H}_1$ and $\mathbf{M}_1$.

$$\nabla \times \mathbf{H}_1 = \mathbf{0} \tag{9.66}$$

$$\nabla \cdot \mathbf{B}_1 = 0 \tag{9.67}$$

$$\mathbf{B}_1 = \mu_0(\mathbf{M}_1 + \mathbf{H}_1) \tag{9.68}$$

Eliminating $\mathbf{B}_1$ reduces this set of equations to the pair

$$\nabla \times \mathbf{H}_1 = \mathbf{0} \tag{9.69}$$

$$\nabla \cdot \mathbf{H}_1 = -\nabla \cdot \mathbf{M}_1 \tag{9.70}$$

Next the magnetization must be expressed in terms of the perturbation quantities. Figure 9.6 introduces some additional notation with reference to the magnetization curve. The bed solids saturate magnetically at a value denoted $M_{p,s}$. For the unperturbed bed the quantity M_0, representing the magnetization per unit volume of the bed, is $M_0 = (1 - \varepsilon_0)M_{p,0} = (1 - \varepsilon_0)\chi_0 H_0$. The magnetic parameters of interest, the tangent slope and the chord slope, are defined by

$$\text{tangent susceptibility} = \hat{\chi} \equiv (\partial M_p/\partial H)_{H_0} \tag{9.71}$$

$$\text{chord susceptibility} = \chi_0 \equiv M_{p,0}/H_0 \tag{9.72}$$

From the definition of $\hat{\chi}$ a small variation δH produces a small change $\delta M_{p,0} = \hat{\chi}\,\delta H$. According to (9.63), however, $\delta H = H - H_0 = \mathbf{H}_1 \cdot \mathbf{i}_0$, and so the magnitude of M_p has the form

$$M_p(H) = M_{p,0}(H_0) + \hat{\chi}(\mathbf{H}_1 \cdot \mathbf{i}_0) \tag{9.73}$$

It will be assumed throughout that the bed magnetization $\mathbf{M}$ is parallel

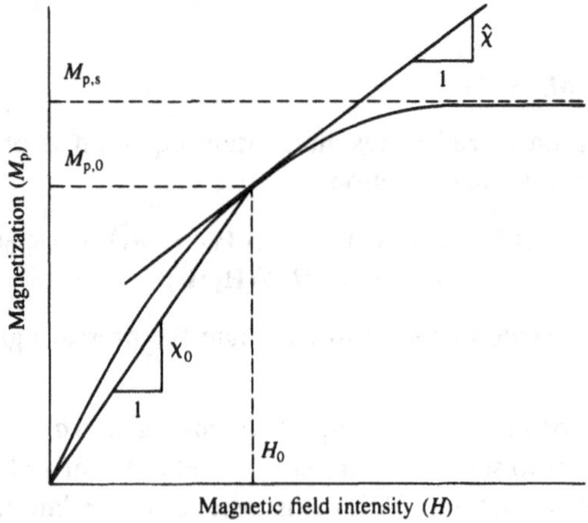

9.6 **Magnetic parameters of bed solids for a material that magnetizes nonlinearly and exhibits saturation.**

to the local field **H** so that using Eq. (9.44) with $\mathbf{M}_f = \mathbf{0}$ gives $\mathbf{M} = \varepsilon_p M_p \mathbf{H}/H$, corresponding to a bed that is ferromagnetically soft. Eliminating M_p with the use of (9.73) and H with (9.63) and collecting perturbation terms of first order only gives the following relationship for the vector perturbation of magnetization $\mathbf{M}_1$:

$$\frac{\mathbf{M}_1}{M_{p,0}(H_0)} = \frac{1-\varepsilon_0}{H_0}\left[\mathbf{H}_1 - (\mathbf{i}_0 \cdot \mathbf{H}_1)\left(1 - \frac{\hat{\chi}}{\chi_0}\right)\mathbf{i}_0\right] - \varepsilon_1 \mathbf{i}_0 \qquad (9.74)$$

This relationship shows that, even though **M** and **H** remain parallel, $\mathbf{M}_1$ and $\mathbf{H}_1$ are not. Now taking the divergence of (9.74) and combining with (9.70) yields an equation that relates the field $\mathbf{H}_1$ to the presence of the voidage perturbation ε_1:

$$\left(\frac{1}{\chi_0} + 1 - \varepsilon_0\right)\nabla \cdot \mathbf{H}_1$$
$$= (1-\varepsilon_0)\left(1 - \frac{\hat{\chi}}{\chi_0}\right)\mathbf{i}_0 \cdot \nabla(\mathbf{i}_0 \cdot \mathbf{H}_1) + \mathbf{H}_0 \cdot \nabla \varepsilon_1 \qquad (9.75)$$

As usual, any arbitrary disturbance may be represented as a sum of Fourier modes. Thus, obtaining a general solution of (9.75) valid for any distribution of voidage can be circumvented because only harmonic disturbance modes need be considered. [The solution obtained finally appears as (9.84).] It may be assumed that initially a train of waves of solids concentration having infinitesimally small amplitude is suddenly produced in the fluid mixture. The development of a single Fourier component may be considered in isolation owing to the linearity of the equations. Then, the motion of a wave through the mixture and its change in amplitude with respect to time may be computed analytically. Again, if the amplitude increases exponentially, the uniform density distribution is said to be unstable to that disturbance mode.

Thus, consider plane waves in the bulk of the form

$$\varepsilon_1 = \hat{\varepsilon}_1 E \tag{9.76}$$

$$\mathbf{H}_1 = \hat{\mathbf{H}}_1 E \tag{9.77}$$

where

$$E = \exp(st)\exp(i\boldsymbol{k}\cdot\mathbf{x}) \tag{9.78}$$

with $\hat{\varepsilon}_1$ the amplitude of the perturbation in voidage, $\hat{\mathbf{H}}_1$ a constant vector to be established by analysis, $\boldsymbol{k}$ the wave vector of the plane-wave disturbance, and $\mathbf{x}$ the position vector ($\boldsymbol{k}\cdot\mathbf{x} = k_x x + k_y y + k_z z$).

Imposing Ampère's law in the form of (9.69) on the solution yields

$$\mathbf{0} = \nabla \times \mathbf{H}_1 = E\nabla \times \hat{\mathbf{H}}_1 - \hat{\mathbf{H}}_1 \times \nabla E = -iE\hat{\mathbf{H}}_1 \times \boldsymbol{k} \tag{9.79}$$

so

$$\hat{\mathbf{H}}_1 \times \boldsymbol{k} = \boldsymbol{0} \tag{9.80}$$

This shows that $\hat{\mathbf{H}}_1$ is a vector parallel to the wave-number vector $\boldsymbol{k}$. If $\boldsymbol{k}_0$ denotes the unit vector in the direction of the wave-number vector $\boldsymbol{k}$ then the term $\mathbf{i}_0\cdot\mathbf{H}_1$ in (9.74) may be replaced by $E\hat{H}_1\cos\theta$, where $\hat{H}_1$ is the magnitude of $\mathbf{H}_1$ and θ is defined as

$$\theta \equiv \cos^{-1}(\mathbf{i}_0\cdot\boldsymbol{k}_0) \tag{9.81}$$

The following solution for $\hat{H}_1$ may now be found by substituting for ε_1 and $\mathbf{H}_1$ from (9.76) and (9.77) into (9.75):

$$\hat{H}_1/M_{\mathrm{p},0}\hat{\varepsilon}_1 = \alpha/\cos\theta \tag{9.82}$$

$$\alpha = \frac{\cos^2\theta}{1 + (1-\varepsilon_0)(1-\cos^2\theta)\chi_0 + \hat{\chi}(1-\varepsilon_0)\cos^2\theta} \tag{9.83}$$

With (9.82), the magnetic term of (9.65) becomes

$$M_0\nabla(\mathbf{H}_1\cdot\mathbf{i}_0) = \mu_0\alpha(1-\varepsilon_0)M_{\mathrm{p},0}^2\nabla\varepsilon_1 \tag{9.84}$$

It is now possible to formulate a partial-differential equation in the voidage perturbation ε_1. Equation (9.65) is divided by ρ_p, its divergence is computed, and substitution for $\nabla\cdot\mathbf{w}_1$ and $\nabla\cdot\mathbf{v}_1$ is made from (9.60) and (9.61). Evaluating each term in turn gives

$$\nabla\cdot\left[(1-\varepsilon_0)\frac{\partial\mathbf{w}_1}{\partial t}\right] = \frac{\partial}{\partial t}[(1-\varepsilon_0)\,\nabla\cdot\mathbf{w}_1] = \frac{\partial^2\varepsilon_1}{\partial t^2} \tag{9.85}$$

$$\nabla\cdot(\mathbf{i}g\varepsilon_1) = \mathbf{i}g\cdot\nabla\varepsilon_1 = g\frac{\partial\varepsilon_1}{\partial x} \tag{9.86}$$

$$\begin{aligned}\nabla\cdot\left[\frac{\beta_0}{\rho_p}(\mathbf{v}_1-\mathbf{w}_1)\right] &= \frac{\beta_0}{\rho_p}(\nabla\cdot\mathbf{v}_1-\nabla\cdot\mathbf{w}_1)\\ &= \frac{\beta_0}{\rho_p}\left(-\frac{1}{\varepsilon_0}\frac{\partial\varepsilon_1}{\partial t}-\frac{1}{\varepsilon_0}v_0\frac{\partial\varepsilon_1}{\partial x}-\frac{1}{1-\varepsilon_0}\frac{\partial\varepsilon_1}{\partial t}\right)\\ &= -\frac{\beta_0}{\rho_p\varepsilon_0(1-\varepsilon_0)}\frac{\partial\varepsilon_1}{\partial t}-\frac{\beta_0 v_0}{\rho_p\varepsilon_0}\frac{\partial\varepsilon_1}{\partial x}\\ &= -\frac{g}{\varepsilon_0 v_0}\frac{\partial\varepsilon_1}{\partial t}-\frac{g(1-\varepsilon_0)}{\varepsilon_0}\frac{\partial\varepsilon_1}{\partial x}\end{aligned} \tag{9.87}$$

$$\begin{aligned}\nabla\cdot\left(\frac{\beta_0'}{\rho_p}\mathbf{v}_0\varepsilon_1\right) &= \frac{\beta_0'}{\rho_p}\mathbf{v}_0\cdot\nabla\varepsilon_1 = \frac{v_0\beta_0'}{\rho_p}\frac{\partial\varepsilon_1}{\partial x}\\ &= g(1-\varepsilon_0)\frac{\beta_0'}{\beta_0}\frac{\partial\varepsilon_1}{\partial x}\end{aligned} \tag{9.88}$$

$$\nabla\cdot\left[\frac{M_{p,0}}{\rho_p}\nabla(\mathbf{H}_1\cdot\mathbf{i}_0)\right] = \mu_0\alpha(1-\varepsilon_0)\frac{M_{p,0}^2}{\rho_p}\nabla^2\varepsilon_1 \tag{9.89}$$

where in going from the third to the fourth line in (9.87) and from the first line to the second line in (9.88) use was made of the steady solution (9.54). Putting all the parts together yields the partial-differential equation for the voidage perturbation ε_1:

$$\begin{aligned}&\frac{\partial^2\varepsilon_1}{\partial t^2}+\frac{g}{\bar{u}_0}\frac{\partial\varepsilon_1}{\partial t}+\frac{g}{\varepsilon_0}\left[1-2\varepsilon_0-\varepsilon_0(1-\varepsilon_0)\frac{\beta_0'}{\beta_0}\right]\frac{\partial\varepsilon_1}{\partial x}\\ &\qquad -\mu_0\alpha(1-\varepsilon_0)\frac{M_{p,0}^2}{\rho_p}\nabla^2\varepsilon_1 = 0\end{aligned} \tag{9.90}$$

where $\bar{u}_0 = \varepsilon_0 v_0$ denotes the fluidizing velocity on a superficial basis. This equation duplicates equation (19) of Anderson and Jackson (1968) with vanishing fluid density and substitution of the magnetic factor for the coefficient of $\nabla^2\varepsilon_1$.

Equation (9.90) is applicable to general plane-wave voidage perturbations in three spatial dimensions in which mean flow is directed along the x axis and applied field is oriented at angle θ relative to the direction of the disturbance mode. This equation is the key governing relationship in this theory.

Formally, with $g = 0$, the second and third terms disappear, and what remains is the classical wave equation, with solutions yielding no growth or decay of the wave amplitude. Neglecting the first term of (9.90) yields the convective diffusion equation. Thus, if ε_1 is regarded as the concentration of a diffusive species, the resulting equation describes the dispersion of the substance injected as a marker into a flow of a streamtube. In the absence of the third term in (9.90) the equation reduces to the form of the telegrapher's equation. Then ε_1 represents the current or voltage in a lossless electrical transmission line having distributed capacitance and reluctance, and there is no decay or growth of the transmitted signal. When magnetization is absent, the last term disappears, and the resultant equation describes ordinary fluidized solids. With all terms present the equation appears not to have an analog to any well-known physical problem, and so the equation must be solved to gain any further physical understanding.

General solution for the voidage perturbation

In general, s in equation (9.78) is a complex number and may be written

$$s \equiv \xi - i\eta \tag{9.91}$$

Thus equation (9.78) may be rewritten

$$E = \exp(\xi t)\exp[i(\boldsymbol{k}\cdot\mathbf{x} - \eta t)] \tag{9.92}$$

The quantity $\Omega = \boldsymbol{k}\cdot\mathbf{x} - \eta t$ is the *phase*, and for plane-wave solutions the phase surfaces Ω = const are parallel planes. The gradient of Ω in space is the wave number $\boldsymbol{k}$, whose direction is normal to these planes and whose magnitude $|\boldsymbol{k}| = k$ is the number of crests per 2π units of distance in that direction. The wavelength λ for these plane waves is then $\lambda = 2\pi/k$. Writing

$$\Omega = k[\boldsymbol{k}_0\cdot\mathbf{x} - (\eta/k)t] = \boldsymbol{k}\cdot[\mathbf{x} - (\eta/k)t\boldsymbol{k}_0] \tag{9.93}$$

shows that the *phase velocity* is given by

$$\mathbf{V}_p = (\eta/k)\boldsymbol{k}_0 \tag{9.94}$$

If $\mathbf{V}_p$ is not independent of k, different wave numbers lead to different phase speeds, which again is the phenomenon of dispersion.

The real part ξ of s determines the rate of growth or decay of the wave with time. If ξ is positive, the disturbance grows and the state of uniform fluidization is unstable, whereas if ξ is negative, the disturbance is negative and the uniform state is stable. ξ is termed the *growth factor*.

A growth distance can be defined:

$$\Delta x \equiv \eta/\xi k \tag{9.95}$$

This gives the distance of travel in which the amplitude of the wave grows or decays by a factor of e, the growth or decay being distinguished by the sign of ξ.

Substitution of the plane-wave expression (9.76) for ε_1 into (9.90) then yields an algebraic equation:

$$\hat{\varepsilon}_1[As^2 + Ps + (f + ibPC)] = 0 \tag{9.96}$$

where

$$A = \varepsilon_0/(1 - \varepsilon_0) \tag{9.97}$$

$$P = \frac{\varepsilon_0}{1 - \varepsilon_0}\frac{g}{\bar{u}_0} \tag{9.98}$$

$$C = 1 - 2\varepsilon_0 - \varepsilon_0(1 - \varepsilon_0)(\beta_0'/\beta_0) \tag{9.99}$$

$$b = (\bar{u}_0/\varepsilon_0)k\cos\gamma \tag{9.100}$$

$$f = \mu_0\alpha(\varepsilon_0/\rho_p)M_{p,0}^2k^2 \tag{9.101}$$

where γ is the angle between the direction of flow and the wave vector $\boldsymbol{k}$, as indicated in Figure 9.5:

$$\gamma \equiv \cos^{-1}(\mathbf{i}\cdot\boldsymbol{k}) \tag{9.102}$$

Note that A, P, and C depend on the properties of the unperturbed system, whereas b and f depend on the magnitude of the wave vector $\boldsymbol{k}$.

Equation (9.96) is satisfied if either

$$\hat{\varepsilon}_1 = 0 \tag{9.103}$$

or

$$As^2 + Ps + (f + ibPC) = 0 \tag{9.104}$$

If (9.104) is satisfied, there exists a nontrivial solution $\varepsilon_1 \neq 0$ of equations (9.60)–(9.62) and (9.65). A nontrivial solution of these equations may also exist when (9.104) is not satisfied, provided that $\hat{\varepsilon}_1 = 0$, corresponding to a particle distribution that remains uniform and unperturbed. From the continuity equations (9.60) and (9.61) it then follows that

$$\nabla \cdot \mathbf{v}_1 = \nabla \cdot \mathbf{w}_1 = 0 \tag{9.105}$$

or, representing $\mathbf{v}_1$ and $\mathbf{w}_1$ in terms of their normal modes,

$$\mathbf{v}_1 = \hat{\mathbf{v}}_1 \exp(st) \exp(i\boldsymbol{k} \cdot \mathbf{x}) \tag{9.106}$$
$$\mathbf{w}_1 = \hat{\mathbf{w}}_1 \exp(st) \exp(i\boldsymbol{k} \cdot \mathbf{x}) \tag{9.107}$$

it follows that

$$\boldsymbol{k} \cdot \hat{\mathbf{v}}_1 = \boldsymbol{k} \cdot \hat{\mathbf{w}}_1 = 0 \tag{9.108}$$

showing that the velocity perturbations are transverse to the wave vector, as was pointed out for the nonmagnetic flow by Anderson and Jackson (1968, p. 15). These transverse waves generate no disturbances in the voidage and hence are not of immediate interest. Accordingly, attention will be limited to solutions corresponding to the roots of (9.104).

Equation (9.104) is quadratic in s, and hence the real and imaginary parts of its two roots can be readily obtained. Multiplying both sides of (9.104) by $4A/P^2$ and using the abbreviations

$$a \equiv 2A/P \tag{9.109}$$
$$z \equiv 4Af/P^2 \tag{9.110}$$
$$q \equiv 4AbC/P \tag{9.111}$$

allows this equation to be rewritten

$$(as + 1)^2 = 1 - z - iq \tag{9.112}$$

The real and imaginary parts of the two roots of this equation can now be obtained with the use of complex-variable algebra; they are

$$\xi = \frac{1}{a}\left[-1 \pm \sqrt{\frac{r + (1 - z)}{2}}\right] \tag{9.113}$$

$$\eta = \pm\frac{1}{a}\sqrt{\frac{r - (1 - z)}{2}} \tag{9.114}$$

where $r \equiv [(1 - z)^2 + q^2]^{1/2}$. In (9.113) the choice of the negative sign before the radical corresponds to waves that decay and hence need not be considered further here.

Before further development the quantity C of (9.99) will be made definite with the aid of (9.56), which reduces in the limit of a massless gas to

$$\beta(\varepsilon_0) = K(1 - \varepsilon_0)^2/\varepsilon_0^2 \tag{9.115}$$

With any influence of the magnetic-field bed history on K ignored, simple differentiation shows that

$$\beta_0'/\beta_0 = -2/\varepsilon_0(1 - \varepsilon_0) \tag{9.116}$$

and consequently

$$C = 3 - 2\varepsilon_0 \tag{9.117}$$

Magnetic stabilization is achieved if $\xi < 0$, neutral stability when $\xi = 0$. Taking the positive choice of sign in (9.113) shows that stability requires

$$z > \tfrac{1}{4}q^2 \tag{9.118}$$

Substituting for z and q in (9.118) and then for A, C, b, and f in the resulting expression gives the following criterion of stability, a central result of this theory:

$$N_m N_v \begin{cases} > 1 & \text{(unstable)} \\ = 1 & \text{(neutrally stable)} \\ < 1 & \text{(stable)} \end{cases} \tag{9.119}$$

where

$$N_m \equiv \rho_p \bar{u}_0^2/\mu_0 M_{p,0}^2 \tag{9.120}$$

is a dimensionless group representing a ratio of kinetic energy to magnetic energy, and

$$N_v \equiv \frac{(3 - 2\varepsilon_0)^2}{\varepsilon_0^2(1 - \varepsilon_0)}[1 + (1 - \varepsilon_0)\chi_0 - (1 - \varepsilon_0)(\chi_0 - \hat{\chi})\cos^2(\gamma - \phi)]\frac{\cos^2\gamma}{\cos^2(\gamma - \phi)} \tag{9.121}$$

where $\phi \equiv \gamma - \theta$ is the angle between the direction of flow and the applied field.

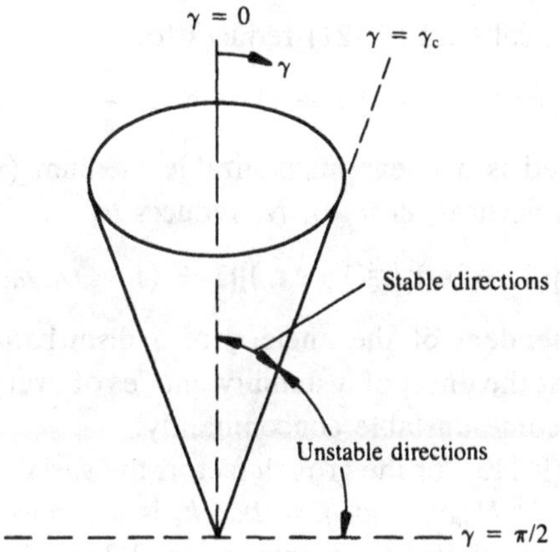

9.7 Stability cone of propagating modes in an unbounded bed.

In retrospect, the stability arises because when a voidage gradient appears in the mixture, the force $\mathbf{f}_{mp} = \mu_0 M \nabla H$ of equation (9.40) ceases to remain zero. In the one-dimensional case this force is proportional to the gradient $\nabla \varepsilon_p$ and has the opposite sign. This comes about because in one-dimensional flow $B = \mu_0(H + M) = \mu_0(H + \varepsilon_p M_p) = \text{const}$, so for a saturated sample with M_p constant, $\nabla H = -M_p \nabla \varepsilon_p$ and $\mathbf{f}_{mp} = -\mu_0 M_p^2 \varepsilon_p \nabla \varepsilon_p$. The presence of this force tends to smooth dilative or compressive perturbations of the void fraction and with it perturbations of the velocity and all other parameters.

Nature of the predicted behavior

Refer now to Eq. (9.121); if $\gamma - \phi = \pi/2$ (wave is perpendicular to field) and $\gamma \neq \pi/2$ (wave is other than transverse to the flow), N_v is infinite and no stabilization is possible. Since disturbance waves of all orientations can be present, apparently an oblique field cannot stabilize, and so the applied field must be vertically oriented ($\phi = 0$) to achieve stabilization.

When the field is vertically oriented and $\chi_0 > \hat{\chi}$, N_v is greatest when $\gamma = \pi/2$ and decreases as γ decreases to 0. As N_m is increased (by increasing the velocity or decreasing the magnetization) in a stable bed, a critical value of γ appears, $\gamma = \gamma_c$, such that outside a cone of half angle γ_c the propagating modes are unstable; see Figure 9.7 (Hagan

1981). In a vertical field (9.121) reduces to

$$N_v = [(3 - 2\varepsilon_0)^2/\varepsilon_0^2(1 - \varepsilon_0)][1 + (1 - \varepsilon_0)\chi_0 - (1-\varepsilon_0)(\chi_0 - \hat{\chi})(\cos^2\gamma)]$$

When the bed is a linear magnetizable medium ($\chi_0 = \hat{\chi}$) and the applied field is vertical ($\phi = 0$), N_v reduces to

$$N_v = [(3 - 2\varepsilon_0)^2/\varepsilon_0^2(1 - \varepsilon_0)][1 + (1 - \varepsilon_0)\chi_0]$$

which is independent of the angle γ of a disturbance wave. In this circumstance, at the onset of instability, modes of every orientation and wavelength become unstable concomitantly.

In equation (9.113) for the growth factor, the variables z and q can be expressed $z = k_r^2/N_m N_v$ and $q = 2k_r$; k_r is a reduced wave number, defined as $k_r \equiv k/(g/\varepsilon_r \bar{u}_0^2)$, where $\varepsilon_r = 2(3 - 2\varepsilon_0)(\cos\gamma)/\varepsilon_0$. This permits the generalized plot of Figure 9.8 to be constructed, showing that the longest-wavelength disturbances are stable and the shortest-wavelength disturbances reach an asymptotic growth rate that is constant at a given value of $N_m N_v$. Again it is seen that all modes are stable when $N_m N_v = 1$.

The dependence of the transition velocity on the magnetization of the bed can be readily developed. Eliminating β_0 from (9.54) and (9.115), letting m indicate the minimum fluidization condition, and taking a ratio of terms to eliminate constants give

$$\frac{\bar{u}_0}{\bar{u}_m} = \frac{\varepsilon_0^3(1 - \varepsilon_m)}{(1 - \varepsilon_0)\varepsilon_m^3} \tag{9.122}$$

where $\bar{u}_0 = \varepsilon_0 v_0$ and $\bar{u}_m = \varepsilon_m v_m$. Next, with $\cos\gamma = \cos\theta = 1$ equation (9.121) reduces to

$$N_v = \frac{(3 - 2\varepsilon_0)^2}{\varepsilon_0^2(1 - \varepsilon_0)}[1 + (1 - \varepsilon_0)\hat{\chi}] \tag{9.123}$$

Combining this expression for N_v with the neutral-stability criterion $N_m N_v = 1$ of (9.119) and the definition of N_m from (9.120) gives the normalized magnetization, which appears as the abscissa in Figure 9.9:

$$\frac{M_{p,0}}{\rho_p^{1/2}\mu_0^{-1/2}\bar{u}_m} = \frac{3 - 2\varepsilon_0}{\varepsilon_0}\frac{[1 + (1 - \varepsilon_0)\hat{\chi}]^{1/2}}{(1 - \varepsilon_0)^{1/2}}\frac{\bar{u}_0}{\bar{u}_m} \tag{9.124}$$

The upper curve in Figure 9.9 is then prepared by assigning a value of ε_m, then choosing values of ε_0, and finally plotting the corresponding values of $\bar{u}_0/\bar{u}_m$ from (9.122) and the normalized magnetization from (9.124). Arbitrarily assigned values of $\hat{\chi} = 1$ and $\varepsilon_m = 0.35$ are used in

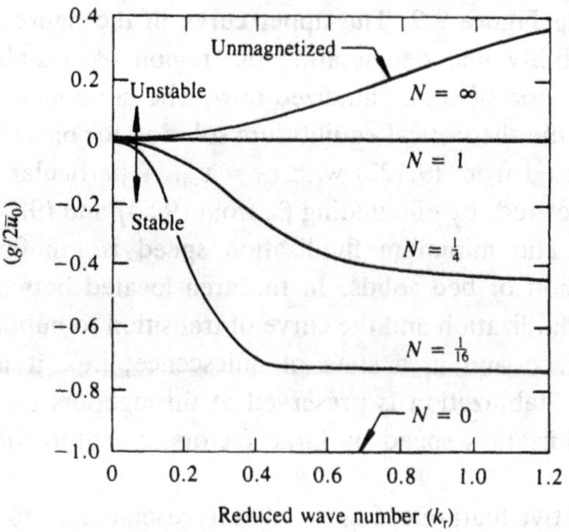

9.8 Normalized growth factor in beds of infinite extent. (*After Rosensweig 1979c.*)

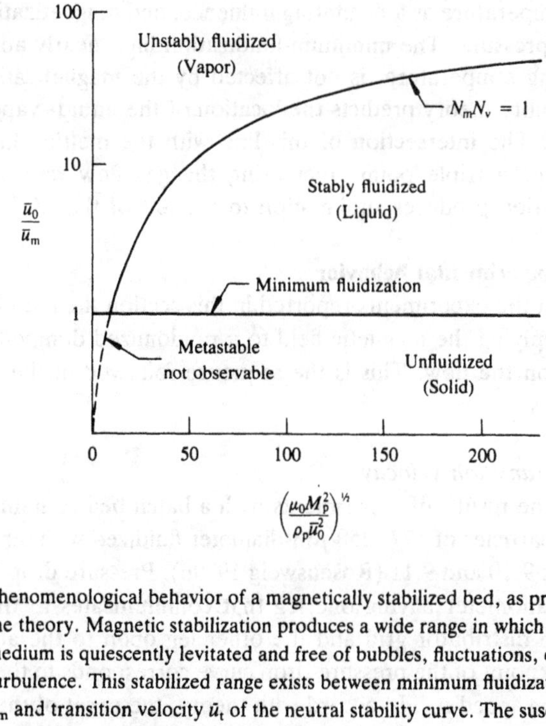

9.9 Phenomenological behavior of a magnetically stabilized bed, as predicted by the theory. Magnetic stabilization produces a wide range in which the bed medium is quiescently levitated and free of bubbles, fluctuations, or turbulence. This stabilized range exists between minimum fluidization velocity $\bar{u}_m$ and transition velocity $\bar{u}_t$ of the neutral stability curve. The susceptibility χ is taken to be constant at 1 with $\varepsilon_d = 0.35$.

constructing Figure 9.9. The upper curve in the figure represents the neutral-stability line demarcating the region of unstable fluidization from the region of stably fluidized flow. The horizontal line $\bar{u}_0/\bar{u}_m = 1$ represents the theoretical equilibrium solution for onset of fluidization, as determined from (9.122) with $\varepsilon_0 = \varepsilon_m$. A particular value of $\bar{u}_m$ is found, if desired, by eliminating β_0 from (9.54) and (9.115) and setting $\varepsilon_0 = \varepsilon_m$. The minimum fluidization speed is uninfluenced by the magnetization of bed solids. In the area located between the line of minimum fluidization and the curve of transition to bubbling, the bed is both fluidized and in a state of quiescence; i.e., it is magnetically stabilized. Stabilization is preserved at throughputs exceeding the incipient fluidization speed by large factors, a feature of technological interest.

A suggestive feature of Figure 9.9 is its resemblance to a thermodynamic phase diagram for a pure substance. As indicated in Figure 9.9, the unfluidized bed is analogous to a solid phase, the stabilized region to a liquid phase, and the bubbling bed to a vapor. The gas velocity plays the role of temperature as an agitating influence, and magnetization corresponds to pressure. The minimum-fluidization line, nearly analogous to the melting temperature, is not affected by the magnetization (pressure). Stability theory predicts the location of the liquid–vapor equilibrium line. The intersection of this line with the melting line has the topology of a triple point. Increasing the gas flow rate at constant magnetization produces sublimation to the left of the triple point.

9.8 Experimental behavior

In the experiments reported in this section the path is followed of first applying the magnetic field to a randomized dumped bed, then bringing on the flow. This is the sequence followed in the theoretical derivations.

Transition velocity

The results of experiments with a batch bed containing spherical steel particles of 177–250-μm-diameter fluidized with air are shown in Figures 9.10 and 9.11 (Rosensweig 1979b). Pressure drop is detected with a manometer having one leg that communicates to the bed just above the distributor grid and the other leg open to the atmosphere. The breakpoint of the pressure-drop curve corresponds to the minimum fluidization speed $\bar{u}_m$. In accord with theory, the onset of the minimum fluidization speed is independent of the presence of a magnetic field.

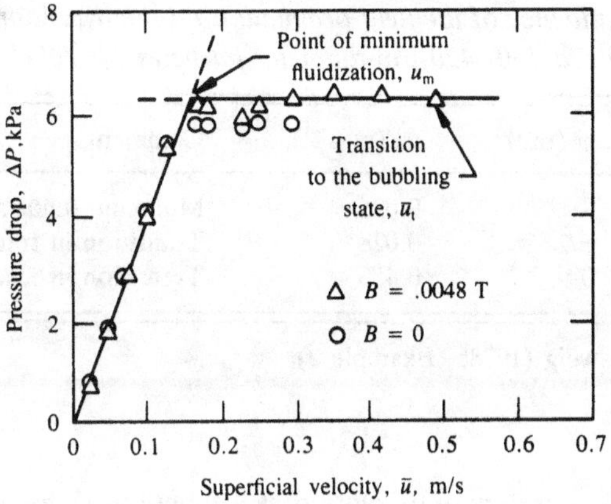

9.10 Experimental pressure in unstabilized and magnetically stabilized fluidized beds of 177–250-μm steel particles. (*After Rosensweig et al. 1981.*)

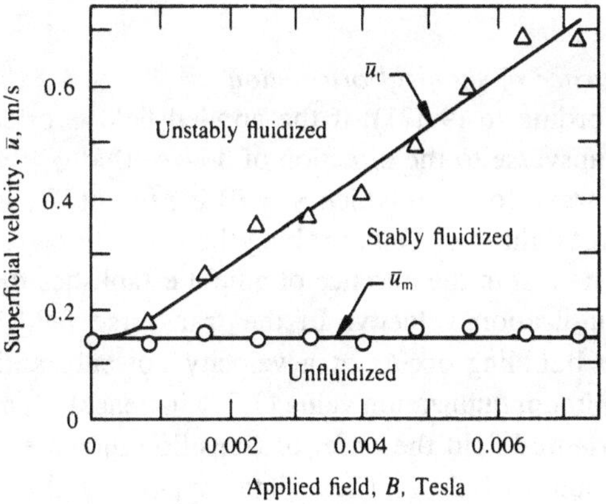

9.11 Experimental phase diagram of a magnetized fluidized bed containing 177–250-μm steel particles. (*After Rosensweig 1979.*)

For increasing superficial velocity of the air flow, the bed expands as pressure remains constant. Transition to the bubbling state occurs sharply at a velocity $\bar{u}_t$. The transition may be detected optically, magnetically, or with pressure sensing as a sudden onset of fluctuation. A pair of corresponding values $\bar{u}_m$ and $\bar{u}_t$ are determined in this manner for each given intensity of the applied magnetic field. The corresponding

Table 9.2 *Influence of the field orientation on the transition to bubbling for air-fluidized 150–420-μm-diameter particles of nickel on alumina*

B_0 (T)	ϕ (rad)	$\bar{u}_0$ ($m \cdot s^{-1}$)	Comment
0	—	0.026	Minimum fluidization velocity
0.057	$\pi/2$	0.028	Transition in transverse field
0.057	0	0.455	Transition in axial field

From Rosensweig (1978b, Example 2).

values of $\bar{u}_m$ and $\bar{u}_t$ may be plotted as a function of the field intensity, producing the experimental phase diagram plotted in Figure 9.11. Comparison of Figure 9.11 with the appearance of the theoretical phase diagram shown in Figure 9.9 provides qualitative confirmation of the general features predicted by the theory.

Influence of the field orientation

According to (9.121), if the applied field is oriented horizontally, i.e., transverse to the direction of flow so that $\phi = \pi/2$, an axially propagating wave (one for which $\gamma = 0$) is present, N_v is infinite, and the bed is unstabilizable. Data probing the prediction are abstracted in Table 9.2. The test in the absence of a field establishes the value of the minimum fluidization velocity. In the transverse field, the onset of transition to bubbling occurs at a velocity not substantially different from the minimum fluidization value (7.7% increase). In comparison, in an axially oriented field the onset of transition increased by the extraordinary amount of 1650%. These data support the theoretical prediction that a transverse field is unable to produce stabilization.

Although a number of points of agreement exist between theory and experiments concerning the onset of instability in magnetically stabilized fluidized beds, the quantitative predictions of theory are found to underpredict the value of transition velocity found in experiments. A more disturbing discrepancy concerns the influence of variables such as the particle size. The theory predicts a decrease in the transition velocity with increasing particle size, whereas experiment invariably yields an increase. One aspect of phenomenology that bears on the adequacy of the model is the description of the solids rheology, as discussed next.

Rheology of magnetically stabilized fluidized solids

The constitutive relation (9.46b) is a drastic idealization of the rheological (stress–strain) behavior of magnetized stabilized fluidized solids. The medium was assumed to be inviscid, and in fact in lightly magnetized beds, or in more highly magnetized beds with fluid velocity that approaches the transition velocity, the bed gives this impression; e.g., when stirred with a blade the surrounding medium remains undisturbed except in the immediate vicinity of the blade. Under higher magnetization, or at a fluid velocity that is well below the transition velocity, the beds develop a resistance to deformation due to a more effective magnetic attraction of the particles for each other. Accordingly, it has become of interest to characterize experimentally the rheology of bed media, and a result that has been established in several ways is that these media possess properties of a *viscoplastic fluid*. That is, in common with certain types of paints and greases, the media will not flow at all unless acted upon by at least some critical stress, called the *yield stress*. The magnitude of the yield stress may be large or small depending on conditions, but it is always present to some degree. In motion, the resultant flow can differ from Bingham flow in exhibiting wall slip.

The yield stress in a magnetically stabilized bed has been measured by determining the break-away force needed to withdraw a vertical flat immersed plate; see Figure 9.12 (Lee 1983). The plate was coated with an adherent layer of bed particles to prevent wall slip.

Values determined for a bed of iron particles, plotted in Figure 9.13, display a dependence on the immersed depth and particle magnetization. The yield stress increases with magnetization at a given bed depth. Sufficiently far below the surface, the yield stress reaches an asymptotic, limiting value, consistent with the assumption in the stability model of bed homogeneity. Near the surface, however, a transition zone is present, and at the surface the yield stress disappears. This end effect has been analyzed within the framework of the inviscid model by applying as a surface boundary condition the occurrence of unit fluid fraction ($\varepsilon_f = 1$), necessitated from a surface-force balance (Rosensweig et al. 1983).

Any model for the yield stress must be found outside the framework of the continuum model, in as much as the phenomenon results from the distribution of the magnetic field on a microscale. However, the number of relevant variables responsible for the behavior is not excessive, and a dimensional analysis is valuable to simplify the relationship. Now, the

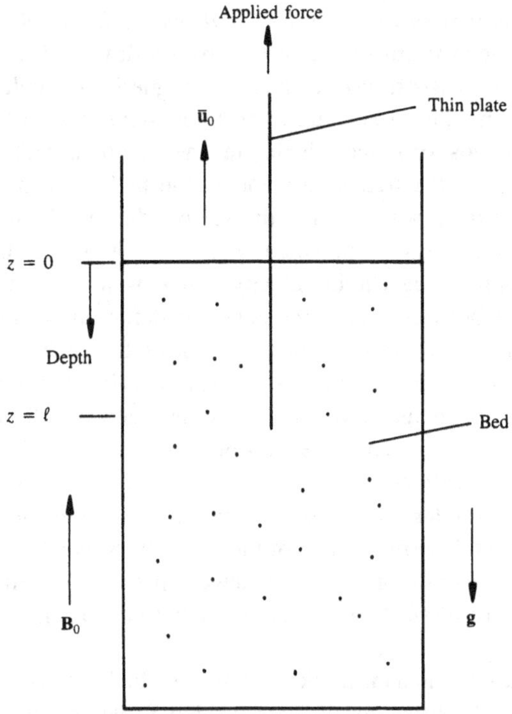

9.12 Drawn-plate experiment to determine the yield stress of a stably fluidized magnetized-particle bed. (*After Lee 1983.*)

yield stress may be attributed to the breaking of contacts. The strength of a contact will depend on the particle magnetization M_p and particle size d_p, whereas the number of contacts will depend on the void fraction ε_0 and the particle size. In addition, M_p moderates the nature of contacts in affecting the bed configuration, e.g., the clustering or chaining of particles. However, the fluidizing velocity $\bar{u}_0$, particle density ρ_p, fluid viscosity η_f, and gravitational constant g determine the value of ε_0 but should have no direct influence on the measured yield stress τ_y. Thus a functional relationship is postulated with the following choice of variables:

$$\tau_y = f(\mu_0, M_p, d_p, \varepsilon_0) \tag{9.125}$$

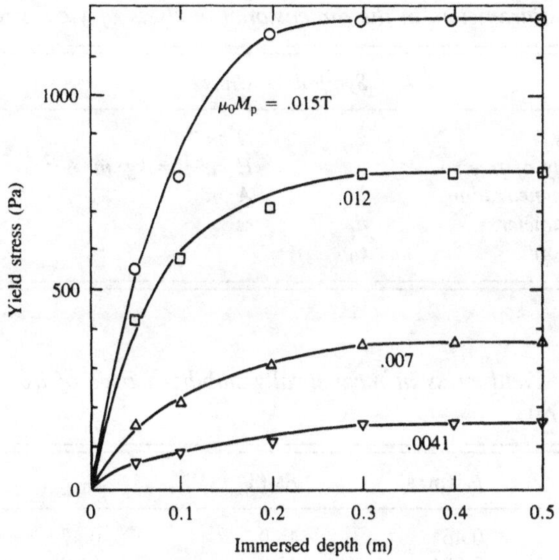

9.13 Measured yield stress in beds of 250–297-μm iron particles, with $\bar{u}_0 = 0.23\ \mathrm{m \cdot s^{-1}}$ ($\bar{u}_0/\bar{u}_m = 2$). (*After Lee 1983.*)

and Table 9.3 may be derived. Because ε_0 carries no units, it survives as a group. If a power-law relationship is assumed between the remaining variables, then so far as units are concerned

$$\tau_y = \mu_0^a M_p^b d_p^c \tag{9.126}$$

and the exponents a, b, and c are to be determined such that

$$N/m^2 = (N/A^2)^a (A/m)^b (m)^c$$

There are three independent units, and therefore three linear equations may be written for the exponents:

$$\begin{aligned} 1 &= a && \text{(for } N\text{)} \\ 0 &= -2a + b && \text{(for } A\text{)} \\ -2 &= -b + c && \text{(for } m\text{)} \end{aligned}$$

The number of equations equals the number of exponents, showing that only one group can result from the combination of these variables. Solving the equations gives

$$a = 1, \qquad b = 2, \qquad c = 0$$

Table 9.3 *Parameters in the dimensional analysis of yield stress*

Quantity	*Symbol*	*Units*
Yield stress	τ_y	$N \cdot m^{-2}$
Permeability of free space	μ_0	$H \cdot m^{-1} = kg \cdot m \cdot A^{-2} \cdot s^{-2} = N \cdot A^{-2}$
Particle magnetization	M_p	$A \cdot m^{-1}$
Particle diameter	d_p	m
Void fraction	ε_0	—

Table 9.4 *Yield stress in magnetically stabilized beds of iron particles (z = 0.3 m)*

d_p (μm)	$\bar{u}_0$ ($m \cdot s^{-1}$)	B_0 ($T \times 10^4$)	ε_0	τ_y (Pa)
250–297	0.46	46.8	0.479	229
74–105	0.094	45.9	0.485	275
250–297	0.46	29.5	0.479	130
74–105	0.081	34.8	0.485	126

Thus, the particle diameter d_p drops out of the relationship, and the group that is determined is $\tau_y/\mu_0 M_p^2$. The net result of this analysis can be expressed as

$$\tau_y/\mu_0 M_p^2 = f(\varepsilon_0) \tag{9.127}$$

Now the data of Figure 9.13 are obtained at constant fluid velocity, and hence ε_0 is constant. Hence, the expectation, from (9.127), is that τ_y should vary directly with the square of the asymptotic values of M_p. A plot of the experimental asymptotic yield-stress values, shown in Figure 9.14, supports this deduction.

Additional data show the yield stress in beds of iron particles of different sizes. Table 9.4 compares yield stresses in beds having nearly the same void fraction. Even though the fluidizing gas velocities differ by a factor of 5, the yield-stress values are close to one another when compared for similar solid magnetization and void fraction. This supports the further prediction of the dimensional analysis that particle size is not a controlling variable.

An empirical equation using the modulus that has been derived to fit

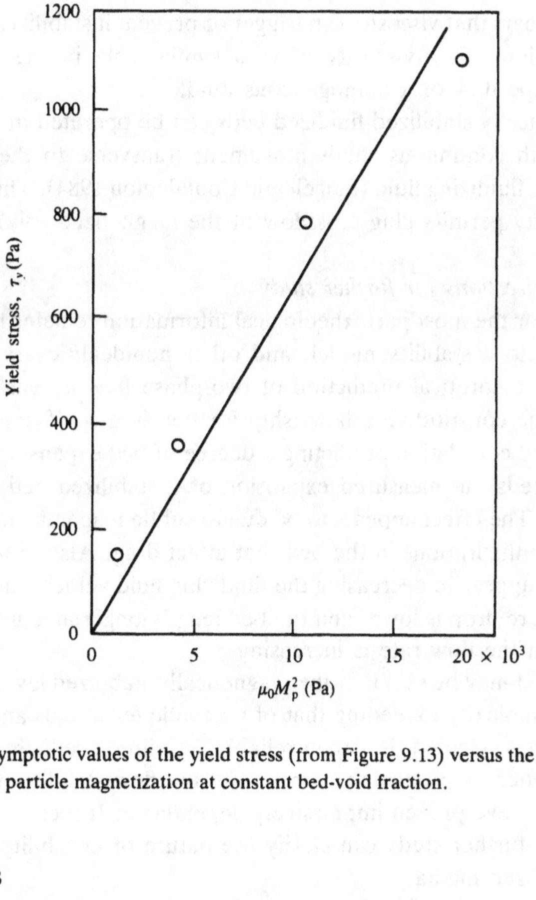

9.14 Asymptotic values of the yield stress (from Figure 9.13) versus the square of the particle magnetization at constant bed-void fraction.

the data is

$$\tau_y/\mu_o M_p^2 = A/\varepsilon_0^{12}, \qquad 0.48 < \varepsilon_0 < 0.59 \tag{9.128}$$

where $A = 3.3 \times 10^{-5}$. It can be seen that the yield stress is very highly dependent on the void fraction; by the same token, the uncertainty in the exponent (12) is great. When an alternating field (60 Hz) is used to magnetize the beds, the value of A is decreased considerably, to about $A = 0.5 \times 10^{-5}$, and the bed fluidity is increased correspondingly. Near the transition speed $\bar{u}_t$ it is found experimentally that the yield stress decreases abruptly to nearly zero. Accordingly, yield-stress effects may not be critical in determining stability. Resistance to rate of deformation may play a role in determining stability; some aspects of assuming a constant viscosity in the magnetized medium have been studied in a theoretical model (see the Comments and Supplemental References), and it appears that viscosity can trigger or prevent instability, depending on conditions. The viscosity plays a similar role in the stability of parallel-pipe flow of a homogeneous fluid.

Magnetically stabilized fluidized beds can be operated in a crossflow mode, with continuous solids movement transverse to the ascending flow of the fluidizing fluid (Siegell and Coulaloglou 1984). The viscoplastic property permits plug crossflow of the magnetized solids phase.

Directions for further study

For the most part, rheological information remains to be incorporated into a stability model, and other nonidealities exist as well. Thus, the theoretical prediction of two-phase flow is sensitive to the form of the constitutive relationship for the drag coefficient, with the packed-bed correlation predicting a degree of bed expansion that gradually exceeds the measured expansion of a stabilized bed as velocity increases. The effect appears to be due to subtle magnetic influences on particle configurations in the bed that affect drag. Also, a hysteresis is known to appear in decreasing the fluidizing fluid velocity such that the bed pressure drop is lower and the bed length longer at a given velocity than when the flow rate is increasing.

Overall it may be said that the magnetically stabilized levitated media have a complexity exceeding that of magnetic ferrofluids and such that analysis has yielded less-than-reliable stability predictions for the former, whereas predictions concerning the flow stability of magnetic ferrofluids have proven impressively dependable. It seems likely, however, that further study can clarify the nature of instability in magnetized fluidized media.

Comments and supplemental references

The earliest studies revealing aspects of the magnetic stabilization of fluidized beds are those of

Filippov (1960, 1961)
Kirko and Filippov (1960)

The use of an alternating field concomitant with particles (magnetite) having appreciable coercivity and the use of cylindrically shaped particles (iron) at other times lead to more complex behavior than is treated in this chapter. It may be recognized that the coercive particles are subject to body torque and spin motion, and that the cylindrical iron particles are apt to magnetize to a high intensity because their demagnetization coefficient is small in the direction of the long axis. The beds of coercive particles fail to stabilize completely and exhibit spouting motions. The cylindrical particle beds agglomerate, to yield solid-like rheology. In comparison, the bed state emphasized in the text is quiescent but flowable, corresponding to magnetically soft particles that are not too strongly magnetized.

Numerous publications of Ivanov and co-workers in Bulgaria emphasize the use of magnetic field to prevent loss of particles from fluidized agitated beds of ferromagnetic catalyst, as in ammonia synthesis; see, for example,

Ivanov and Zrunchev (1969)

Studies of heat transfer are reported by

Bologa and Syutkin (1977)

Techniques for averaging in multiphase media to obtain the continuum description of these discrete systems is a still-evolving research area. Illustrative complementary approaches may be found in

Whitaker (1966)
Ishii (1975)
Buyevich and Shchelchkova (1978)
Drew (1983)

The approach adopted in this chapter employs volume averaging, but it is also possible to employ ensemble, time, or mixed averaging. Features common to all approaches are the need to account for discontinuities at the fluid–particle interface and the need to choose terms to be represented with empirical constitutive expressions.

Mathematical solutions of conventional fluidized-bed stability are reviewed in

Jackson (1971)

The same governing equations are used to describe the flow field associated with the motion of an isolated bubble in a fluidized bed.

The nonlinear growth of waves is numerically computed using the method of characteristics by

Fanucci, Ness, and Yen (1979)

Formation of a shock wave is hypothesized as the onset of bubbling.

The stability and void fraction distribution associated with magnetized, fluidized beds of finite extent, as well as the influence of solids-phase viscosity on bed stability, are topics examined by

Rosensweig et al. (1983)

where the abscissa in Fig. 7B should read $\varepsilon_0 M_p$, making the high-through-put region inaccessible.

Experimental characterization of magnetically stabilized fluidized-solids flowability, the fluid-tracer radial-dispersion coefficient, and the particle–fluid heat-transfer coefficient are given by

Rosensweig et al. (1981)

The bed structure has been probed using computer simulation, magnetic permeability, and photomicrographs by

Rosensweig, Jerauld, and Zahn (1981)

Magnetically stabilized fluidized beds are flowable, so it is possible to circulate the bed solids and to carry out continuous processes of chemical conversion, adsorption separation, particulate filtration, and other operations of technological interest. A reference describing applications and characteristics of the beds for use in processes is

Lucchesi et al. (1979)

The stabilized bed has attractive features as a particulate filter of high-temperature high-pressure operability, with the possibility of continuous regeneration. Collection efficiency studies are reported by

Albert and Tien (1983)

Use of the bed for countercurrent solvent–solid contact in continuous affinity chromatography is treated by

Burns and Graves (1984)

Confirming studies of magnetized stabilized bed behavior very similar to the work reported in this chapter are found in the work of

Arnaldos, Casal, and Puigjaner (1983)

Stabilization of electrically polarizable two-phase suspensions of particles, drops, or bubbles is analyzed by

Naletova (1982)

and analysis and experiments of electrically stabilized fluidized solids are reported in a study by

Zahn and Rhee (1983)

APPENDIXES

App. 1 Vector and tensor notation

The *symbolic* or *Gibbs* notation is utilized, in which vectors are designated by boldface Roman or Greek letters, such as $\mathbf{v}$, $\mathbf{H}$, and $\boldsymbol{\xi}$, and second-order tensors by boldface sans serif letters, such as T or D.

Unit Cartesian coordinate vectors are denoted $\mathbf{i}$, $\mathbf{j}$, $\mathbf{k}$ or $\mathbf{e}_1$, $\mathbf{e}_2$, $\mathbf{e}_3$. $\mathbf{n}$ is the unit normal to a surface and faces outward if the surface is closed. $\mathbf{t}$ denotes a unit tangential vector.

The *scalar product* of two vectors is denoted by $\mathbf{a}\cdot\mathbf{b}$ and the *vector product* by $\mathbf{a}\times\mathbf{b}$. The latter is a *pseudovector*, as its sense of direction is reversed in mirror reflection; a vector such as velocity that is unchanged in mirror reflection is known as a *polar vector*.

A *dyad* or indeterminate product of $\mathbf{a}$ and $\mathbf{b}$ is formed by writing the vectors side by side as $\mathbf{ab}$. If $\mathbf{a} = a_x\mathbf{i} + a_y\mathbf{j} + a_z\mathbf{k}$, and $\mathbf{b} = b_x\mathbf{i} + b_y\mathbf{j} + b_z\mathbf{k}$, then $\mathbf{ab} = a_xb_x\mathbf{ii} + a_xb_y\mathbf{ij} + \cdots$. Dyads in general are not commutative: $\mathbf{ab} \neq \mathbf{ba}$.

The scalar product $\mathbf{a}\cdot\mathbf{b}$, the vector product $\mathbf{a}\times\mathbf{b}$, and the indeterminate product $\mathbf{ab}$ represent the three kinds of products that may be formed from two vectors.

A *dyadic* D is a finite sum of dyads: $\mathsf{D} = \mathbf{a}_1\mathbf{b}_1 + \mathbf{a}_2\mathbf{b}_2 + \cdots + \mathbf{a}_n\mathbf{b}_n$. The *transpose* or *conjugate dyadic* is given as $\mathsf{D}^{\mathrm{T}} \equiv \mathbf{b}_1\mathbf{a}_1 + \mathbf{b}_2\mathbf{a}_2 + \cdots + \mathbf{b}_n\mathbf{a}_n$. A dyadic D is *symmetric* if $\mathsf{D} = \mathsf{D}^{\mathrm{T}}$ and *antisymmetric* if $\mathsf{D} = -\mathsf{D}^{\mathrm{T}}$.

Any dyadic D can be decomposed into the sum $\mathsf{D} = \mathsf{D}_{\mathrm{s}} + \mathsf{D}_{\mathrm{a}}$, where $\mathsf{D}_{\mathrm{s}} = \frac{1}{2}(\mathsf{D} + \mathsf{D}^{\mathrm{T}})$ is a symmetric part and $\mathsf{D}_{\mathrm{a}} = \frac{1}{2}(\mathsf{D} - \mathsf{D}^{\mathrm{T}})$ is antisymmetric.

The *scalar of the dyadic* is written sca $\mathsf{D} \equiv \mathbf{a}_1\cdot\mathbf{b}_1 + \mathbf{a}_2\cdot\mathbf{b}_2 + \cdots$. The *vector of the dyadic* is defined by vec $\mathsf{D} \equiv \mathbf{a}_1\times\mathbf{b}_1 + \mathbf{a}_2\times\mathbf{b}_2 + \cdots$.

The dot products of a vector $\mathbf{v}$ with the dyadic D are also defined: $\mathbf{v}\cdot\mathsf{D} \equiv (\mathbf{v}\cdot\mathbf{a}_1)\mathbf{b}_1 + (\mathbf{v}\cdot\mathbf{a}_2)\mathbf{b}_2 + \cdots$ and $\mathsf{D}\cdot\mathbf{v} \equiv \mathbf{a}_1(\mathbf{b}_1\cdot\mathbf{v}) + \mathbf{a}_2(\mathbf{b}_2\cdot\mathbf{v}) + \cdots$. The cross products $\mathbf{v}\times\mathsf{D}$ and $\mathsf{D}\times\mathbf{v}$ are defined analogously.

The dyadic $\mathbf{I}$ is called the *unit dyadic* and is written as $\mathbf{I} = \mathbf{ii} + \mathbf{jj} + \mathbf{kk}$. The unit dyadic has the property $\mathbf{I}\cdot\mathbf{v} = \mathbf{v}\cdot\mathbf{I} = \mathbf{v}$ for any vector $\mathbf{v}$. $\mathbf{I}$ plays a role in tensor analysis similar to that of the number 1 in arithmetic.

The dot product of the dyads $\mathbf{ab}$ and $\mathbf{cd}$ is defined, according to the *nesting convention*, as $\mathbf{ab}\cdot\mathbf{cd} \equiv (\mathbf{b}\cdot\mathbf{c})\mathbf{ad}$. In evaluating multiple internal products, the nesting convention is followed, as indicated by the example $(\mathbf{ab}) \dot{\times} (\mathbf{cd}) = (\mathbf{a}\cdot\mathbf{d})(\mathbf{b} \times \mathbf{c})$.

From a tensor $\mathbf{T}$ a scalar can be formed by taking the *trace* of $\mathbf{T}$, that is, by summing the diagonal elements of $\mathbf{T}$. Thus, $\operatorname{tr}\mathbf{T} \equiv T_{ii} = T_{11} + T_{22} + T_{33}$. The trace of a tensor is an *invariant*, independent of the choice of orientation of the coordinate system. It is common practice to write $\mathbf{T}\cdot\mathbf{T}$ as $\mathbf{T}^2$, $\mathbf{T}\cdot\mathbf{T}^2$ as $\mathbf{T}^3$, and so on. A tensor has three invariants including $\operatorname{tr}\mathbf{T}$. The other two may be expressed as $\operatorname{tr}\mathbf{T}^2$ and $\operatorname{tr}\mathbf{T}^3$. The *magnitude of a tensor* is

$$|\mathbf{T}| \equiv \sqrt{\tfrac{1}{2}\mathbf{T}:\mathbf{T}^{\mathrm{T}}}$$

For a symmetric tensor the magnitude is the invariant $\sqrt{\frac{1}{2}\operatorname{tr}\mathbf{T}^2}$.

The *alternator* is the polyadic $\boldsymbol{\varepsilon} \equiv \mathbf{e}_i\mathbf{e}_j\mathbf{e}_k\,\varepsilon_{ijk}$, where ε_{ijk} is defined in Section 8.6. Certain products of the alternator are called *duals*. For a vector $\mathbf{A}$, dual $\mathbf{A} \equiv \boldsymbol{\varepsilon}\cdot\mathbf{A} = -\mathbf{I} \times \mathbf{A}$; for a tensor $\mathbf{D}$, dual $\mathbf{D} \equiv -\frac{1}{2}\boldsymbol{\varepsilon}:\mathbf{D} = \frac{1}{2}\mathbf{I} \dot{\times} \mathbf{D}$.

Vector identities

$$(\mathbf{A} \times \mathbf{B})\cdot\mathbf{C} = \mathbf{A}\cdot(\mathbf{B} \times \mathbf{C}) = (\mathbf{C} \times \mathbf{A})\cdot\mathbf{B}$$

$$\mathbf{A} \times (\mathbf{B} \times \mathbf{C}) = \mathbf{B}(\mathbf{A}\cdot\mathbf{C}) - \mathbf{C}(\mathbf{A}\cdot\mathbf{B})$$

$$(\mathbf{A} \times \mathbf{B})\cdot(\mathbf{C} \times \mathbf{D}) = (\mathbf{A}\cdot\mathbf{C})(\mathbf{B}\cdot\mathbf{D}) - (\mathbf{A}\cdot\mathbf{D})(\mathbf{B}\cdot\mathbf{C})$$

$$\nabla \times \nabla\phi = \mathbf{0}$$

$$\nabla\cdot(\nabla \times \mathbf{A}) = 0$$

$$\nabla(\phi_1\phi_2) = \phi_1\nabla\phi_2 + \phi_2\nabla\phi_1$$

$$\nabla\cdot(\phi\mathbf{A}) = \phi\nabla\cdot\mathbf{A} + \mathbf{A}\cdot\nabla\phi$$

$$\nabla \times (\phi\mathbf{A}) = \phi\nabla \times \mathbf{A} + \nabla\phi \times \mathbf{A}$$

$$\nabla(\mathbf{A}\cdot\mathbf{B}) = \mathbf{A}\cdot\nabla\mathbf{B} + \mathbf{B}\cdot\nabla\mathbf{A} + \mathbf{A} \times (\nabla \times \mathbf{B}) + \mathbf{B} \times (\nabla \times \mathbf{A})$$

$$\nabla\cdot(\mathbf{A} \times \mathbf{B}) = \mathbf{B}\cdot(\nabla \times \mathbf{A}) - \mathbf{A}\cdot(\nabla \times \mathbf{B})$$

$$\nabla \times (\mathbf{A} \times \mathbf{B}) = \mathbf{A}(\nabla\cdot\mathbf{B}) - \mathbf{B}(\nabla\cdot\mathbf{A}) + \mathbf{B}\cdot\nabla\mathbf{A} - \mathbf{A}\cdot\nabla\mathbf{B}$$

$$\nabla \times (\nabla \times \mathbf{A}) = \nabla(\nabla\cdot\mathbf{A}) - \nabla^2\mathbf{A}$$

$$\mathbf{A}\cdot\nabla\mathbf{A} = \tfrac{1}{2}\nabla(\mathbf{A}\cdot\mathbf{A}) - \mathbf{A} \times (\nabla \times \mathbf{A})$$

$$\nabla\cdot(\nabla\phi_1 \times \nabla\phi_2) = 0$$

Tensor identities

$$\mathbf{AB}\cdot\mathbf{C} = \mathbf{A}(\mathbf{B}\cdot\mathbf{C})$$
$$\mathbf{A}\cdot\mathbf{BC} = (\mathbf{A}\cdot\mathbf{B})\mathbf{C}$$
$$\nabla\cdot(\mathbf{AB}) = \mathbf{A}\cdot\nabla\mathbf{B} + \mathbf{B}(\nabla\cdot\mathbf{A})$$
$$\mathbf{A}\cdot\mathbf{T} = \mathbf{T}^{\mathrm{T}}\cdot\mathbf{A}$$
$$\mathbf{I}\cdot\mathbf{A} = \mathbf{A}\cdot\mathbf{I} = \mathbf{A}$$
$$\nabla\cdot(\phi\mathbf{I}) = \nabla\phi$$
$$\nabla\cdot(\phi\mathbf{T}) = \phi\nabla\cdot\mathbf{T} + \nabla\phi\cdot\mathbf{T}$$
$$\nabla\cdot\mathbf{T}_{\mathrm{a}} = -\tfrac{1}{2}\nabla\times\mathbf{A}, \text{ where } \mathbf{T}_{\mathrm{a}} = \tfrac{1}{2}(\mathbf{T}-\mathbf{T}^{\mathrm{T}}) \text{ and } \mathbf{A} = \operatorname{vec}\mathbf{T}$$

Integral theorems

1. *Divergence theorem:*

(a) For vectors, $\int_V \nabla\cdot\mathbf{A}\,dV = \oint_S \mathbf{A}\cdot d\mathbf{S}$

(b) For tensors, $\int_V \nabla\cdot\mathbf{T}\,dV = \oint_S d\mathbf{S}\cdot\mathbf{T}$

(c) Corollaries are $\int_V \nabla\psi\,dV = \oint_S \psi\,d\mathbf{S}$

$$\int_V \nabla\times\mathbf{A}\,dV = \oint_S d\mathbf{S}\times\mathbf{A}$$

2. *Stokes's theorem:*

$$\int_S (\nabla\times\mathbf{A})\cdot d\mathbf{S} = \oint_L \mathbf{A}\cdot d\mathbf{l}$$

3. *Reynolds' transport theorem for differentiating a volume integral:*

$$\frac{D}{Dt}\int_V \psi\,dV = \int_V \frac{\partial\psi}{\partial t}\,dV + \oint_S \psi\mathbf{v}\cdot d\mathbf{S}$$

4. *Transport theorem for differentiating a surface integral:*

$$\frac{D}{Dt}\int_S \mathbf{A}\cdot d\mathbf{S} = \int_S \left[\frac{D\mathbf{A}}{Dt} - \mathbf{A}\cdot\nabla\mathbf{v} + \mathbf{A}(\nabla\cdot\mathbf{v})\right]\cdot d\mathbf{S}$$
$$= \int_S \left[\frac{\partial\mathbf{A}}{\partial t} + \nabla\times(\mathbf{A}\times\mathbf{v}) + \mathbf{v}(\nabla\cdot\mathbf{A})\right]\cdot d\mathbf{S}$$

5. *Transport theorem for line integrals:*

$$\frac{D}{Dt}\int_L \mathbf{A}\cdot d\mathbf{l} = \int_L \left(\frac{D\mathbf{A}}{Dt} + \mathbf{A}\cdot\nabla\mathbf{v}\right)\cdot d\mathbf{l}$$

Comments and supplementary references

The seminal reference is the textbook founded upon the lectures of J. Willard Gibbs:

Gibbs and Wilson (1909)

A clear introduction to vectors and tensors is given by

Mase (1970)

A similar treatment is that of

Bird, Stewart, and Lightfoot (1960)

For study at a more advanced level, see

Aris (1962)

Borisenko and Tarapov (1979)

App. 2 **Application of the Maxwell stress tensor: analysis of a sheet jet**

This problem of a sheet jet is of interest to illustrate a technique of integral analysis using the Maxwell stress tensor. Figure A2.1 illustrates the appearance of a sheet jet. Such jets are of common occurrence in atomizing nozzles, fountains, coating flows, and elsewhere. Their appearance is dominated by the sheet's edges, which tend to gobble up the jet as it advances unless restrained by some means – for example, with wire guides.

A cross section taken perpendicular to the axial velocity is shown in Figure A2.2. G. I. Taylor (1959c), using a momentum balance over the control volume bounded by the sides 1–2–3–4, derived an expression for the velocity of the sheet edge *when the applied field is zero*. Taylor's model was extended by Melcher (1978) to predict the rate of this inward advance if an *electric field* is used to inhibit this process. Here an analysis is developed for the flow of a sheet jet in a *magnetic field*. In Figure A2.2 magnetic flux, emanating from opposed slab magnets is shown being conducted into the sheet, which is considered highly permeable. Note that to the right, or in the absence of the sheet jet, no magnetic flux is present between the magnets, as they are opposed to each other. The field surrounding the sheet is distributed in a complex pattern whose details, fortunately, can be ignored here.

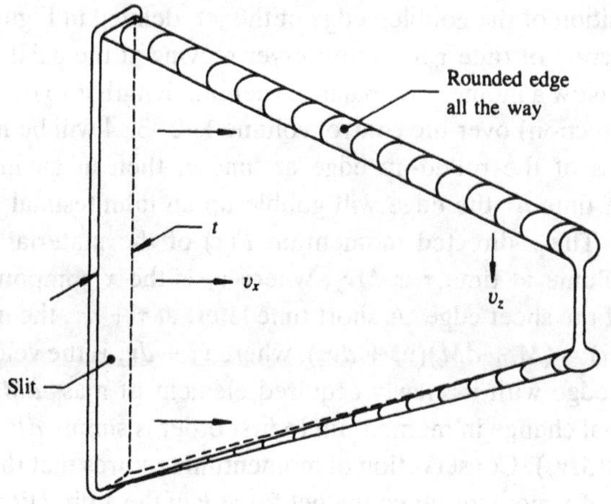

A2.1 An unobstructed sheet jet of uniform thickness t and axial velocity v_x issuing from a slit. The forces due to surface tension contract the jet width.

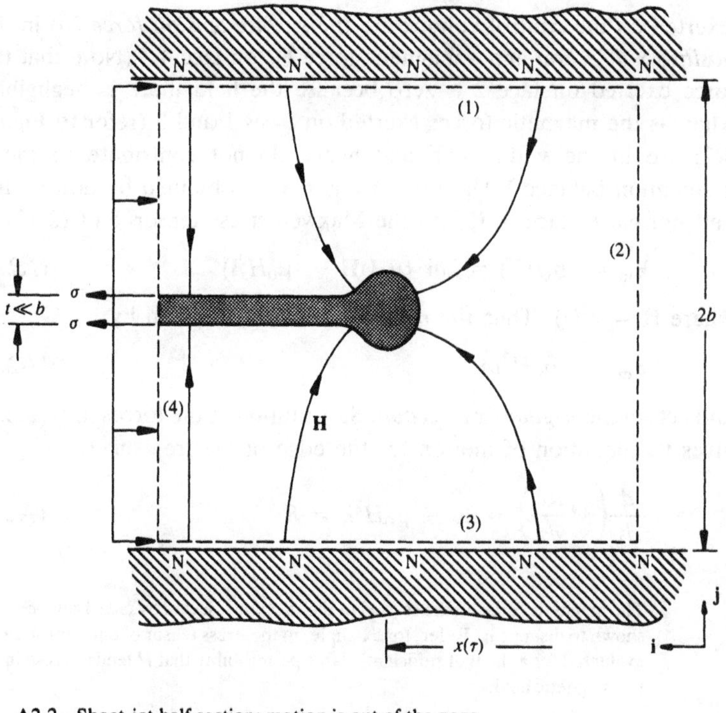

A2.2 Sheet-jet half section; motion is out of the page.

The position of the gobbled edge of the jet, defined in Figure A2.2 by x, is a function of time τ for an observer moving at the axial velocity v_z of the jet. Now a momentum balance (per unit length into the paper and in the x direction) over the control volume 1–2–3–4 will be made. If M is the mass of the rolled-up edge at time τ, then in an infinitesimal amount of time $d\tau$ the edge will gobble up an infinitesimal amount of mass dM. The x-directed momentum $P(\tau)$ of the material inside the control volume at time τ is Mv_x, where v_x is the x component of the velocity of the sheet edge. A short time later, at $\tau + d\tau$, the momentum is $P(\tau + d\tau) = (M + dM)(v_x + dv_x)$, where $v_x + dv_x$ is the velocity of the rolled-up edge with its newly acquired element of mass dM. Thus the infinitesimal change in momentum to first order is simply $dP = M\,dv_x + v_x\,dM = d(Mv_x)$. Conservation of momentum requires that the time rate of change of x momentum be the net force F in the x direction acting on the control volume:

$$F = \frac{dP}{d\tau} = \frac{d(Mv_x)}{d\tau} = \frac{d}{d\tau}\left(M\frac{dx}{d\tau}\right) \tag{A2.1}$$

Exerted on the control volume are the *surface tension force* $2\sigma\mathbf{i}$ in the *positive x direction* and a magnetic force $\mathbf{F}_m$ on face 4.* (Note that the force exerted on face 2 is zero because the field there is negligible, whereas the magnetic forces exerted on faces 1 and 3 (refer to Figure 3.7) are in the y direction and hence do not contribute to the x momentum balance.) The force $\mathbf{F}_m$ is readily obtained by dotting the unit normal to face 4, $\mathbf{i}$, into the Maxwell stress tensor $\mathbf{T}$ of (3.15):

$$\mathbf{F}_m = 2b(\mathbf{i}\cdot\mathbf{T}) = 2b\mathbf{i}\cdot(\mu_0\mathbf{HH} - \tfrac{1}{2}\mu_0H^2\mathbf{I}) \tag{A2.2}$$

where $\mathbf{H} = \pm H\mathbf{j}$. Thus the *magnetic force* $\mathbf{F}_m$ is given by

$$\mathbf{F}_m = -\mu_0H^2b\mathbf{i} \tag{A2.3}$$

and acts in the *negative x direction*. Substitution of the forces into (A2.1) gives the equation of motion for the edge of the free sheet:

$$\frac{d}{d\tau}\left(M\frac{dx}{d\tau}\right) = 2\sigma - \mu_0H^2b = F \tag{A2.4}$$

*The magnetic force on the section of the sheet intersected by face 4 can be shown to disappear. Refer, for example, to the stress tensor of equation 4.29 evaluated for a sheet of sufficiently large permeability that H tends to zero in the magnetic field.

It is readily seen from (A2.4) that surface tension pulls the edges of the sheet jet together while the magnetic force tries to suppress this tendency. Thus, as long as $F > 0$, i.e., as long as the surface-tension force is greater than the magnetic force, the motion of the rolled-up edge is inward. The analysis predicts that it is possible to stop this inward motion of the edge ($F = 0$) and indeed to reverse it ($F < 0$).

Equation (A2.4) can be integrated once to give

$$Mv_x - (Mv_x)_0 = F\tau \tag{A2.5}$$

where the subscript 0 indicates that the quantities inside the parentheses are evaluated at time $\tau = 0$. Equation (A2.5) is the statement that the change in momentum (the left side) is equal to the impulse (the right side).

If at time $\tau = 0$ the free edge is formed at $x = 0$ and initially $M = M_0 = 0$, mass conservation, for inward motion, requires that $M = \rho x t$, where ρ is the mass density and t is sheet thickness. Thus (A2.5) becomes

$$\frac{1}{2}\frac{dx^2}{d\tau} = \frac{F}{\rho t}\tau \tag{A2.6}$$

Integrating (A2.6) with the condition $x = 0$ at $\tau = 0$ gives

$$x = \sqrt{\frac{F}{\rho t}}\,\tau = \sqrt{\frac{2\sigma - \mu_0 H^2 b}{\rho t}}\,\tau \tag{A2.7}$$

The velocity of the sheet edge is constant and equal to

$$v_x = \frac{dx}{d\tau} = \sqrt{\frac{F}{\rho t}} = \frac{x}{\tau} \tag{A2.8}$$

This problem concerning a sheet jet has used the concept of integral momentum balance in a manner similar to the use of integral momentum analysis in ordinary fluid mechanics, such as in the approximate analysis of boundary-layer growth or orifice discharge.

Comments and supplemental references

Photographs of sheet jets are found in Taylor's papers; see

Taylor (1959 a–c)

and a recent finite-element numerical simulation of the full Navier–Stokes equation governing the dynamics of a sheet jet is described by

Kistler (1983)

An application of integral analysis using the stress tensor of a magnetic fluid to deduce the support force and stiffness of a ferrohydrostatic bearing is presented by

Rosensweig (1978a)

REFERENCES

Abraham, M., and Becker, R. 1937. *The classical theory of electricity and magnetism*. Blackie and Son, London.
Abramowitz, M., and Stegun, I. A. 1964. *Handbook of mathematical functions with formulas, graphs, and mathematical tables*. NBS Appl. Math. Ser. 55, U.S. Government Printing Office, Washington, D.C.
Adelstein, O., Lamotte, A., Baddour, R. F., Cotton, C. K., and Whiteside, G. M. 1979. Preparation and magnetic filtration of polyacrylamide gels containing immobilized proteins and a ferrofluid. *J. Mol. Catal.* 6 (3), 199–225.
Albert, R. J., and Tien, C. 1983. Filtration in magnetically stabilized fluidized beds. Paper 45e. AIChE Annu. Meet., Washington, D.C.
Anderson, T. B., and Jackson, R. 1967. A fluid mechanical description of fluidized beds – equation of motion. *Ind. Eng. Chem.* 6, 527–39.
1968. Fluid mechanical description of fluidized beds. *Ind. Eng. Chem. Fundam.* 7 (1), 12–21.
Andres, U. 1976. Magnetic liquids. *Mater. Sci. Eng.* 26, 269–75.
Aris, R. 1962. *Vectors, tensors, and the basic equations of fluid mechanics*. Prentice-Hall, Englewood Cliffs, New Jersey.
Arkhipenko, V. I., Barkov, Yu. D., and Bashtovoi, V. G. 1978. Shape of a drop of magnetized fluid in a homogeneous magnetic field. *Magnetohydrodynamics* 14 (3), 373–5.
Arnaldos, J., Casal, J., and Puigjaner, L. 1983. Magnetically stabilized fluidized bed. Characteristics and possible applications (in French). *Powder Technol.* 36 (1), 33–8.
Ausloos, M., Clippe, P., Kowalski, J. M., and Pekalski, A. 1980. Phase diagrams of model magnetofluids. *IEEE Trans. Magnetics* MAG-16 (2), 233–6.
Bacri, J.-C., and Salin, D. 1983. Dynamics of the shape transition of a magnetic ferrofluid drop. *J. Phys. Lett.* 44, 415–20.
1984. First order transition in the instability of a magnetic fluid interface. *J. Phys. Lett.* 45 (11), 558–64.
Bailey, R. L. 1983. Lesser known applications of ferrofluids. *J. Magnetism Magnetic Mater.* 39 (1, 2), 178–82.
Barclay, J. A. 1982. Use of a ferrofluid as the heat-exchange fluid in a magnetic refrigerator. *J. Appl. Phys.* 53 (4), 2887–94.
Barton, A. F. M. 1983. *Handbook of solubility parameters and other cohesion parameters*. CRC Press, N. W. Boca Raton, Florida.

Basaran, O. A. 1984. *Electrohydrodynamics of drops and bubbles.* Ph.D. thesis, University of Minnesota.
Bashtovoi, V. G., and Krakov, M. S. 1978. Surface instability in the nonisothermal layers of magnetized fluids. *Magnetohydrodynamics* 14 (3), 285–90.
Batchelor, G. K. 1970. *An introduction to fluid dynamics.* Cambridge University Press, London.
Bates, L. F. 1963. *Modern magnetism.* Cambridge University Press, New York.
Bean, C. P., and Livingston, J. D. 1959. Superparamagnetism, *J. Appl. Phys.* 30 (4), 120S–9S.
Berkovsky, B. M., and Bashtovoi, V. 1980. Instabilities of magnetic fluids leading to a rupture of continuity. *IEEE Trans. Magnetics* MAG-16, 288–97.
Berkovsky, B. M., and Orlov, L. P. 1973. Investigation of free-surface forms in magnetic liquids' analogue of pinch effect. *Magnetohydrodynamics* 4 (9), 468–73.
Berkovsky, B. M., Vislovich, A. N., and Kashevsky, B. E. 1980. Magnetic fluid as a continuum with internal degrees of freedom. *IEEE Trans. Magnetics* MAG-16 (2), 329–42.
Berkowitz, A. E., Lahut, J. A., Jacobs, I. S., Levinson, L. M., and Forester, D. W. 1975. Spin pinning at ferrite–organic interfaces. *Phys. Rev. Lett.* 34 (10), 594–7.
Bibik, E. E., Matgullin, B. Y., Raikher, Y. L., and Shliomis, M. I. 1973. The magnetostatic properties of colloidal magnetite. *Magnetohydrodynamics*, 9 (1), 58–62.
Bird, R. B., Stewart, W. E., and Lightfoot, E. N. (1960) *Transport phenomena.* Wiley, New York.
Birnbaum, S., and Larsson, P. O. 1982. Application of magnetic immobilized microorganisms: Ethanol production by *Saccharomyces cerevisiae. Appl. Biochem. Biotechnol.* 7, 55–7.
Bitter, F. 1931. On inhomogeneities in the magnetization of ferromagnetic materials. *Phys. Rev.* 38, 1903.
1937. *Introduction to ferromagnetism.* McGraw-Hill, New York.
Blums, E. 1980. Some aspects of heat and mass transfer in magnetic fluids. *IEEE Trans. Magnetics* MAG-16 (2), 347–51.
Bologa, M. K., and Syutkin, S. V. 1977. The influence of an electromagnetic field on the structural hydrodynamic properties of a fluidized bed. *Electron. Obrab. Mater.* (*USSR*) 68 (1), 37–42.
Borisenko, A. I., and Tarapov, I. E. 1979. *Vector and tensor analysis with applications*, transl. by R. A. Silverman, Ed. Dover, New York.
Bozorth, R. M. 1951. *Ferromagnetism.* Van Nostrand, New York.
Brancher, J. P. 1978. Waves and instabilities on a plane interface between ferrofluids and nonmagnetic fluids. *Thermomechanics of Magnetic Fluids*, B. Berkovsky, Ed. Hemisphere, Washington, D.C. pp. 181–94.
Brenner, H. 1970. Rheology of a dilute suspension of dipolar spherical particles in an external field. *J. Colloid Interface Sci.* 32 (1), 141–58.
1984. Antisymmetric stresses induced by the rigid-body rotation of dipolar suspensions: Vortex flows. *Int. J. Eng. Sci.* 22, 645–82.
Brenner, H., and Weissman, M. H. 1972. Rheology of a dilute suspension of dipolar spherical particles in an external field. II. Effects of rotary Brownian motion, *J. Colloid Interface Sci.* 41 (3), 499–531.
Brevik, I. 1982. Fluids in electric and magnetic fields: pressure variation and stability. *Canad. J. Phys.* 60, 449–55.
Brown, R., and Horsnell, T. S. 1969. The wrong way round. *Electr. Rev.* (14 February), 235–6.

Brown, W. F., Jr. 1951. Electric and magnetic forces: A direct calculation, I, II. *Amer. J. Phys.* 19 (5), 290–304; 19 (6), 333–50.

Burns, M. A., and Graves, D. J. 1984. Continuous affinity chromatography using a magnetically-stabilized fluidized bed. Paper 96b. AIChE Annu. Meet. San Francisco.

Buyevich, Yu. A., and Shchelchkova, I. N. 1978. Flow of dense suspensions. *Progr. Aerospace Sci.* 18, 121–50.

Byrne, J. V. 1977. Ferrofluid hydrostatics according to classical and recent theories of stresses. *Proc. IEE (London)* 124 (11), 1089–97.

Cabannes, H. 1970. *Theoretical magnetofluiddynamics.* Academic, New York.

Calugaru, G., Badescu, R., and Luca, E. 1976a. Magnetoviscosity of ferrofluids. *Rev. Roum. Phys.* 21 (3), 305–8.

1976b. A new aspect of the movement of ferrofluids in a rotating magnetic field. *Rev. Roumaine Phys.* 21 (4), 439–40.

Caroli, C., and Pincus, P. 1969. Response of an isolated magnetic grain suspended in liquid to a rotating field. *Phys. Kondens. Mat.* 9, 311.

Chandrasekhar, S. 1961. *Hydrodynamic and hydromagnetic stability.* Oxford University Press, London.

Chantrell, R. W., Bradbury, A., Popplewell, J., and Charles, S. W. 1982. Agglomeration formation in magnetic fluid. *J. Appl. Phys.* 53 (3), 2742–4.

Chantrell, R. W., Popplewell, J., and Charles, S. W. 1978. Measurements of particle size distribution parameters in ferrofluids, *IEEE Trans. Magnetics* MAG-14 (5), 975–7.

Charles, S. W., and Popplewell, J. 1980. Ferromagnetic liquids. In *Ferromagnetic Materials*, Vol. 2, E. P. Wohlfarth, Ed. North-Holland, Ams., pp. 509–59.

Charles, S. W., and Rosensweig, R. E. 1983. Magnetic fluids bibliography. *J. Magnetism Magnetic Mater.* 39, 192–220.

Chu, B. T. 1959. Thermodynamics of electrically conducting fluids. *Phys. Fluids* 2 (5), 473–84.

Chung, D. Y., and Isler, W. E. 1977. Sound velocity measurements in magnetic fluids. *Phys. Lett. A* 61 (6), 373–4.

Condiff, D. W., and Dahler, J. S. 1964. Fluid mechanical aspects of antisymmetric stress. *Phys. Fluids* 7 (6), 842–54.

Cowley, M. D., and Rosensweig, R. E. 1967. The interfacial stability of a ferromagnetic fluid. *J. Fluid Mech.* 30 (4), 671–88.

Curtis, R. A. 1971. Flows and wave propagation in ferrofluids. *Phys. Fluids* 14 (10), 2096–102.

Dahler, J. S., and Scriven, L. E. 1961. Angular momentum of continua. *Nature* 192, 36–7.

1963. Theory of structured continua. I. General consideration of angular momentum and polarization. *Proc. Roy. Soc. London Ser. A* 275, 504–27.

deGennes, P. G., and Pincus, P. A. 1970. Pair correlations in a ferromagnetic colloid. *Phys. Kondens. Mat.* 11, 189.

de Groot, S. R., and Mazur, P. 1962. *Non-equilibrium thermodynamics.* North-Holland, Amsterdam.

Drazin, P., and Reid, W. 1981. *Hydrodynamic stability.* Cambridge University Press, New York.

Drew, D. A. 1981. Personal communication.

1983. Mathematical modeling of two-phase flow. *Annu. Rev. Fluid Mech.* 15, 261.

Edison, T. 1887. Improvements in electrical generators. British patent 16709, 13 June.

Einstein, A. (1906, 1911). On the movement of small particles suspended in a stationary

liquid demanded by the molecular kinetic theory of heat. *Annalen der Physik* 17, 549; 19, 371; 34, 591. English transl.: In *A. Einstein: Investigations on the theory of the Brownian movement*, R. Furth, Ed. Dover, New York, 1956.
Elmore, W. C. 1938a. Ferromagnetic colloid for studying magnetic structures. *Phys. Rev.* 54 (4), 309–10.
1938b. The magnetization of ferromagnetic colloids. *Phys. Rev.* 54 (12), 1092–5.
Essmann, U., and Trauble, H. 1967. The direct observation of individual flux lines in type II superconductors. *Phys. Lett. A* 24 (10), 526–7.
Ezekiel, F. D. 1975. Uses of magnetic fluids in bearings, lubrication and damping. *Mech. Eng.* 97 (7), 94ff.
Faber, T. E. 1958. The intermediate state in superconducting plates. *Proc. & Soc. London* A248, 460–81.
Fanucci, J. B., Ness, N., and Yen, R. H. 1979. On the formation of bubbles in gas-particulate fluidized beds. *J. Fluid Mech.* 94 (Pt. 2), 353–67.
Fay, H., and Quets, J. M. 1980. Density separation of solids in ferrofluids with magnetic grids. *Separation Sci. Technol.* 15 (3), 339–69.
Filippov, M. V. 1960. The effect of a magnetic field on a ferromagnetic particle suspension bed. *Prikl. Magnitogidrodin. Tr. Inst. Fiz. Akad. Nauk. Latvia SSR* (*USSR*) (12), 215–36.
1961. Resistance and expansion of a fluidized bed of magnetite in a magnetic field. *Izv. Akad. Nauk. Latv. SSR* (*USSR*) 12 (173), 47–51.
Finlayson, B. A. 1970. Convective instability of ferromagnetic fluids. *J. Fluid Mech.* 40 (4), 753–67.
Frenkel, J. 1955. *The kinetic theory of liquids*. Dover, New York.
Gailitis, A. 1969. Form of surface stability of a ferromagnetic fluid. *Magnetohydro-dynamics* 5 (1), 44–5.
1977. Formation of the hexagonal pattern on the surface of a ferromagnetic fluid in an applied magnetic field. *J. Fluid Mech.* 82 (3), 401–13.
Gibbs, J. W. 1906. *The Scientific Papers of J. Willard Gibbs*, Vol. 2. Longmans, Green. Republished: Dover, New York, 1961.
Gibbs, J. W., and Wilson, E. B. 1909. *Vector analysis*, 2nd ed. Scribner's, New York. Republished by Dover, New York, 1960.
Glazov, O. A. 1976. Inducing motion of a ferromagnetic liquid by a travelling magnetic field. *Magnetohydrodynamics* 12 (4), 400–3.
Goldberg, P., Hansford, J., and van Heerden, P. J. 1971. Polarization of light in suspensions of small ferrite particles in a magnetic field. *J. Appl. Phys.* 42 (10), 3874–6.
Gotoh, K., Isler, W. E., and Chung, D. Y. 1980. Theory of ultrasonic attenuation in magnetic fluids. *IEEE Trans. Magnetics*. MAG-16 (2), 211–3.
Gubanov, A. I. 1960. Quasi-classical theory of amorphous ferromagnetics. *Soviet Phys. Solid State* 2, 468.
Hagan, P. S. 1981. Personal communication.
Hakim, S. S., and Higham, J. B. 1962. An experimental determination of the excess pressure produced in a liquid dielectric by an electric field. *Proc. Phys. Soc.* 80, 190–8.
Hall, W. F., and Busenberg, S. N. 1969. Viscosity of magnetic suspension. *J. Chem. Phys.* 51 (1), 137–44.
Hassett, K. L., Stecher, L. C., and Hendrickson, D. N. 1980. Polymer-anchored metal oxide particles. I. Superparamagnetic magnetite microcrystals stabilized by lignosulfonate. *Inorg. Chem.* 19 (2), 416–22.
Hathaway, D. B. 1979. Use of ferrofluid in moving-coil loudspeakers. *dB-Sound Eng.*

Mag. 13 (2), 42–4.
Hayes, C. F. 1975. Association in a ferromagnetic colloid. *J. Colloid Interface Sci.* 52 (2), 239–43.
Heck, C. 1974. *Magnetic materials and their applications*. Crane, Russak, New York.
Helmholtz, H. 1882. Abhandlungen I. *Ann. Phys. Chem.* 13, 798.
Hess, P. H., and Parker, P. H., Jr. 1966. Polymers for stabilization of colloidal cobalt particles. *J. Appl. Polymer Sci.* 10, 1915–27.
Higgins, B. G., and Scriven, L. E. 1979. Interfacial shape and evolution equations for liquid films and other viscocapillary flows. *Ind. Eng. Chem. Fundam.* 18 (3), 208–15.
Hsieh, R. K. T. 1980. Continuum mechanics of a magnetically saturated fluid. *IEEE Trans. Magnetics* MAG-16 (12), 207–10.
Hughes, W. F., and Young, F. J. 1966. *Electromagnetodynamics of fluids*. Wiley, New York.
Ishii, M. 1975. *Thermo-fluid dynamic theory of two-phase flow*. Eyrolles, France.
Ivanov, D. G., and Zrunchev, I. A. 1969. A new method of ammonia synthesis at high pressures involving fluidization of the catalyst in a magnetic field. *C. R. Acad. Bulgare Sci.* 22 (12), 1405–8.
Jackson, J. D. 1975. *Classical electrodynamics*. Wiley, New York.
Jackson, R. 1971. Fluidmechanical theory. In *Fluidization*, J. F. Davidson and D. Harrison, Eds. Academic, London, pp. 65–119.
Jacobs, I. S., and Bean, C. P. 1963. Fine particles, thin films, and exchange anisotropy. In *Magnetism*, Vol. III, G. T. Rado and H. Suhl, Eds. Academic Press, New York.
Jeans, J. 1925. *The mathematical theory of electricity and magnetism*. Cambridge University Press, London.
Jenkins, J. T. 1971. Some simple flows of a para-magnetic fluid. *J. Physique* 32, 931.
Jones, T. B. 1978. Theory and application of ferrofluid seals. In *Thermomechanics of magnetic fluids*, B. Berkovsky, Ed. Hemisphere, Washington, D.C. pp. 255–98.
Jones, T. B., and Bliss, G. W. 1977. Bubble dielectrophoresis. *J. Appl. Phys.* 48 (4), 1412–7.
Jordan, P. C. 1973. Association phenomena in a ferromagnetic colloid. *Molecular Phys.* 25 (4), 961–73.
Kagan, I. Ya., Rykov, V. G., and Yantovskii, E. I. 1973. Flow of a dielectric ferromagnetic suspension in a rotating magnetic field. *Magnetohydrodynamics*. 9 (2), 259–60.
Kaiser, R., and Miskolczy, G. 1970a. Magnetic properties of stable dispersions of subdomain magnetite particles. *J. Appl. Phys.* 41 (3), 1064–72.
1970b. Some applications of ferrofluid magnetic colloids. *IEEE Trans. Magnetics* MAG-6 (3), 694–8.
Kamiyama, S., Koike, K., and Oyama, T. (1983). Pipe flow resistance of magnetic fluid in a nonuniform transverse magnetic field. *J. Magnetism Magnetic Mater.* 39 (1,2), 23–6.
Kaplan, B. Z., and Jacobson, D. M. 1976. Electrical properties of magnetizable liquids. *Nature* 259 (5545), 654–6.
Karim, S. M., and Rosenhead, L. 1952. The second coefficient of viscosity of liquids and gases. *Rev. Mod. Phys.* 24 (2), 108–16.
Kellogg, O. D. 1953. *Foundations of potential theory*. Dover, New York.
Kelvin, Lord. 1910. *Mathematical and physical papers*. Dover, New York.
Kendra, M. 1982. Personal communication.
Khalafalla, S. E., and Reimers, G. W. 1973a. Separating non-ferrous fluid metals in incinerator residue using magnetic fluids. *Separation Sci.* 8, 161–78.

1973b. Magnetofluids and their manufacture. U. S. Patent 3,764,540.
1974. Production of magnetic fluids by peptization techniques. U. S. Patent 3,843,540.
Kirko, I. M., and Filippov, M. V. 1960. Special features of a fluidized bed of ferromagnetic particles in a magnetic field. *Z. Tekhn. Fiz.* 30 (9), 1081–4.
Kistler, S. F. 1983. The fluid mechanics of curtain coating and related viscous free surface flows, Ph.D. thesis, University of Minnesota.
Korteweg, D. J. 1880. *Ann. Phys. Chem.* 9, 48–61.
Kuhn, L., and Myers, R. A. 1979. Ink-jet printing. *Sci. Am.* 240 (4), 162ff.
Kunii, D., and Levenspiel, O. 1977. *Fluidization engineering.* Wiley, New York.
Kruyt, H. R. 1952. *Colloid Science.* Elsevier, New York, Vol. I.
Lalas, D. P., and Carmi, S. 1971. Thermoconvective stability of ferrofluids. *Phys. Fluids* 14 (2), 436–7.
Lamb, H. 1945. *Hydrodynamics.* Dover, New York.
Landau, L. D., and Lifshitz, E. M. 1959. *Fluid mechanics.* Pergamon, New York.
1960. *Electrodynamics of continuous media.* Pergamon, London.
Lee, W.-K. 1983. The rheology of magnetically stabilized fluidized solids. *AIChE Symp. Ser.* 222, 79–87.
Levi, A. C., Hobson, R. F., and McCourt, F. R. 1973. Magnetoviscosity of colloidal suspensions. *Canad. J. Phys.* 51, 180–94.
Lewis, D. J. 1950. The instability of liquid surfaces when accelerated in a direction perpendicular to their planes. II. *Proc. Roy. Soc. London Ser. A* 117, 81–96.
Lucchesi, P. J., Hatch, W. H., Mayer, F. X., and Rosensweig, R. E., 1979. Magnetically stabilized beds – new gas solids contacting technology. Paper SP-4 in *Proc. 10th World Petroleum Congr.* Heyden, Philadelphia Pa., Vol. 4, pp. 419–425 (discussion).
Mackor, E. L. 1951. A theoretical approach of the colloidal-chemical stability of dispersions in hydrocarbons. *J. Colloid Sci.* 6, 492–5.
Mailfert, R., and Martinet, A. 1973. Flow regimes for a magnetic suspension under a rotating magnetic field. *J. Phys.* 34 (203), 197–202.
Malik, S. K., and Singh, M. 1984. Nonlinear dispersive instabilities in magnetic fluids. *Quart. Appl. Math.* 42 (3), 359–71.
Martinet, A. 1974. Birefringence and linear dichroism of ferrofluids in magnetic field. *Rheology Acta* 13 (2), 260–4.
1978. Experimental evidences of static and dynamic anisotropies of magnetic colloids. In *Thermomechanics of magnetic fluids*, B. Berkovsky, Ed. Hemisphere, Washington, D.C., pp. 97–114.
1982. *Ferrofluids.* 16-mm film, 25 min. Serdav, 27 rue Paul Bert 94200, Ivry, France (French and English versions).
1983. The case of ferrofluids. In *Aggregation Processes in Solution.* Elsevier, New York, Chapter 18, pp. 1–41.
Martsenyuk, Yu., Raikher, L., and Shliomis, M. I. 1974. On the kinetics of magnetization of suspensions of ferromagnetic particles. *Sov. Phys. JETP* 38 (2), 413–6.
Maruno, S., Yubakami, K., and Soga, S. 1983. Plain paper recording process using magnetic fluids. *J. Magnetism Magnetic Mater.* 39 (1, 2), 187–9.
Mase, G. E. 1970. *Theory and problems of continuum mechanics.* McGraw-Hill, New York.
Massart, R. 1981. Preparation of aqueous magnetic liquids in alkaline and acidic media. *IEEE Trans. Magnetics* MAG-17 (2), 1247–8.
Matsuki, H., Yamasawa, K., and Murakami, K. 1977. Experimental considerations on a new automatic cooling device using temperature-sensitive magnetic fluid. *IEEE*

Trans. Magnetics MAG-13 (5), 1143–5.
McNab, T. K., Fox, R. A.,, and Boyle, J. F. 1968. Some magnetic properties of magnetite (Fe_3O_4) microcrystals. *J. Appl. Phys.* 39 (12), 5703–11.
McTague, J. P. 1969. Magnetoviscosity of magnetic colloids. *J. Chem. Phys.* 51 (1), 133–6.
Mehta, R. V. 1978. Optical properties of certain magnetic fluids. In *Thermomechanics of Magnetic Fluids*, B. Berkovsky, Ed. Hemisphere, Washington, D.C. pp. 139–48.
Melcher, J. R. 1963. *Field-coupled surface waves*. M.I.T. Press, Cambridge, Massachusetts.
1978. The electrohydrodynamics of G. I. Taylor. *J. Electrostatics* 5, 1–9.
1981. *Continuum Electromechanics*, M.I.T. Press, Cambridge, Massachusetts.
Melcher, J. R., and Taylor, G. I. 1969. Electrohydrodynamics: a review of the role of interfacial shear stresses. In *Annual Reviews of Fluid Mechanics*, Vol. I, W. R. Sears and M. Van Dyke Ed. Annual Reviews, Palo Alto, California, pp. 111–46.
Mellilo, L., and Raj, K. 1981. Ferrofluids as a means of controlling woofer design parameters. *J. Audio Eng. Soc. (USA)* 29 (3), 132–8.
Moon, P., and Spencer, D. E. 1961. *Field theory for engineers*. Van Nostrand, Princeton, New Jersey.
Morimoto, Y., Akimoto, M., and Yotsumoto, Y. 1982. Dispersion state of protein-stabilized magnetic emulsions. *Chem. Pharm. Bull.* 30 (8), 3024–7.
Moskowitz, R. 1975. Dynamic sealing with magnetic fluids. *ASLE Trans.* 18 (2), 135–43.
Moskowitz, R., and Rosensweig, R. E. 1967. Nonmechanical torque-driven flow of a ferromagnetic fluid by an electromagnetic field. *Appl. Phys. Lett.* 11 (10), 301–3.
Naletova, V. A. 1982. Stabilization of bubble-liquid processes by an electric field. *Izv. Akad. Nauk. SSR Meh. Zhidkosti i. Gaza* 4, (July–August), 5–12.
Neal, J. A. 1977. Aryl sulfonate-aldehyde composition and process for its preparation. U. S. Patent 4,018,691.
Néel, L. 1949. Effect of thermal fluctuations on the magnetization of small particles. *C. R. Acad. Sci. Paris* 228, 664.
1953. *Rev. Mod. Phys.* 25, 293.
Neuringer, J. L., and Rosensweig, R. E. 1964. Ferrohydrodynamics. *Phys. Fluids* 7 (12), 1927–37.
Newbower, R. S., 1972. A new technique for circulatory measurements employing magnetic fluid tracers. In *Proc. 1972 Biomedical Symp.*, San Diego.
O'Grady, K., Bradbury, A., Charles, S. W., Menear, S., Popplewell, J., and Chantrell, R. W. 1983. Curie–Weiss behavior in ferrofluids. *J. Magnetic Mater.* (February), 31–4.
Panofsky, W. K., and Phillips, M. 1962. *Classical electricity and magnetism*. Addison-Wesley, Reading, Massachusetts.
Papell, S. S. 1965. Low viscosity magnetic fluid obtained by the colloidal suspension of magnetic particles. U.S. Patent 3,215,572.
Papell, S. S., and Faber, O. C., Jr. 1966. Zero and reduced gravity simulation of magnetic colloid pool boiling system. Rep. NASA TN-D-3288.
Papirer, E., Horny, P., Balard, H., Anthore, R., Petipas, R., and Martinet, A. 1983. The preparation of a ferrofluid by decomposition of dicobalt octacarbonyl. I. Experimental parameters. II. Nucleation and growth of particles. *J. Colloid Interfacial Sci.* 94 (1), 207–19, 220.
Pearson, J. R. A. 1958. On convection cells induced by surface tension. *J. Fluid Mech.* 4, 489–500.
Penfield, P., and Haus, H. A. 1967. *Electrodynamics of moving media*. M.I.T. Press,

Cambridge, Massachusetts.
Perry, M. P., and Jones, T. B. 1976. Dynamic loading of a single-stage ferromagnetic liquid seal. *J. Appl. Phys.* 49 (4), 2334–8.
Peterson, E. A., and Krueger, D. A. 1977. Reversible field induced agglomeration in magnetic colloids. *J. Colloid Interface Sci.* 62 (1), 22–34.
Pimbley, G. H., Jr. 1976. Stationary solutions of Rayleigh–Taylor instability. *J. Math. Anal. Appl.* 55, 170–206.
Popplewell, J., Charles, S. W., and Chantrell, R. 1977. The long term stability of magnetic liquids for energy conversion devices. *Energy Convers.* 16 (3), 133–8.
Popplewell, J., Al-Qenaie, A., Charles, S. W., Moskowitz, R., and Raj, K. 1982. Thermal conductivity measurements on ferrofluids. *Colloid Polymer Sci.* 260 (3), 333–8.
Raj, K. 1981. Personal communication.
Rayleigh, Lord 1878. On the instability of jets. *Proc. London Math. Soc.* 10, 4. Republished: *Theory of Sound*, Dover, New York.
1883. Investigation of the character of the equilibrium of an incompressible heavy fluid of variable density. *Proc. London Math. Soc.* 14, 170–7. Republished: *Scientific Papers*. Cambridge University, Cambridge, England, 1900, Vol. 2, pp. 200–7.
1916. On convective currents in a horizontal layer of fluid when the higher temperature is on the under side. *Philos. Mag.* 32, 529–46. Reprinted in *Scientific Papers*. Cambridge, England, 1920, Vol. 6, pp. 432–46.
Reitz, J. R., Milford, F. J., and Christy, R. W. 1979. *Foundations of electromagnetic theory*. Addison-Wesley, Reading, Massachusetts.
Resler, E. L., Jr., and Rosensweig, R. E. 1964. Magnetocaloric power. *AIAA J.* 2 (8), 1418–22.
1967. Regenerative thermomagnetic power. *J. Eng. Power* 89, 399–406.
Romankiw, L. T., Slusarczak, M., and Thompson, D. A. 1975. Liquid magnetic bubbles. *IEEE Trans. Magnetics* MAG-11 (1), 25–8.
Rosensweig, R. E. 1966a. Magnetic fluids. *Int. Sci. Tech.* (July), 48–56.
1966b. Fluidmagnetic buoyancy. *AIAA J.* 4 (10), 1751–8.
1966c. Buoyancy and stable levitation of a magnetic body immersed in a magnetizable fluid. *Nature (London)* 210 (5036), 613–614.
1969. Material separation using ferromagnetic liquid techniques. U.S. Patent 3,700,595.
1970. Method of substituting one ferrofluid solvent for another. U.S. Patent 3,531,413.
1971. Magnetic fluid seals. U.S. Patent 3,620,584.
1975. Ferrofluid compositions and method of making same. U.S. Patent 3,917,538.
1978a. Phenomena and relationships of magnetic fluid bearings. In *Thermomechanics of magnetic fluids*. B. Berkovsky, Ed. Hemisphere, Washington, D.C., pp. 231–54.
1978b. Process for operating a magnetically stabilized fluidized bed. U.S. Patent 4,115,927. Reissue Patent 31,439, 15 November 1983.
1979a. Fluid dynamics and science of magnetic liquids. In *Advances in Electronics and Electron Physics*, Vol. 48. L. Marton, Ed. Academic Press, New York, pp. 103–99.
1979b. Fluidization: Hydrodynamic stabilization with a magnetic field. *Science* 204 (6 April), 57–60.
1979c. Magnetic stabilization of the state of uniform fluidization. *Ind. Eng. Chem. Fundam.* 18 (3), 260–9.
1982a. Magnetic fluids. *Sci. Am.* 247 (4), 136–45, 194.
1982b. Pattern formation in magnetic fluids. In *Evolution of Order and Chaos in Physics, Chemistry, and Biology*, H. Haken, Ed. Springer-Verlag, Berlin,

pp. 52–64.
Rosensweig, R. E., Jerauld, G. R., and Zahn, M. 1981. Structure of magnetically stabilized fluidized solids. In *Continuum models of discrete systems*, Vol. 4. O. Brulin and R. K. T. Hsieh, Eds. North-Holland, Amsterdam, pp. 137–44.
Rosensweig, R. E., and Kaiser, R. 1967. *Study of Ferromagnetic Liquid, Phase I.* NTIS Rep. No. NASW-1219; NASA Rep. NASA-CR-91684. NASA Office of Advanced Research and Technology, Washington, D.C.
Rosensweig, R. E., Kaiser, R., and Miskolczy, G. 1969. Viscosity of magnetic fluid in a magnetic field. *J. Colloid Interface Sci.* 29 (4), 680–6.
Rosensweig, R. E., Miskolczy, G., and Ezekiel, F. D. 1968. Magnetic fluid seals. *Mach. Des.* 40 (28 March), 145–50.
Rosensweig, R. E., Nestor, J. W., and Timmins, R. S. 1965. Ferrohydrodynamic fluids for direct conversion of heat energy. In *Mater. Assoc. Direct Energy Convers. Proc. Symp. AIChE-I. Chem. Eng. Ser.* 5, 104–18; discussion, 133–7.
Rosensweig, R. E., Siegell, J. H., Lee, W. K., and Mikus, T. 1981. Magnetically stabilized fluidized solids. *AIChE Symp. Series* 77 (205), 8–16.
Rosensweig, R. E., Zahn, M., Lee, W. K., and Hagan, P. S. 1983. Theory and experiments in the mechanics of magnetically stabilized fluidized solids. In *Theory of dispersed multiphase flow*, R. E. Meyer, Ed. Academic, New York, 359–84.
Rosensweig, R. E., Zahn, M., and Shumovich, R. 1983. Labyrinthine instability in magnetic and dielectric fluids. *J. Magnetism Magnetic Mater.* 39 (1, 2), 127–32.
Rosensweig, R. E., Zahn, M., and Vogler, T. 1978. Stabilization of fluid penetration through a porous medium using a magnetizable fluid. In *Thermomechanics of magnetic fluids*, B. Berkovsky, Ed. Hemisphere, Washington, D.C. pp. 195–211.
Roth, J. R., Rayk, W. D., and Reinman, J. J. 1970. Technological problems anticipated in the application of fusion reactors to propulsion and power generation. Rep. NASA-TMX-2106.
Russakoff, G. 1970. A derivation of the macroscopic Maxwell equations. *Amer. J. Phys.* 38, 1188–95.
Saffman, P. G., and Taylor, G. I. 1958. The penetration of a fluid into a porous medium or Hele-Shaw cell containing a more viscous liquid. *Proc. Roy. Soc. London Ser. A* 245, 312–29.
Sano, K., and Doi, M. 1983. Theory of agglomeration of ferromagnetic particles in magnetic fluids. *J. Phys. Soc. Jpn.* 52 (8), 2810–5.
Schaufeld, J. 1982. Ferrofluid inertia dampers enhance stepper motor performance. *Design News* (5 April).
Scholten, P. C. 1978. Colloid chemistry of magnetic fluids. In *Thermomechanics of magnetic fluids*, B. Berkovsky, Ed. Hemisphere, Washington, D.C., pp. 1–26.
1983. How magnetic can a magnetic fluid be? *J. Magnetism Magnetic Mater.* 39 (1,2), 99–105.
Scriven, L. E. 1960. Dynamics of a fluid interface. *Chem. Eng. Sci.* 12, 98–108.
Scriven, L. E., and Sternling, C. V. 1964. On cellular convection driven by surface-tension gradients: effects of mean surface tension and surface viscosity. *J. Fluid. Mech.* 19, 321–40.
Serrin, J. 1959. Mathematical principles of classical fluid mechanics. In *Handbuch der Physik* Vol. 8:1, Fluid Dynamics 1, S. Flugge, Ed. Springer-Verlag, Berlin, pp. 125–263.
Shaposhnikov, I. G., and Shliomis, M. I. 1975. Hydrodynamics of magnetizable media. *Magnetohydrodynamics* 11 (1), 37–46.
Sharma, V. K., and Waldner, F. 1977. Superparamagnetic and ferrimagnetic resonance of ultrafine magnetite (Fe_3O_4) particles in ferrofluids. *J. Appl. Phys.* 48 (10),

4298–302.

Shimoiizaka, J., Nakatsuka, K., and Chubachi, R. 1978. Rheological characteristics of water base magnetic fluids. In *Thermomechanics of Magnetic Fluids*, B. Berkovsky, Ed. Hemisphere, Washington, D.C. pp. 67–76.

Shimoiizaka, J., Nakatsuka, K., Fujita, T., and Kounosu, A. 1980. Sink-float separators using permanent magnets and water based magnetic fluid. *IEEE Trans. Magnetics* MAG-16 (2), 368–71.

Shliomis, M. I. 1972. Effective viscosity of magnetic suspensions. *Soviet Phys. JETP* 34 (6), 1291–4.

1974a. Magnetic fluids. *Soviet Phys. Uspekhi* (Engl. transl.) 17 (2), 153–169.

1974b. Certain gyromagnetic effect in a liquid paramagnetic. *Soviet Phys JEPT* 39 (4), 701–4.

1975. Nonlinear effects in suspension of ferromagnetic particles under action of a rotating magnetic field. *Soviet Phys. Dokl.* 19 (10), 686–687.

Shliomis, M. I., and Raikher, Yu. L. 1980. Experimental investigations of magnetic fluids. *IEEE Trans. Magnetics* MAG-16 (2), 237–50.

Shulman, Z. P., Kordonskii, V., and Demchuk, S. A. 1977. Effect of a heterogeneous rotating magnetic field on the flow and heat exchange in ferrosuspensions. *Magnetohydrodynamics* 13 (4), 406–9.

Siegell, J. H., and Coulaloglou, C. A. 1984, Crossflow magnetically stabilized fluidized beds. Paper 106d. AIChE Annu. Meet. San Francisco.

Skjeltorp, A. T. 1983a. One- and two-dimensional crystallization of magnetic holes. *Phys. Rev. Lett.* 51 (25), 2306–9.

1983b. Studies of two-dimensional lattices using ferrofluid. *J. Magnetism Magnetic Mater.* 37, 253–6.

Slepian, J. 1950. Electromagnetic pondermotive force within material bodies. *Proc. Natl. Acad. Sci.* (*US*) 36, 485–97.

Sternling, C. V., and Scriven, L. E. 1959. Interfacial turbulence: hydrodynamic instability and the Marangoni effect. *AIChE J.* 6, 514–23.

Stratton, J. A. 1941. *Electromagnetic theory*. McGraw-Hill, New York.

Suyazov, V. M. 1982. Theory of the effective viscosity of magnetofluids; I. Fundamental equations of the theory. *Magnetohydrodynamics* 18 (2), 105–18.

1983. Theory of the effective viscosity of magnetofluids; II. Contribution of anisotropy to the effective viscosity. *Magnetohydrodynamics* 18 (3), 211–20.

Taketomi, S. 1983. Magnetic fluid anomalous pseudo-Cotton Mouton effects about 107 times larger than that of nitrobenzene. *Jap. J. Appl. Phys.* 22 (7), 1137–43.

Taktarov, N. G. 1975. Breakup of magnetic liquid jets. *Magnetohydrodynamics* 11 (2), 156–8.

Tarapov, I. Ye., and Patsegon, N. F. 1980. Nonlinear waves in conductive magnetizable fluid. *IEEE Trans. Magnetics* MAG-16 (2), 309–16.

Taylor, G. I. 1950. The instability of liquid surfaces when accelerated in a direction perpendicular to their planes. *Proc. Roy. Soc. London Ser. A* 201, 192–6.

1959a. The dynamics of thin sheets of fluid I. Water bells. *Proc. Roy. Soc. London Ser. A* 253, 289–95.

1959b. The dynamics of thin sheets of fluid II. Waves on fluid sheets. *Proc. Roy. Soc. London Ser. A* 253, 296–312.

1959c. The dynamics of thin sheets of fluid III. Disintegration of fluid sheets. *Proc. Roy. Soc. London Ser. A* 253, 313–21.

Taylor, G. I., and McEwan, A. D. 1965. The stability of a horizontal fluid interface in a vertical electric field. *J. Fluid Mech.* 22, Pt. 1, 1–16.

Tesla, N. 1890. Pyromagnetic-electric generator. U.S. Patent 428,057, 13 May.

Thiele, A. A. 1969. The theory of cylindrical magnetic domains. *Bell Syst. Tech. J.* 48 (10), 3287–335.

Thomas, J. R. 1966. Preparation and magnetic properties of colloidal cobalt particles. *J. Appl. Phys.* 37 (7), 2914–5.

Tritton, D. J. 1977. *Physical fluid dynamics.* Van Nostrand Reinhold, New York.

Truesdell, C. 1954. *The kinematics of vorticity.* Indiana University Press, Bloomington.

Truesdell, C., and Toupin, R. 1960. Charge and magnetic flux. In *Handbuch der Physik 3:1*, The classical field theories, S. Flugge, Ed. Springer-Verlag, Berlin, p. 660.

Tsebers, A. O., and Maiorov, M. M. 1980. Magnetostatic instabilities in plane layers of magnetizable liquids. *Magnetohydrodynamics* 16 (1), 21–8.

Tsvetkov, V. N. 1939. Motion of anisotropic fluids in a rotating magnetic field. *Z. Eksper. i. Theoret. Fiziki.* 9, 602–5 (in Russian); *Acta Physicochima USSR* 10, 555–68 (in German).

Twombly, E., and Thomas, J. W. 1980. Mathematical theory of nonlinear waves on the surface of a magnetic fluid. *IEEE Trans. Magnetic* MAG-16, 214–20.

Van der Voort, E. 1969. Ideal magnetocaloric conversion. *Appl. Sci. Res.* 20 (2, 3), 98–114.

Whitaker, S. 1966. The equations of motion in porous media. *Chem. Eng. Sci.* 21, 291–300.

Wolfe, R., and North, J. C. 1974. Planar domains in ion-implanted magnetic bubble garnets revealed by ferrofluid. *Appl. Phys. Lett.* 25 (2), 122–4.

Zahn, M. 1979. *Electromagnetic field theory.* Wiley, New York.

Zahn, M., and Rhee, S.-W. 1982. Electric field effects on the equilibrium and small signal stabilization of electrofluidized beds. Conf. record, *IEEE Trans. Ind. Appl. Soc.* 82-CH1817-6, 1027–34.

Zahn, M., and Rosensweig, R. E. 1980. Stability of magnetic fluid penetration through a porous medium with uniform magnetic field oblique to the interface. *IEEE Trans. Magnetics* MAG-16 (2), 275–82.

Zahn, M., and Shenton, K. E. 1980. Magnetic fluids bibliography. *IEEE Trans. Magnetics*, MAG-16 (2), 387–415.

Zaitsev, V. M., and Shliomis, M. I. 1969. Entrainment of ferromagnetic suspension by a rotating field. *J. Appl. Mech. Tech. Phys.* 10 (5), 696–700.

1970. Nature of the instability of the interface between two liquids in a constant field. *Soviet Phys.-Dokl.* 14 (10), 1001–2.

Zelazo, R. E., and Melcher, J. R. 1969. Dynamics and stability of ferrofluids: surface interactions. *J. Fluid Mech.* 39 (1), 1–24.

CITATION INDEX

Abraham, M., 122
Abramowitz, M., 210
Adelstein, O., 4
Akimoto, M., 4
Albert, R. J., 314
Anderson, T. B., 276, 289, 296, 299
Andres, U., 44
Aris, R., 318
Arkhipenko, V. I., 131
Arnaldos, J., 314
Ausloos, M., 73

Bacri, J.-C., 198
Badescu, R., 67, 270
Bailey, R. L., 2, 152
Barclay, J. A., 176
Barkov, Yu. D., 131
Barton, A. F. M., 49
Basaran, O. A., 131
Bashtovoi, V. G., 131, 198, 222, 224, 225
Batchelor, G. K., 32
Bates, L. F., 138
Bean, C. P., 58, 60
Becker, R., 122
Berkovsky, B. M., 134, 198, 222, 224, 225, 270, 271
Berkowitz, A. E., 61
Bibik, E. E., 60
Bird, R. B., 318
Birnbaum, S., 4
Bitter, F., 32, 33
Bliss, G. W., 159
Blums, E., 176
Bologa, M. K., 313
Borisenko, A. I., 318
Boyle, J. F., 61
Bozorth, R. M., 32, 62
Brancher, J. P., 198
Brenner, H., 66, 260, 270, 271
Brevik, I., 119
Brown, R., 270
Brown, W. F., Jr., 31
Burns, M. A., 314
Busenberg, S. N., 65, 263
Buyevich, Yu. A., 313
Byrne, J. V., 100, 159

Cabannes, H., 31
Calugaru, G., 67, 270
Carni, S., 231
Caroli, C., 255
Casal, J., 314
Chandrasekhar, S., 193, 235, 236
Chantrell, R. W., 43, 68, 70, 175
Charles, S. W., 43, 45, 62, 72, 73, 175, 236
Christy, R. W.,32, 98
Chu, B. T., 116
Chung, D. Y., 72, 123
Condiff, D. W., 253, 270
Cowley, M. D., 100, 103, 111, 116, 122, 193, 194, 195, 197, 198
Coulaloglou, C., 312
Curtis, R. A., 231

Dahler, J. S., 31, 252, 253, 269, 270
deGennes, P. G., 68
deGroot, S. R., 271
Demchuk, S. A., 271
Doi, M., 73
Drazin, P., 235
Drew, D. A., 282, 285, 313

Edison, T., 161
Einstein, A., 63
Elmore, W. C., 33, 58
Essmann, U., 215

Ezekiel, F. D., 158, 159

Faber, O. C., Jr., 236
Faber, T. E., 215
Fanucci, J. B., 313
Fay, H., 4
Filippov, M. V., 312
Finlayson, B. A., 231
Fox, R. A., 61
Frenkel, J., 61

Gailitis, A., 196, 197
Gibbs, J. W., 180, 318
Glazov, O. A., 263
Goldberg, P., 70
Gotoh, K., 123
Graves, D. J., 314
Gubanov, A. I., 72

Hakim, S. S., 118, 158
Hall, W. F., 65, 263
Hassett, K. L., 43
Hathaway, D. B., 2, 146
Haus, H. A., 100, 116, 122
Hayes, C. F., 67
Heck, C., 165
Helmholtz, H., 100, 236
Hendrickson, D. N., 43
Hess, P. H., 43, 67, 69
Higgins, B. G., 159
Higham, J. B., 118, 158
Hobson, R. F., 66
Horsnell, T. S., 270
Hsieh, R. K. T., 270
Hughes, W. F., 32

Ishii, M., 285, 313
Isler, W. E., 72, 123
Ivanov, D. G., 313

Jackson, J. D., 87, 98
Jackson, R., 276, 289, 296, 299, 313
Jacobs, I. S., 58
Jacobson, D. M., 72
Jeans, J., 98
Jenkins, J. T., 270
Jerauld, G. R., 314
Jones, T. B., 132, 133, 145, 147, 148, 159
Jordan, P. C., 68

Kagan, I. Ya., 270
Kaiser, R., 38, 45, 60, 66, 156, 237, 266, 267
Kamiyama, S., 271
Kaplan, B. Z., 72
Karim, S. M., 120
Kashevsky, B. E., 270, 271
Kellog, O. D., 98
Kelvin, Lord, 236
Kendra, M., 198
Khalafalla, S. E., 40, 157
Kirko, I. M., 312
Kistler, S. F., 321
Koike, K., 271
Kordonskii, V., 271
Korteweg, D. J., 100
Krakov, M. S., 222
Krueger, D. A., 69
Kruyt, H. R., 44, 64
Kuhn, L., 4
Kunii, D., 274

Lalas, D. P., 231
Lamb, H., 190
Landau, L. D., 98, 122, 148, 190
Larsson, P. O., 4
Lee, W. -K., 307, 308, 309
Levenspiel, O., 274
Levi, A. C., 66
Lewis, D. J., 236
Lifshitz, E. M., 98, 122, 148, 190
Lightfoot, E. N., 318
Livingston, J. D., 60
Luca, E., 67, 270
Lucchesi, P. J., 4, 314

McCourt, F. R., 66
McEwan, A. D., 235
Mackor, E. L., 46
McNab, T. K., 40, 61
McTague, J. P., 65
Mailfert, R., 43
Maiorov, M. M., 210, 211
Malik, S. K., 198
Martinet, A., 43, 60, 62, 69, 73, 158
Martsenyuk, Yu., 61
Maruno, S., 4, 133
Mase, G. E., 32, 318
Massart, R., 43
Matsuki, H., 175
Mazur, P., 271
Mehta, R. V., 70
Melcher, J. R., 32, 99, 116, 123, 202, 203, 206, 207, 235, 236, 318
Melillo, L., 146
Milford, F. J., 32, 98
Miskolczy, G., 60, 66, 156, 158, 240, 266, 267
Moon, P., 98
Morimoto, Y., 4
Moskowitz, R., 2, 159, 237, 239, 262, 263
Murakami, K., 175
Myers, R. A., 4

Naletova, V. A., 314
Neal, J. A., 43
Néel, L., 61
Ness, N., 313
Nestor, J. W., 47, 63, 72, 175
Neuringer, J. L., 31, 119, 137, 158
Newbower, R. S., 4
North, J. C., 4

O'Grady, K., 69
Orlov, L. P., 134
Oyama, T., 271

Panofsky, W. K., 98, 123
Papell, S. S., 38, 236
Papirer, E., 73
Parker, P. H., Jr., 43, 67, 69
Patsegon, N. F., 123
Pearson, J. R. A., 231
Penfield, P., 100, 116, 123
Perry, M. P., 145, 147, 148
Peterson, E. A., 69
Phillips, M., 98, 122
Pimbley, G. H., Jr., 236
Pincus, P. A., 68, 255
Popplewell, J., 43, 45, 62, 72, 175
Puigjaner, L., 314

Quets, J. M., 4

Raikher, L., 61, 66
Raj, K., 64, 146
Rayk, W. D., 175
Rayleigh, Lord, 217, 230, 236
Reid, W., 235
Reimers, G. W., 40, 157
Reinman, J. J., 175
Reitz, J. R., 32, 98
Resler, E. L., Jr., 161, 168, 175
Rhee, S. -W., 314
Romankiw, L. T., 208
Rosenhead, S. M., 120
Rosensweig, R. E., 2, 4, 31, 32, 38, 45, 47, 50, 51, 55, 63, 66, 72, 73, 100, 103, 111, 116, 118, 122, 131, 135, 137, 144, 149, 150, 152, 157, 158, 159, 161, 168, 175, 193, 194, 195, 197, 198, 207, 208, 213, 214, 215, 225, 228, 236, 237, 239, 262, 263, 266, 267, 274, 291, 303, 304, 305, 307, 313, 314, 322
Roth, J. R., 175
Russakoff, G., 99
Rykov, V. G., 270

Saffman, P. G., 225
Salin, D., 198
Sano, K., 73
Schaufeld, J., 152
Scholten, P. C., 73
Scriven, L. E., 31, 159, 231, 252, 269, 270
Serrin, J., 32
Shaposhnikov, I. G., 31, 118
Sharma, V. K., 63
Shchelchkova, I. N., 313
Shenton, K. E., 73, 236
Shimoiizaka, J., 39, 157
Shliomis, M. I., 31, 60, 61, 62, 65, 66, 118, 196, 198, 231, 255, 256, 260, 263, 266
Shulman, Z. P., 271
Shumovich, R., 213, 214, 215
Siegell, J. H., 312
Singh, M., 198
Skjeltorp, A. T., 157, 199, 236
Slepian, J., 100
Slusarczak, M., 208
Soga, S., 4, 133
Spencer, D. E., 98
Stecher, L. C., 43
Sternling, C. V., 231
Stewart, W. E., 318
Stratton, J. A., 98, 123, 146
Suyazov, V. M., 257
Syutkin, S. V., 313

Taketomi, S., 70
Taktarov, N. G., 221
Tarapov, I. E., 123, 318
Taylor, G. I., 32, 225, 235, 236, 318, 321
Tesla, N., 161
Thiele, A. A., 215
Thomas, J. R., 43
Thomas, J. W., 198
Thompson, D. A., 208
Tien, C., 314
Timmins, R. S., 47, 63, 72, 175
Toupin, R., 99
Trauble, H., 215
Tritton, D. J., 32, 158
Truesdell, C., 99, 158
Tsebers, A. O., 210, 211
Tsvetkov, V. N., 270
Twombly, E., 198

van der Voort, E., 174, 175
Vislovich, A. N., 270, 271
Vogler, T., 225, 228

Waldner, F., 63
Weissman, M. H., 66, 271
Whitaker, S., 313
Wilson, E. B., 318
Wolfe, R., 4

Yamasawa, K., 175

Yantkovskii, E. I., 270
Yen, R. H., 313
Yotsumoto, Y., 4
Young, F. J., 32
Yubakami, K., 4, 133
Zahn, M., 32, 73, 98, 213, 214, 215, 225, 228, 236, 314
Zaitsev, V. M., 196, 198, 260, 263
Zelazo, R. E., 116, 118, 202, 203, 206, 207
Zrunchev, I. A., 313

SUBJECT INDEX

adiabatic temperature change, 167
magnetocaloric effect, 161
agglomeration of colloidal particles, 36
see also clustering and chaining
alternating tensor, 316
definition, 246
uses, 247, 248
Ampère's law, 77, 86
application, 91, 92, 137
averaged, 282, 288
angular momentum,
balance, 252, 253
density, 249
anisotropy constant, 61
antiferromagnetism, 6
antisymmetric dyadic, 247, 315
antisymmetric stress, 254
as component of asymmetric stress, 254
in rotating field, 258–63
vorticity and, 254
applications of ferrohydrodynamics,
accelerometer, 152
bearings, 159, 322
damper, 159
enzyme and microorganism support, 4
fluidized beds, 272–314
loudspeaker coolant, 2
lubricant retention, 159
magnetic-domain detection, 4
magnetocaloric power, 161–75
nuclear fusion, 175
pressure seals, 144, 159
printing, 134
pumping, 164
sink–float separation, 157
targeted drug delivery, 4
vacuum feedthroughs, 2
aqueous ferrofluid,
charge-stabilized, 42, 43
electron micrograph of, 41
sterically stabilized, 71
asymmetric stress, 237–71
constitutive relations for, 254
in data correlation, 267
in torque-driven flow, 267
vector of, 248
atomization, 318
sheet jet, 318–21
averaging,
Darcy's law, 289
Maxwell's equations, 280, 281, 282
in two-phase flow, 276–89
weighting function, 276

Bénard convection, 228
Bernoulli equation, generalized
for conical meniscus, 137
derivation, 23, 124–6
energy conversion and, 164
families of, 130
Gouy experiment, 136
for magnetic nozzle, 134
for plug seal, 144–5
in printing, 134
Quincke problem, 132
for rotary shaft seal, 142
surface elevation and, 133
Bessel equation,
for round jet, 218
for torque-driven flow, 261
bifurcation,
of circular drop, 211
labyrinthine instability and, 208
normal-field instability and, 198
body couple, 14, 249

body force,
from antisymmetric stress, 254, 258
from gradient field, 110
in other disciplines, 1
boiling, 236
Bond number,
magnetic, 213, 223
boundary conditions,
fluidized solids in a field, 307
for magnetic field, 77
spin, 260
velocity, 260
boundary layer,
flat plate, 240
torque-driven flow, 262
Brownian-orientational motion,
influence on magnetization, 255
influence on viscosity, 263
relaxation time and, 255
Brownian translational motion, 35
colloidal segregation and, 48
bubbles,
in ferrofluids, 212
in fluidized beds, 274
bulk pressure, 130
composite, 121
fluid-magnetic, 112
magnetostrictive, 112
thermodynamic, 110
buoyancy, 154, 157

capillary constant, 201
capillary pressure, 127, 130, 131, 181
capillary wave, 190
Carnot cycle, 171
Cauchy equation of motion, 245
Cauchy stress principle, 26
cavitation, 110
cgs units, 10
charge stabilization of colloid, 43
chemical precipitation, colloid preparation, 40
clustering and chaining
analytical description, 67
numerical experiments, 68
coating flows, 318
cobalt particles
ferrofluid containing, 43
intrinsic superparamagnetism of, 62
coercivity, 14
coexistence curve
of ferrofluid phases, 73
of particle bed, 304
colloidal stability
attraction of magnetic particles and, 36
in gravitational field, 35
ionic-charge mechanism, 43
in magnetic-field gradient, 34
of metallic base ferrofluid, 45
steric mechanism of, 46
van der Waals forces and, 37
conjugate dyadic, 315
constitutive laws,
for angular-momentum conversion, 254
body-couple, 255
couple-stress, 254
of electromagnetic field vectors, 90
fluidized-bed, 285, 286, 291
viscous-stress, 120
contacting, use of particle beds in, 272, 314
continuity equation
averaged in two-phase flow, 279
Eulerian formulation, 19, 20
Lagrangian formulation, 21
and Reynolds' transport theorem, 245
correlation phenomena in particle clustering and chaining, 66, 68
Coulomb's law
as force at a distance, 8, 74
surface-stress description and, 81–4
couple stress tensor, 250
constitutive relationship for, 254
coupling coefficient, 59
crystal anisotropy, superparamagnetism and, 61
Curie temperature
of ferrofluid relative to melting point, 33
of ferromagnetic solids, 165, 166
cycle
magnetocaloric, 168
thermodynamic analysis, 168–74
turbine, 174
cylindrical jet, acceleration of, 135
cylindrical-jet stability
in radial gradient field, 222
in uniform axial field, 217

Darcy's law,
in porous media flow, 226, 289
demagnetization factor,
definition, 136
interpreting data with, 136
labyrinthine spacing and, 212
diamagnetism, 6, 7
dielectric fluids,
dielectrophoretic force and, 160
labyrinthine instability of, 215–16
dimensionless group,
Bond number and, 213, 223
coupling coefficient and, 59
droplet elongation and, 129
in fluidized-bed stability, 300
Rayleigh number and, 229–31
Reynolds' number and , 291

yield stress of fluidized bed and, 310–11
dipole
far-field, 13
interaction energy of, 15
moment of, 13
near-field, 12
dispersing agent
chemical structure of, 51
entropic repulsion by, 48
interchange and, 54
size reduction and, 38
dispersion relationship
for cylinder in gradient field, 222
for cylindrical jet, 216
definition, 189
for fluid interfaces, 199
Kelvin–Helmholtz, 202, 204
for porous medium, 226
Rayleigh–Taylor, 200
for two-phase flow, 298
displacement current
as field source, 92
in Maxwell's equation, 84
disturbance wave, mathematical representation, 232
divergence theorem, 317
domain magnetization, 5
droplets
circular, 211, 212
in multiphase flow, 314
round, elongation of, 129
dual
of tensor, 316
of vector, 316
dyad, 315
dyadic, 315

Earnshaw's theorem,
circumvention, 149
proof of, 147
efficiency
of Carnot cycle, 171
of magnetocaloric cycles, 171, 172, 174
eigenvalue problem, 190
electrical polarization
of fluidized solids, 314
of labyrinthine fluids, 214
of suspensions, 314
electrodeposition, 45
electrohydrodynamics, 1
electromotive force, 89
energy conversion, 161
energy density,
of electromagnetic field, 94
magnetostatic, 95, 98
entropic repulsion, 48
equation of motion
averaged in two-phase flow, 282, 288
of ferrofluid with antisymmetric stress, 258
in Neuringer–Rosensweig model, 118–22
equation of state (magnetic)
Curie–Weiss, 69
linear, 166
Weiss, 175
Ergun equation, 291
Eulerian observer, 22, 23
exchange of stability in normal-field instability, 191
experimental data
for ferrofluid plug seal, 148
for fluid-cylinder stability, 225
for labyrinthine spacing, 214
for magnetic fluidization, 305, 309
for normal-field instability, 195
for surface-waves resonance, 203

ferroelectrics in energy conversion, 174
ferric induction, 11
ferrimagnetism, 6
ferrofluid
colloidal stabilization of, 46–50
compressibility, 111
density of, 71
electrical conductivity of, 72
magnetization of, 55–63, 255–7
modification of, 50–4
pour point of, 71
preparation of, 38–46
pyromagnetic coefficient of, 167
specific heat of, 71
surface tension of, 71, 229
thermal conductivity of, 71, 72
thermal-expansion coefficient of, 71
viscosity of, 63–77, 263–9
ferrohydrodynamic stability problems
of circular drop, 211
of cylindrical column, 216–25
of fluidized bed, 291–306
of fluid jet, 217–22
Kelvin–Helmholtz, 202–6
labyrinthine instability in, 207–15
normal-field, 178–99
of porous media flow, 225–8
Rayleigh–Taylor, 200–2, 206–7
thermoconvective, 228–32
ferrohydrodynamics, scope of, 1–4
ferromagnetism, 4–6, 62
field variables, units of, 8–11, 17
fingering instability, 225–8
prevention of, 227–8
fixed bed, 273
flocculation, 50

flocculation (cont'd)
  irreversible, 50–2
  reversible, 50–2
fluid-cylinder stability, 215–25
fluidization, 272–314
  applications of, 273, 314
  fluid-particle interaction in, 285, 291
  magnetically stabilized, 276, 291–314
  minimum or incipient, 273, 290
  ordinary, 273, 276
  stress of phase in, 285
fluid jet, 217–22
fluid-magnetic pressure, 112
flushing, 42
flux, 253
force density,
  electric, 2
  interfacial capillary, 127, 181
  interfacial magnetic, 129
  Kelvin, 13, 111, 118
  Lorentz, 1
  from stress tensor, 110
Fourier analysis, 178, 234, 295

Galilean transformation, 89
Gauss's laws, 86, 90, 281
Gibbs's notation, 315
gradient-field stabilization
  cylindrical column, 222
  Rayleigh–Taylor, 207
gravity wave, 190
growth factor, 298

Hamaker constant, 37
harmonics, 178
heat and mass transfer, 2, 175, 228, 240
Helmholtz free energy, 101
hysteresis
  in fluidized bed, 312
  for hard magnet, 14
  lag angle in rotating field and, 257, 258
  normal-field instability and, 198

identities
  integral-theorem, 317
  tensor, 317
  vector, 316
immersed-body force, 153
inertial dampers, 159
interfacial force balance, 127–9
internal angular momentum
  diffusion of, 260
  equation of change of, 253, 259
interpenetrating continua, 276
irreversible thermodynamics, 271
irrotational flow, 125
isothermal process, 168

Kelvin–Helmhotz instability, 202–6
kinematic condition, 182
Kortweg–Helmholtz force density, 114, 118, 119
Kronecker delta, 108

labyrinthine instability, 208–15
  analogs of, 214
  in dielectric fluids, 214
  horizontal, 212, 215
  vertical, 209
Lagrangian observer, 22, 23, 162
Langevin function, 58
Laplace's equation
  in cylindrical coordinates, 217
  of magnetic perturbation potential, 185
  of pressure perturbation, 183
  in spherical coordinates, 80
  velocity potential and, 216
Leibniz formula, 112
levitation
  density separation in, 156–7
  of nonmagnetic body, 150–7
  self-, of magnet, 150–2, 159
light polarization, 67, 70
lignosulfate, 42
linear analysis, 178, 291
linear-momentum balance, 253
liquid crystals, 240, 270
Lorentz transformation, 87

magnetic Bond number, 213, 223
magnetic clutch fluid, 7
magnetic domains, 4, 5
magnetic energy
  of dipole, 15, 56
  of field, 95, 98
  of magnetized matter, 162
magnetic field, 9
magnetic-field solutions
  dipole, 12–13
  electric capacitor, 92
  for long, straight conductor, 91
  for magnetized sphere, 80, 81, 82
  for perturbed cylindrical jet, 219
  for perturbed planar interface, 185, 187–8, 189
magnetic fluid
  nonexistence of, in nature, 7, 33
  paramagnetic salt solution and, 44
  see also ferrofluid
magnetic force
  in ferrofluid bulk, 110–19
  in ferrofluid surface, 129
  magnetostrictive component of, 113
  see also force density
magnetic induction, 9, 76

magnetic material
hard, 14, 15
soft, 14, 15
magnetic permeability
definition, 78, 91
of fluidized bed, 314
magnetic poles
in isolation, 8, 9
surface density, 31
volume density, 31
magnetic saturation, 56, 58
magnetic torque, 14
magnetic work, *see* magnetic energy
magnetically stabilized bed, 274, 276
magnetism, *see* specific type
magnetite
grinding of, 38
particle size of, 40, 41
precipitation, 40–4
magnetization curve
of Curie–Weiss solids, 164
of ferrofluid, 58, 59
magnetization relaxation
Brownian, 61–3, 255–7, 265
Néel, 61–3
magnetization vector, 11, 13, 86
magnetobirefringence, 70
magnetocaloric cycle efficiency
nonregenerative, 171
regenerative, 172, 174
magnetocaloric heat pump, 174
magnetocaloric heating and cooling effects, 161, 167
magnetohydrodynamics, 1
magnetostriction, 111–13, 118, 123
mass conservation, *see* continuity equation
maxwell, 11
Maxwell's equations, 85, 86
charge conservation and, 90
see also specific equation
Maxwell stress tensor, 81–4
in analysis of sheet jet, 318–22
portrait of, 85
mean curvature, 127
capillary pressure and, 127, 180
obtaining, 180
metallic ferrofluid, 45
momentum equation, *see* equation of motion
Monge representation, 180
Monte Carlo computation, 68–69, 70
Mössbauer recoilless emission
particle-surface dead layer and, 60–61
Néel relaxation and, 61
multiphase flow, 272

Navier–Stokes equation, 122
and antisymmetric flow, 257–8
and two-phase flow, 276
nesting convention, 316
nonlinear analysis of normal-field instability, 196–8
normal-field instability
experiments and, 193–5
linear analysis of, 178–92
nonlinear analysis of, 196–8
thin-film, 198
two-dimensional melting and, 199
normal mode, definition, 233
normal-stress boundary condition, 179

oersted, 11
orifice discharge, 321

paramagnetic salt solution, 44–5
paramagnetism, 6
peptization, 42
permanent magnet, 14
permeability, 78, 91
chord, 192
geometric mean of, 192
tangent, 192
phase, 297
phase change, 110, 236
phase diagram of magnetized fluidized solids, 304, 305
phase velocity, 298
polar vector, 315
polarized matter, 10, 13
polyadic, 316
polymer
dispersing agent, 47, 51
flocculating agent, 50
Poynting vector, 95
pressure
of fluidized bed, 275, 305
generated by heated ferrofluid, 165–6
interfacial capillary, 127, 181
interfacial magnetic normal, 129
see also bulk pressure
pseudovector
in antisymmetric flow field, 247
definition, 315
pyromagnetic coefficient, 167

radial dispersion, 314
rare-earth magnets, 175
Rayleigh number, 229
in magnetic gradient field, 230
in uniform field with temperature gradient, 230
Rayleigh–Taylor instability, 200
relaxation time constant,
Brownian orientational, 61–3

relaxation time constant (cont'd)
  magnetization of ferrofluid and, 255–7, 265
  Néel magnetization of solid and, 61–3
remanence, 14
repulsion, 46–9, 152
Reynolds number, 291
Reynolds stresses, 284
Reynolds' transport theorem, 243–5, 279, 317
rheology
  of ferrofluid couple stress, 254
  of ferrofluid viscosity, 263–9
  of fluidized magnetized medium, 307–12
  of Newtonian fluid, 120
  vortex viscosity in ferrofluid and, 254

saturation magnetization, 56, 58
scalar potential
  of magnetic field, 77
  of velocity field, 216
shaft seal, 144
sheet jet, 318
SI units, 9, 87
small-disturbance analysis, 178
specific heat, 71, 162
spin viscosity
  bulk coefficient of, 254
  shear coefficient of, 254
standing wave, 234, 235
steric hindrance, 35
steric repulsion, 46
steric stabilization, 48
Stokes's theorem, 317
stress dyadic, *see* stress tensor
stress tensor, 25–30
  ferrohydrostatic, 108
  of nonpolar fluid, 246
  of polar fluid, 254
stream tube, 18
striction pressure, 112
substantial derivative, 22, 119
superconductor
  field-intensity, 167
  type-I, 215
  type-II, 215
superficial velocity, 273
superparamagnetism
  extrinsic, 62
  intrinsic, 61
superposition, 234
surface, mean curvature of, 127, 181
surface adsorption, 38, 39, 46, 54
surface couple density, 249
surface waves classified, 235
susceptibility, 44
  chord, 293, 294
  tangent, 293, 294
symbolic notation, 315
symmetric dyadic, 315
symmetry breaking, 177

Taylor wavelength, 201
thermal energy, 34, 46, 56, 266
thermoconvective instability, 228
thermodynamics
  first law of, 163
  second law of, 163
  see also specific terms
total angular momentum, 249
  equation of change and, 251
traction vector, 27
  in magnetic field, 116–17
transpose dyadic, 315
turning point, 198
two-phase flow, 276

unit dyadic, 316

vacuum feedthrough, 2
van der Waals attraction
  integration for spheres and, 37
  London model of, 37
vector of a dyadic, 315
velocity, 18
  averaged, in two-phase region, 280
  interstitial, 291, 302
  perturbation, 183, 291
  superficial, 273
velocity potential, 216
  and Bernoulli equation, 125
  of round jet, 217
viscoplastic, 46
viscosity
  first and second coefficients, 120
  of magnetized ferrofluid, 64–7, 263–9
  of unmagnetized ferrofluid, 63, 71
voidage, 277
vortex viscosity, 254
vorticity, 125, 158

wavelength, 233
wave number, 233
work
  as a cyclic process, 162
  of displacement, 15
  of magnetization of material, 101